# A CENTURY OF MARITIME SCIENCE

## The St Andrews Biological Station

# A Century of Maritime Science

## *The St Andrews Biological Station*

EDITED BY JENNIFER HUBBARD,
DAVID J. WILDISH, AND ROBERT L. STEPHENSON

UNIVERSITY OF TORONTO PRESS
Toronto Buffalo London

Toronto Buffalo London
www.utppublishing.com

ISBN 978-1-4426-4858-6

**Library and Archives Canada Cataloguing in Publication**

A century of maritime science : the St. Andrews Biological Station/edited by Jennifer Hubbard, David Wildish, and Rob Stephenson.

Includes bibliographical references and index.
ISBN 978-1-4426-4858-6 (bound)

1. Canada. Department of Fisheries and Oceans. Biological Station (St. Andrews, N.B.) – History. 2. Marine biology – Research – Canada – History. 3. Marine biologists – Canada – History. 4. Research institutes – Canada – History. I. Hubbard, Jennifer Mary, 1960–, author, editor II. Wildish, D. J., 1939–, author, editor III. Stephenson, Robert (Robert L.), author, editor

GC59.15.C45 2016 551.46'132 C2015-906995-5

Unless otherwise noted all illustrations are courtesy of the St Andrew's Biological Station.

University of Toronto Press acknowledges the financial assistance to its publishing program of the Canada Council for the Arts and the Ontario Arts Council, an agency of the Government of Canada.

Canada Council for the Arts
Conseil des Arts du Canada

Funded by the Government of Canada
Financé par le gouvernement du Canada

# Contents

# Acknowledgments

The editors wish to thank the Department of Fisheries and Oceans, St Andrews Biological Station for its ongoing support of this project, including financial contributions to the publication of this volume. Publication of this book was also enabled by a grant provided by the Office of the Dean of Arts, Ryerson University. We thank Charlotte McAdam and Joanne Cleghorn of the St Andrews Biological Station library for their assistance with archival material and photographs, and Benjamin Allen Stevens for his help in preparing the index. Finally, we thank our two readers, Dr Brian Payne and Dr Carmel Finley, for their patience in reading and critiquing earlier versions of this book and for their valuable comments.

A CENTURY OF MARITIME SCIENCE

The St Andrews Biological Station

# Introduction

JENNIFER HUBBARD

Scientists of the recent past must be the first to write their own history. They are the only individuals who have an intimate and deep understanding of their aims and techniques and the obstacles – fiscal, technological, and conceptual – that have had to be overcome in the development of their science. The material in this volume grew out of presentations made at the October 2008 "Workshop on the Evolution of Marine Science in Canada" celebrating one hundred years of science at the St Andrews Biological Station. Most chapters presented here have been written by the scientists who have experienced at least a part of the stories they tell. There are also two chapters contributed by historians of science who focus on the station's historical background and early significance; but the main body of the work tells the story of several of the most important strands of research at St Andrews. Most also give insights into the work that was done since the Second World War, which is a period for which historical treatments in the history of marine science remain sparse. This compilation of articles provides an invaluable introduction to a range of scientific specializations that either originated in St Andrews or that its scientists helped to develop, and shows how science is shaped not only by individuals, but also by the institution in which they work and the problems encountered in local environments surrounding them.

This book provides an addition to histories of science in Canada, and the general history of marine science.[1] Tim D. Smith, Eric Mills, Helen Rozwadowski, Jennifer Hubbard, and more recently Carmel Finley have made contributions to the historical understanding of the scientific, institutional, and political development of biological and physical oceanography and fisheries science.[2] In addition, former Department of Fisheries and Oceans communications director for the Maritimes region Joseph Gough and scientist George A. Rose have recently contributed histories of Canadian fisheries management and the ecological history of the North Atlantic cod fisheries.[3] The main focus in histories of fisheries science has been the study of population dynamics and approaches to fisheries management. This volume is different. It instead offers the perspective of working scientists on these and other important research streams, including marine toxicology, marine environmental science, benthic ecology and flow studies, scallop and salmon studies, and the development of fish culture or farming. In addition there is a contribution by a marine technologist who helped to provide scientists with the esoteric tools needed to conduct their research, and a contribution by a woman scientist, Dr Mary Arai, who represents the third generation of women in the scientifically eminent Needler family to hold a Ph.D. in science and to conduct research in Canada's marine stations. Each chapter shows how the institution in which science is situated influences research choices, and how that institutional setting encourages collaboration across defined disciplinary boundaries, as practical and fiscal concerns have led scientists to forge common interests and overlapping research programs. One of the chapters, by John Caddy, on the history of scallop research, is both a history and a contribution to continuing research, with suggestions for necessary directions in future research and scallop management. There are many stories and research programs undertaken at the St Andrews Biological Station that do not get the attention they deserve owing to limitations of space and time; the major herring research program, for example, and the use of submersibles in invertebrate research. Some of this will be alluded to in this introduction.

Shining through many of the scientists' chapters is the presence of a philosophy of service and problem solving. This counters my claim in *A Science on the Scales: The Rise of Canadian Atlantic Fisheries Biology 1898–1939* that "the ideal of service that so strongly marked the first four decades of fisheries research in Canada was largely lost in later periods."[4] As Tim Foulkes, a St Andrews technologist, commented in the workshop version of his paper: "From my perspective I saw the biological station at St Andrews as a place where professors used to come to poke about in the mud, catch fish, smoke their pipes, and enjoy the summers collecting esoteric data for their academic publications. Since looking into the history of technology at the station, I have been impressed with how ingenious some of these professors were, how much 'grunt' work they did, and how dedicated they were to fisheries science." This collection of papers also serves as a partial rebuttal of my description of the St Andrews Biological Station – the focus of my book – in the post-war era as a "scientific backwater."[5] Although following the Great Depression the St Andrews Biological Station lost the prominence it enjoyed in its first four decades, its scientists have continued to make vital and original contributions to marine and fisheries science, including publishing several "scientific classics" in their fields, such as Sprague's trilogy in water research (1969–71). Particularly since the 1960s, many have participated in the international research collaborations that are so quintessential in marine science. Later chapters in *A Science on the Scales* focus on mainstream fisheries population studies related to fin-fish fisheries; thus, its historical analysis was supportive of arguments that were put forward in favour of relocating the biological station – following the destructive fire of 1932 – to some Nova Scotia location closer to the important Grand Banks fisheries. The authors of these collected papers demonstrate clearly why that would have been a terrible mistake. There are more fish in the sea than just those that formerly massed on the Grand Banks, and shellfish such as scallops, mussels, and oysters, and crustaceans such as lobster and shrimp also require scientific attention. The Bay of Fundy ecosystem is unique due to its huge tidal range – the world's largest – and biological and oceanographic investigations need to explore conditions in a variety of environments. Universal "laws" in fisheries science have proved elusive, as local conditions shape the life histories of living populations. Also, as some chapters here demonstrate, important new scientific fields such as ecotoxicology were developed in response to destruction or change in local wild populations due to natural and man-made chemicals introduced into the environment from diverse sources. The papers in this volume, then, celebrate the history of a century of local and international science, with the biological station at St Andrews at the nexus of diverse but often intertwining enterprises.

## The Historical Context of the St Andrews Biological Station: The Ideal of Service and the Growth of Science

The first four chapters in this collection provide readers with different aspects of the general background of the creation and development of the St Andrews Biological Station, between them going back to the early nineteenth century and extending to 2008, its centenary year. They deal with, respectively: the Canadian scientific and educational context; the participation of women in science; the influence of German forestry and the ideal of service on fisheries science; and the balancing act between pure and applied science at the St Andrews Biological Station, which faced, following the Second World War, persistent threats of closure, cutbacks, and loss of programs and scientists.

The Canadian marine stations, starting with the Atlantic Biological Station,[6] were founded due to a convergence of interests between biologists at Canadian universities and the then Department of Marine and Fisheries. The scientists wanted facilities to pursue the new, trendy marine biology, while the department recognized that fisheries management required greater scientific understanding. The nascence of this recognition was grounded in the science and scientific education emerging in nineteenth-century British North America, as documented by Eric Mills in this volume's first chapter: "Science in Canada: The Context of the Biological Board of Canada's St Andrews Biological Station." Mills has provided an excellent, concise history of the development of Canadian science in the nineteenth century. He encapsulates the contributions by British naval expeditions from the beginning of the century; the British military's involvement in creating magnetic, geological, astronomical, and hydrological surveys; and the emergence by mid-century of the Geological Survey of Canada, which contributed not only to geology surveys but also to natural history in general. Natural history societies, both amateur and professional, introduced a broad interest in science even before Canada's fledgling universities later added science to their curricula. Mills shows how the railroad construction that followed Confederation enabled St Andrews to be later chosen as the site of the Atlantic Biological Station, by making it accessible to academics from distant locations. Two marine scientific enterprises also prefigured the first marine stations: Canada's hydrographic surveys, linked to the practical need for mapping shipping lanes, and the Canadian Tidal and Current Survey, created in response to a petition by ship owners in 1893. Both were to contribute to investigations initiated by the Biological Board of Canada, created in 1912 to run Canada's marine station. Finally, Mills describes Canada's first, floating marine station, established in response to pressure by academics wanting to emulate the European and American scientists flocking to marine laboratories to study the entrancing morphological variety of marine organisms. Canadian academics also promised to study fishery problems. Mills concludes: "It was the need to develop a Canadian scientific identity uniting utility with scientific understanding that lay behind the origin of Canada's first marine stations. That a small group of Canadian university-based scientists could carry this off in the face of government indifference makes this a particularly interesting case study to historians."

Today it would be impossible to establish a scientific organization with government funding to be run by independent scientists with minimal government oversight. Why, then, was the situation so different in 1898, when the board of management of the Marine Biological Station was founded?

The answer lies in the service ethic that characterized nineteenth-century gentlemanly scientific practice. *A Science on the Scales* highlighted the importance of this tradition but failed to identify the genesis of the service ideal in government science. This sprang from independent but government-funded scientific institutions created to serve the British government in the late eighteenth century, the legacy of Joseph Banks (1743–1820). Banks, the botanical expert on Captain James Cook's first expedition around the world – who built up Kew Gardens in London, and helped found the Royal Institution and other scientific establishments – actually earned his scientific credentials by surveying Newfoundland's cod fisheries for the British government. Banks, like fellow wealthy landowners, believed that, as a privileged class, they were obliged to undertake public office; Banks decided to devote himself to scientific service for the public good. He promoted the landed class's ideal of improvement and preferred science that could be applied to offer practical benefits for the nation.[7]

Banks's friendship with First Lord of the Admiralty John Montagu, fourth Earl of Sandwich, got him appointed to accompany an Oxford friend, Captain Constantine Phipps, to Newfoundland and Labrador, to study this region's natural history. Phipps commanded H.M.S. *Niger* as part of the regular British patrol of Newfoundland's fisheries,[8] and was directed to build fortifications to enable "His Majesty's Subjects to carry on the Fisheries on the Coast,"[9] and to otherwise protect British fishermen, their craft and tackle. From May until October 1766, Banks collected and classified marine and terrestrial plant and animal specimens, and observed the cod fisheries. He established his scientific reputation by writing up the first Linnaean descriptions of Newfoundland and Labrador's plants and animals.[10] Banks, then, conducted perhaps the first scientific study of fish in what were to become, in 1949, Canadian waters.

Banks's fame as naturalist of Captain James Cook's 1768–71 scientific expedition to the South Pacific, together with his unique collections, enabled him to take his place at the heart of London's scientific and social life for over five decades. Elected at the age of thirty-five as president of the Royal Society, Banks held this position with distinction for over forty-one years. Well aware that the English government avoided funding science, Banks developed links between the Royal Society and certain departments. For example, he advised on standardizing Britain's system of weights and measures.[11]

The powerful British aristocracy preferred a small government and bureaucracy, but the expanding British Empire needed similarly expanding governing structures and policies. The landowning oligarchy therefore created non-government institutions to perform functions later filled by government departments. While government-funded, these were run by unpaid volunteers: gentlemen who shared a vision of duty and service to society to repay the benefits of their class – and also to preserve and justify those benefits.[12] Banks himself was consulted on practically all matters related to exploration, science, and the colonies. In 1799 he helped found the Royal Institution, a new scientific society devoted to applying scientific findings to agriculture. While its members improved their estates to bolster their incomes, they also sought to improve the lives of agricultural dependents, and to develop new crops and new industries to benefit the entire British Empire. Banks and the Scottish agricultural reformer Sir John Sinclair founded the Board of Agriculture in 1793 to promote agricultural interests. The Board of Agriculture ended up being a half-public, half-private body, not answerable to the Treasury even though it did receive three thousand pounds per annum from the government.[13] It was similar in its functions to the early Biological Board of Canada.

The British political tradition, then, developed a hands-off approach to scientific institutions; government funding was procured, but scientists worked as volunteers, and set the research agenda in their capacity as experts. This became the model for Canada's first marine biological station, and the board of management of the Marine Biological Station that ran it. This exemplar then also inspired the creation of the permanent Atlantic Biological Station in St Andrews, New Brunswick, and its sister station in Nanaimo, British Columbia, and all later fisheries research and experimental stations in Canada, run first by the Biological Board and then the Fisheries Research Board of Canada. This ideal lasted, even after fisheries biology was professionalized in Canada in the

1930s, until the government finally took control in 1973, and disbanded the Fisheries Research Board of Canada in 1977, events described in some detail by Rob Stephenson in chapter 4. Canadian marine and fisheries biology and oceanography originated because the founding and participating university scientists acted within the gentlemanly tradition of using their expertise to serve government and nation. The Dominion government, thanks to this same tradition, trusted scientists to deliver on their promises to help the fisheries.

These gentlemanly institutions, while bound by British traditions, were at the same time novel spaces free to develop new norms and traditions. As Canada's first oceanographer Harry B. Hachey was later to write,

> When I first arrived at the Atlantic Biological Station in 1927, I was forcibly reminded of the old song, "Old MacDonald had a farm" – you know:– "Here a chick, there a chick, everywhere a chick, chick." There were chicks all over the place, Helen, Emma, Viola, Gene, Nelda and a host of others. A physics laboratory in a University was never like this, and so I was weaned away from confining walls to the great open spaces of the bounding main.[14]

Hachey's quotation is telling: the marine biological stations afforded new spaces where women enjoyed more freedom, albeit unpaid or underpaid, to undertake scientific careers than in traditional spaces such as university science departments, where, for the most part, they were not welcome before the latter decades of the twentieth century. Mary Needler Arai's chapter in this volume, "Some Contributions of Women to the Early Study of Marine Biology of Canadian Waters" increases our understanding of the active role of women in Canadian marine science, showing that they were important contributors who were not barred from doing science by their gender, but who certainly suffered privations: lack of remuneration or in a few cases lack of access to equipment. In every case until the late 1960s, they were not allowed on ship expeditions, a hardship experienced even by such luminaries of oceanography as Mary Sears at Woods Hole, as documented by Kathleen Broome Williams in *The Machine in Neptune's Garden*;[15] Jennifer Martin, a contributor to this volume, was one of the first two women permitted to participate on an ocean cruise of the St Andrews Biological Station research vessel *J.L. Hart* in 1980. Arai also builds on the literature of feminist historians such as Marianne Ainley, whose accounts, incidentally, include Arai, her mother, and grandmother.[16] Arai recounts how some of the women irreverently named above by Hachey either remained single and remained scientists, or lost their income upon marriage. If they married scientists they were free, as in the case of her grandmother, Edith Berkeley, to continue their scientific studies as unpaid "assistants" to their professional husbands. As the third generation of women in her family to hold a Ph.D. in biology, Mary Needler Arai holds a special place in the history of Canadian marine stations, and her insights allow her readers to understand how scientifically important were these women's contributions, and those of turn-of-the-century female taxonomists working – in US museums – to identify specimens from US and Canadian marine expeditions. Their importance is also highlighted by Martin in her chapter on the history of paralytic shellfish poisoning: Alfreda Needler was the first to connect this phenomenon with its causative agent, a microorganism then called *Goniaulax tamarensis*.

One scientist at St Andrews who was most active in supporting women in science was Archibald Gowanlock Huntsman, the director of the Atlantic Biological Station from 1919 to 1934. Huntsman, the father of three daughters – his daughter Elinor also obtained a Ph.D. in biology – was the central figure in the history of the St Andrews station from shortly after he became its curator in 1911 until he was forced to step down in 1934 for rebuilding the station without permission after it was destroyed by a fire. However, Huntsman only features strongly in three chapters.[17] R.H. Peterson tells an irreverent story about him in his chapter on the history of salmon research and Robert Stephenson makes repeated references to his influence. He is, however, a central figure in my own chapter, "The Gospel of Efficiency and the Origins of Maximum Sustainable Yield (MSY): Scientific and Social Influences on Johan Hjort's and A.G. Huntsman's Contributions to Fisheries Science." Chapter 3 looks at outside influences on the development of Canadian fisheries and marine science and also examines how the service ideal influenced the way scientists chose to address fisheries problems. It situates the beginnings of fisheries science within the environmental tradition first highlighted by Samuel P. Hays in his *Conservation and the Gospel of Efficiency*, that is, the wise-use conservation movement that fought to conserve resources so as to allow their efficient use.[18] In looking at the forestry science background of those at the forefront of Gospel of Efficiency conservation, which was shaped by nineteenth-century German forestry science, I discovered parallels between this science and ideals that developed in fisheries science, specifically the goal of maximum sustained yield that dominated fisheries science after the Second World War. Similarities in language and concepts included the forestry goal of optimum sustained yield, methods of aging trees and fish, and other parameters. Huntsman himself consciously borrowed forestry concepts and discussed forestry science with former US chief forester, and fellow University of Toronto professor, Bernhard Fernow. Fernow introduced German forestry science and practices to North America, and influenced the better-known Gifford Pinchot, the primary champion of Gospel of Efficiency conservation. Chapter 3 also explores the consequences of the dual philosophies of the service ideal, and the ideal of maximum sustained yield, for fisheries management and fisheries science in general.

In chapter 4, Robert Stephenson, the former director of the St Andrews Biological Station who organized the historical celebration of its centennial, presents "St. Andrews Biological Station: A Case Study in the History of Public Science in Canada." Because of Stephenson's deep appreciation of the importance of the historical context of science, that 2008 workshop took place and this collection of papers has been drawn together. Stephenson offers a unique survey of the entire century of scientific enterprise at the permanent biological station at St Andrews, adding perspectives drawn from his own experience. Unlike earlier institutional histories, Stephenson addresses the continuing tension between pure and applied science. Together with his account of the pressures experienced by a scientific institution under almost continuous existential as well as fiscal threat following the Second World War, this provides his history with thematic unity. While he mentions the scope of research in each of the major periods delineated by organizations controlling the station, from the Biological Board in 1908 to the Department of Fisheries and Oceans following 1973, his focus is on how the organizational relationships, combined with immediate practical demands, shaped research programs at St Andrews. He addresses a series of institutional historical questions as well. Should volunteer research have been continued after the Great Depression? What has become of the educational role of the St Andrews Biological Station and its role as a "nursery ground for aquatic science"? How have scientists met the contradictory demands to assist in improving and expanding fisheries while at the same time devising methods for conserving the fisheries? He argues that science at St Andrews (at the very least) mirrored evolving concepts and methods in fisheries ecology, and in many cases contributed significantly to "basic concepts and methods in fisheries ecology" and to the "development and the application of these to north-west Atlantic species and populations." Stephenson also refers to three "periods" of the St Andrews Biological Station – the period before 1937, when the station was staffed primarily by academic volunteers, the period of the Fisheries Research Board of Canada (1937–73), and the more recent period in which the station has been within a government department. These periods are indicative of the change in political context for aquatic research in Canada that has taken place over the past century.

During the first period (pre-1937), the government provided a resource for academic "volunteers" who established their own research agendas, and trusted these volunteers to offer assistance to the government owing to the tacit understanding of the service ideal discussed above. Furthermore, the Canadian government had a minimal interest and involvement in science issues in general, except as they assisted with nation building.[19] The federal government was also a reluctant participant in fisheries matters: these held a low priority in comparison to agriculture, mining, and forestry.[20]

The second period, from 1937 to 1973, saw the St Andrews station administered through the Fisheries Research Board of Canada. Like its earlier incarnation, the Biological Board, the Fisheries Research Board was run by scientists and was at arm's length from the political system. This was a time of intensive growth: increasing governmental attention to fisheries development, and to fisheries (and aquatic) research in general, occurred during and after the Second World War. Scientific and other staff at the biological station grew in numbers, to match an expanding research agenda in the post-war period. This reflected the Canadian government's increased interest in, and funding of, all kinds of science and technology research in aid of promoting innovation and commercialization.[21]

The expansion of government science inevitably led to rising concerns about efficacy and accountability. By the early 1960s Canada and other nations belonging to the Organisation for Economic Cooperation and Development (OECD) countries became concerned about the uses to which science funding was being put, and began to undertake a long series of studies concerning science policy and strategies for research and development. In 1960 the Royal Commission on Government Organization was appointed (the Glassco commission), under J. Grant Glassco, a prominent Canadian businessman. In its 1962 report the commission noted that Canadian science was expanding without co-ordination or policy. During the decade that followed, government technocrats devoted considerable attention to the organization of government science. In 1964 a science secretariat was created; in 1966 the Science Council of Canada (1967–72) was established, and so too was the Senate Special Committee on Science Policy. The Ministry of State for Science and Technology was created in 1971 to coordinate Canada's science policy: of major concern was the fact that Canada's private sector trailed most other OECD countries in its contributions to research and development; most science and technology research in Canada was being done or funded by the government.[22]

Senior bureaucrats within the Department of Marine and Fisheries had since 1919 desired to have more control of the scientific operations of the Biological Board and the Fisheries Research Board, to bring it into line with departmental policy. This goal was realized as a result of the science policy debates of the 1960s and 1970s; the Fisheries Research Board was abolished, and fisheries research was absorbed into federal government departments. Intensified government control at the ministerial level was also occurring in Canadian government–supported science elsewhere, such as the environmental sciences, the Canadian Wildlife Service, and the Geological Survey of Canada (the latter coming under ministerial line management in 1972).[23] Since then, Canada's government labs and science-based agencies have been confronted by ongoing pressure to balance basic long-term research with governmental expectations of research and development under often stringent budgetary constraints.[24]

The third phase of the station's history, then, has been characterized by increased attention to political priorities and to governance as part of the government bureaucracy. Stephenson chronicles the particular issues of the biological station over this period, including the challenges of organization, attention to the relevance of science in an increasing climate of fiscal restraint, and the very real issue of the scope and amount of public science within government. Of particular importance is Stephenson's description and analysis of the changes and challenges that occurred when the Fisheries Research Board was ended by federal fiat, and biological station operations were brought into line with more bureaucratic and centralized management by the Department of Fisheries and Oceans. Here Stephenson draws on the reminiscences of former directors who shared a panel at the St Andrews "Workshop on the Evolution of Marine Science in Canada" and highlighted their accomplishments and, with painful honesty in one case, their failures.

These challenges, including such travesties as research teams split across two or more biological stations, and loss of scientists to other Department of Fisheries and Oceans institutions, occurred just as research challenges ballooned with the expanding fishing capacity of Atlantic Canada's fleets. At length, the opportunities created by local conditions, and recognition of the need for research into problems arising from pollution and overfishing, led to a focus on what became areas of excellence, such as aquaculture and environmental research. Stephenson's chapter includes a discussion of the role of public science, and the limitations of government science.

**The St Andrews Biological Station: Perspectives by Working Experts**

The remaining eight chapters in this collection, contributed by experts in the areas of work they describe, highlight the principal developments and most important participants in the history of their fields, albeit from a St Andrews perspective. Some research programs described here are localized contributions to international research on species of world-wide commercial importance; others deal with problems and programs specific to the St Andrews and Bay of Fundy locale. In both cases, St Andrews Biological Station researchers have augmented the frontiers of scientific knowledge, enriched international marine and aquatic science communities, and contributed to both fisheries and environmental management.

The first three of these chapters share a strong focus on the use of technology. The undersea environment, in particular, poses unique challenges for scientists because they must depend on technologies to reveal its features: salinometers, reversing thermometers, Nansen bottles, sonar, current metres, underwater cameras and submersibles equipped with cameras and sampling arms. Recently, historians of science have focussed attention on technology used by marine scientists and the often-ignored role of technologists who build, maintain, and operate this equipment. This was the theme of an important collection of articles in *The Machine in Neptune's Garden*, edited by Helen Rozwadowski and David K. Van Keuren,[25] and particularly articles by Michael S. Reidy, "Gauging Science and Technology in the Early Victorian Era," and by Christine Keiner, "Modeling Neptune's Garden: The Chesapeake Bay Hydraulic Model, 1965–1984."[26]

Tim Foulkes, in chapter 5, "Technology in Marine Science at the St Andrews Biological Station, 1908–2008," provides an unusual and important narrative by a technology expert who participated in building many of the scientific machines required for success in studying marine and aquatic environments. The first part deals with earlier technological challenges at St Andrews, but the meat of Foulkes's chapter is the biological station's workshops, created in 1925, and the projects in which he participated beginning in the 1960s. He describes the expertise of technologists and technicians, and one gets some sense of the flavour of the work atmosphere. The station's fishing-gear engineering research program from 1962 to 1987 focused on the mechanical performance of East Coast otter trawls, but had a number of significant marine technology spin-offs, particularly related to underwater cameras and remote vehicle systems needed by other programs for stock assessment and behaviour studies. For example, the early underwater camera technology developed for Gear Research was adapted by John Caddy for his scallop research, described in a later chapter in this volume. Foulkes lists many technologies developed, along with their applications, giving an idea of the range of contributions. His main story, however, is an in-depth description of the creation of towed underwater camera vehicles, pioneered at St Andrews. These devices compensated to some degree for the elimination of submersibles, used briefly, but deemed too expensive and risky, as Caddy mentions in his chapter. The increasingly sophisticated towed underwater benthic survey vehicles, BRUTIV I to III, were later adapted by engineers at the Bedford Institute of Oceanography (BIO) to the more recent TOWCAM technology, incorporating digital rather than film camera images, and the Grab Acoustic Positioning System (GAPS), a GPS-based benthic sampling system developed for scientists from St Andrews collaborating with engineers from the BIO. This account mirrors the emphasis of Stefan Helmreich's *Alien Ocean*, an exploration of how scientists and other marine research participants use technologies to navigate and investigate the underwater world, which emphasizes the degree to which our knowledge of this environment is technology-mediated.[27]

New technology is also featured in chapter 6, by Blythe D. Chang and Fred H. Page, "An Overview of Physical Oceanographic Research at the St Andrews Biological Station during Its First Century." This is not surprising, as oceanography depends entirely on a series of technologies to "see" the oceans, including tidal gauges, drift bottles, and bathythermographs and other more sophisticated instruments introduced from the late 1940s onward. Chang and Page relate the station's oceanography to earlier programs by William Bell Dawson, whose work is also described by Eric Mills in the first chapter. In his recent *The Fluid Envelope of our Planet*, Mills argued that the Canadian Fisheries Expedition of 1914–15, headed by the Norwegian fisheries biologist Johan Hjort, introduced Canadian scientists to the most advanced methods of dynamic oceanography recently developed in Scandinavia. But at that point Hjort and his assistant Johan Sandström were "evangelizing in the wilderness," as the Biological Board of Canada's scientists – and, he admits, also scientists in non-Scandinavian European countries as well – lacked the proper background and were not prepared for the highly mathematical science being developed in Norway.[28] Rather, scientists based at St Andrews, although limited in their ability to carry out thorough research programs by a lack of equipment, undertook studies of currents, salinity, and temperatures that mostly focused on relating hydrographic conditions to the fisheries. Chang and Page survey the entire history of the biological station's oceanographic programs, highlighting research that received particular emphasis in each period. Much of this research responded to specific problems: proposals to build tidal power dams on the Bay of Fundy; strategic anti-submarine research by the station's Atlantic Oceanographic Group during the Second World War; and more recent fisheries oceanography studies to incorporate environmental and oceanographic data into groundfish stock assessments.

Chang and Page also describe the individuals who developed this science, from early volunteer academics up to their colleagues' and their own personal contributions, from 1984 onward, towards revitalizing the oceanography program at St Andrews. Henry B. Hachey, Canada's first professional Canadian oceanographer – who, as Mills recounts in *The Fluid Envelope of the Planet*, was largely self-taught[29] – had to master dynamic oceanography to research the effects of a proposed tidal power dam on Passamaquoddy Bay and environs. He began using physical models to assist in understanding hydrographic events. Hachey was also the founding leader of the St Andrews–based Atlantic Oceanographic Group (AOG) in 1944, a trans-agency organization that included the Royal Canadian Navy and the National Research Council, which helped train a future generation of oceanographers. Many St Andrews programs responded to specific problems, but long-term research also occurred, including a record of surface water temperatures continuously since 1916. Such data, as Chang and Page point out, offer important insights into issues such as climate change.

In chapter 7, David J. Wildish and S.M.C. Robinson present the history of experimental flow studies at SABS. This chapter focuses on a series of laboratory experiments with flow simulators that augmented field research in two areas. The earliest flow studies here were conducted as part of the successful University of Toronto Ph.D. thesis of C.J. Kerswill, which involved field experiments with the commercially important molluscs of Prince Edward Island. Thus, one stream of research has involved organisms such as clams, oysters, scallops, and mussels that filter feed, catching suspended microorganisms and microscopic detritus as food. The other has concentrated on fish responses to the speed of water flow, with applications in understanding fish responses to pollutants. A strong physiological component of salmon studies began at the biological station when Dr Richard (Dick) L. Saunders was hired in the early 1960s. It was inevitable that he should incorporate flow studies in salmon biology. Because it is difficult to study current flow in nature, technologies, namely, flow simulators or flumes, had to be developed to enable scientists to regulate the speed of an artificial current and simultaneously monitor test animals' responses. Wildish and Robinson describe a series of such devices, each for specific kinds of study, including a flume designed by University of New Brunswick chemical engineering professor Dave Kristmanson, built by technicians at St Andrews, at the beginning of a flourishing and prolific collaboration with Wildish. They also highlight local circumstances which led to specific research programs, including studying the effects of local industrial effluents on filter feeders, and responses of fish at different stages of development to stream currents in clean and polluted water. This chapter in fact provides the historical context of hydrodynamics and boundary-layer studies because St Andrews scientists began to incorporate hydrodynamics into investigations of bottom-dwelling (benthic) organisms just as the field was developing, and made important contributions. Basic information from this research has allowed scientists to predict the effects of environmental conditions on fish and bivalve culture, to better manage and protect the marine environment.

The book's final five chapters each deal with important research programs at St Andrews. Both John Caddy's history of scallop research, which ties in with more mainstream fish population modelling, and Dick Peterson's history of salmon research at St Andrews include personal insights and experience. Jennifer Martin tells the story of a research program pioneered at St Andrews on paralytic shellfish poisoning that made it the world leader in investigating naturally generated biotoxins in the marine environment. Peter Wells describes ecotoxicology research, a field that emerged in the late 1960s at St Andrews and elsewhere in response to growing concern about the environmental effects of human pollutants such as pesticides and industrial runoff. Finally, former St Andrews Biological Station director Robert H. Cook, who oversaw its complex but ultimately successful aquaculture research program, provides a history of aquaculture research in St Andrews and surrounding regions.

John Caddy's account in chapter 8, "A Personal Perspective on the Historical Role of the St Andrews Biological Station in Investigations of Canadian Scallop Fisheries," provides not only a history of the station's scallop research, but a scientific assessment of ongoing management, and suggestions for future research and management measures to assist the scallop fishery's sustainability. Caddy has been in the forefront of scallop research and also collaborated with leading fisheries scientist John Gulland (1923–90), who introduced the widely used method of virtual population analysis. Caddy and Gulland's collaboration resulted in a paper that "started the move away from the assumption of equilibrium conditions of fish stocks to less predictable results of fisheries management,"[30] leading to a more realistic, less mechanistic model for understanding population fluctuations of fished populations. Caddy here presents a history of scallop research at St Andrews: Jack Stevenson's pioneering work in the 1930s; the later work of Carl Medcof; and the shift from the exploratory research phase to science that focused, from the 1960s onward, on conservation. Caddy discusses the changing theories of scallop population dynamics, refined over time as he and others realized that mobile and sedentary populations faced different challenges in reproduction and life histories, and were affected in different ways by harvesting. He pioneered, in this context, the first spatial model for a fishery resource in 1975.[31] The BRUTIV and TOWCAM underwater sledges with cameras and later video cameras, described by Tim Foulkes in chapter 5, were used to take images of scallops on the sea floor and monitor the effects of scallop dredges. They functioned as cheaper substitutes for submarines.

For a period in the late 1960s the station hired submersibles such as the Cubmarine and Perry Oceanographics Shelf Diver (also mentioned by Tim Foulkes in chapter 5); Caddy was able to use these vehicles in his scallop research (as did Derek Iles for herring research). He notes:

"The hire and transport of these machines from Florida and Vancouver cost a fortune – and only for 3–4 weeks a year. I hope however I have made it clear that routine assessments by dredge surveys do not provide much information on scallop life histories – other methodologies using cameras or biologists in person – are needed. (I did enjoy however going down to Florida in winter for the test and hiring studies using cubmarine/shelf diver!)."[32] This submarine research led to an episode that has gone down in SABS lore. It would be a shame to omit it from this history: it did not directly fit into any of the contributed chapters, and so I include it here.

The US research ship *Albatross IV* was used to launch the Canadian research submersible, *Pisces*, with Caddy as the scientist aboard filming the sea floor in the Georges Bank herring spawning grounds. The submersible microphone went dead during the dive, and the submersible's pilot, Jim, brought the vessel to the surface in a fog bank with the research vessel nowhere in sight. Caddy, a poet as well as a scientist, tells the story in "Interuterine Voyage on Georges Bank":

> Much later, as time manifested itself / from the drooping needles / on the air gauge, / Jim calls the surface, / but the microphone is dead; / we surface: no ship in sight.
>
> At this point / the sub's radio sputters, / announcing its dysfunction – / a corroded connection
>
> Jim leaves me / to go up and squat / all six feet of him in the splashing conning tower; the radio on his lap. / He closes the lid: the last I see or hear of him for four hours: / Did you ever try to pray / sick and cold / not knowing if anyone was there?
>
> By later accounts / Jim's boy scout days weren't wasted. / He gets through to the *Albatross*, / who have "seen" us to the north by radar.
>
> In the fog they pick up an old Polish / herring barrel (us, we are too low to "see" / and in the opposite direction).
>
> Jim tells them to sound the foghorn. / He hears them from the north / 13 seconds later. / He straddles the sound / with the hour hand of his watch / pointing at the sun showing / from the lighter haze to the southwest.
>
> He calls: / "Come south! Come south! / You're 10 miles north of us!" / To no avail: professional sailors / have no faith / in those adrift with instruments: / big boats find small ones, / not vice versa!
>
> They head north again / to locate another barrel
>
> "Come south Goddammit!"

> Some time later, they get the message, / and Jim talks the giant to the pick-up point, / at which point, / like I alone had done some time ago, / those on board realize / they had given up hope ...[33]

Because of the submarine pilot's competence, Caddy lived to participate in international negotiations to set the Canadian-US boundaries for scallop and other fisheries on Georges Bank in Washington, DC.[34] Caddy's chapter in this volume situates St Andrews Biological Station scallop research in the context of the scallop fisheries' shift from inshore to offshore locations, and the changing political context as offshore resources changed from being managed through the International Commission for the Northwest Atlantic Fisheries (ICNAF) to being managed by individual nations when their jurisdiction was extended to 200 nautical miles in 1977 under the Law of the Sea Convention, formalized in 1982. In this era the desire for control over offshore oil and other resources on the continental shelf also influenced fisheries resources diplomacy.

Dick Peterson's contribution (chapter 9, "Fifty Years of Atlantic Salmon Field Studies at St Andrews Biological Station") traces the birth and death of a research program at St Andrews, namely, research on wild Atlantic salmon. This program was inaugurated by A.G. Huntsman, the station's first director. Peterson comments:

> In Anderson House of the Huntsman Marine Science Centre there is a bust of Dr. Huntsman with a list of scientists inscribed below. About 30 years ago, during an acid rain meeting, I happened to be standing in front of the bust with Dr. Eville Gorham, a distinguished limnologist with ties to the Maritimes. During the conversation, I mentioned that I had never read the inscribed names. Dr. Gorham replied that I should genuflect before it whenever I passed. Perhaps a historical review could be considered a form of genuflexion.[35]

Peterson's reflections in chapter 9, however, are not particularly reverent, and bring into his account some vivid recollections of his early work at counting fences. Earlier, Huntsman had hired several scientists to conduct salmonid field and other research; their work on migrations, bird predation, and population fluctuations were synthesized by Huntsman into a theory on the causes of Atlantic salmon population fluctuations. During the expanding research program of the 1950s and early 1960s, scientists investigated methods to increase salmon production. Many salmon experts were hired in these years, as was Peterson himself. Peterson also describes the origins and development of research into the effects of pesticides and other pollutants (discussed in greater detail by Peter Wells in chapter 11) on salmon and salmon fry in the Miramichi River, once a legendary salmon fishing river. He situates these stories within the context of declining annual catches, the Maritime Atlantic salmon fisheries' changing nature, the demise of the Fisheries Research Board of Canada, and the Department of the Environment's[36] changing priorities that led to the unfortunate end of the salmon research program at St Andrews in the early 1970s. Continued research in a focused program by St Andrews scientists – rather than dispersed into other institutions with no coherent program – may have offered greater insights into the challenges faced by migrating wild populations which crashed irretrievably due to the brief but disastrously intense offshore salmon fishery near Greenland.

In contrast to Peterson's account of an investigation that was prematurely terminated, Jennifer Martin in chapter 10, "Paralytic Shellfish Poisoning (PSP) Research: Seventy Years in Retrospect," describes an ongoing research program that over time has only gained in recognition and importance. Scientists at the St Andrews Biological Station were the first to recognize and study this phenomenon, caused by a microscopic phytoplankter, present in coastal regions around the world, which appears to present its greatest toxicity in colder waters. This research, centred on the organism *Gonyaulax tamarensis* (now called *Alexandrium fundyense*), was begun in the late 1930s when Carl Medcof became interested in outbreaks of toxin accumulation in clams, mussels, and other bivalves. If eaten, the infected bivalves cause paralysis and, in cases of severe poisoning, death in humans (there is no antidote). Alfreda Berkeley Needler in the 1940s collaborated with Medcof in researching "red tides" (algal blooms of *Alexandrium*), and linked both this phenomenon and PSP with the microorganism responsible for them. Since then, scientists such as D.G. Wilder at SABS have developed methods for testing toxicity in shellfish. Others have investigated the incidence and concentrations of PSP in scallops and other bivalves and filter-feeders and studied how quickly infected shellfish can eliminate the toxins in a clean environment. A research program begun by Alan White in the 1970s, joined by Jennifer Martin at the start of her St Andrews career, showed that this microorganism causes mortality in fish like herring and cod, and even humpback whales. During the 1980s, as the station's aquaculture science became increasingly important – as described by Robert Cook in the final chapter – scientists began monitoring phytoplankton for *Alexandrium* concentrations to assist Bay of Fundy aquaculture programs, and to uncover what conditions favour high concentrations of the organism. As the incidence and impact of this food-borne microorganism have increased in colder waters around the world, scientists around North America began to collaborate with the St Andrews scientists. Martin also describes how research, monitoring, and testing work done at St Andrews has expanded beyond PSP to include the occurrence other shellfish-related toxins that can cause lethal poisonings.

Bio-and ecotoxins are also covered in Peter Wells's chapter, "A History of Research in Environmental Science and Ecotoxicology at the St Andrews Biological Station." Wells, an ecotoxicologist, provides one of the first histories of an ecotoxicology research program. As a first-generation practitioner, he experienced first-hand the parallel development of this science in the United States and Canada. As the environmental problems exposed by Rachel Carson's influential book *Silent Spring* became widely recognized, analytical scientists developed increasingly sensitive methods for detecting tiny concentrations of chemicals in the environment. Wells, in chapter 11, tells how ecotoxicology at St Andrews began in response to DDT spraying of New Brunswick forests to control the spruce budworm. DDT had a strong impact on aquatic insects and young salmon in the Miramichi and other salmon rivers, first investigated by biological station scientists in 1951. The studies of DDT's effects by scientists based at St Andrews strongly influenced Rachel Carson, who lived close by on Southport Island in Maine, and worked for the US Bureau of Fisheries (after the war the Fish and Wildlife Service). Numerous salmon researchers also became concerned about zinc and copper runoff from mining pollution and undertook various studies to document the effects on salmon. By the early 1960s John B. Sprague had started studies exclusively dedicated to aquatic environmental toxicology at St Andrews, and had joined new international committees such as the United Nations' expert panel on marine pollution (GESAMP). Environmental science and ecotoxicology programs drew a diverse group of experts from fish behaviourists to marine ecologists. The credibility and international esteem of the ecotoxicology program at St Andrews was largely due to the leadership of Vlado Zitko, who joined the station in 1968. A gifted analytical chemist, he was involved in most of the investigations at St Andrews dealing with environmental problems related to organic and other pollutants and natural toxins. These investigations originated in response to events demanding attention, such as oil spills, or new environmental concerns such as the effects of PCBs and methyl mercury, which bioaccumulate in the aquatic environment. Scientists at St Andrews also studied the effects of aerial pesticide spraying on lobsters, salmon, and other finfish; metal pollutants' effects on lobsters; and how pulp-mill effluents and dredging affect salmon and benthic species. They developed cost-effective monitoring systems and guidelines for industrial effluents to protect nearby ecosystems, and methods for biomonitoring a wide range of pollutants. Aquaculture wastes have come in for especially intensive study. Wells highlights the importance of international collaboration and outreach, and his chapter shows how scientists at St Andrews exemplified concern with real issues, rather than scientific abstractions, although a great deal of fundamental science was pursued in search of answers to the problems they were encountering: thus, they continued to embody the service ideal in science.

The final chapter, "Aquaculture Research and Development at the St Andrews Biological Station, 1908–2008," is by former St Andrews Biological Station director Robert Cook (1977–92), the first post–Fisheries Research Board director, whose term was second in length only to Huntsman's. Cook describes the slow emergence of aquaculture science, from the first tenuous inquiries into the practices and failings of fish hatcheries and oyster culture in the early twentieth century to its recent position as a dominant research program at the station. He chronicles early, historically important oyster culture and lobster hatchery research programs, then comes to his main focus: salmon culture's difficult development and its continuing challenges. Cook was the station's director when salmon culture investigations began to be pursued seriously, at a time when research management of the marine fish program was now headquartered at the Bedford Institute of Oceanography, and St Andrews Biological Station's mandate had become environmental research (including toxicology, acid rain, organic pollutants, etc.), aquaculture research, and fish health protection. The salmon aquaculture focus was of course a response to the lamentable collapse of natural Atlantic salmon fisheries after the short-sighted fishery free-for-all on the newly discovered Atlantic salmon feeding grounds south of Greenland.

Because Cook brought together a series of outside agencies to work on this project, such as the Atlantic Salmon Federation, and the Huntsman Marine Science Centre, he offers an authoritative account of the origins and success of the Maritime aquaculture industry. Early aquaculture was considered only an academic pursuit in Canada: in the 1970s government bureaucrats believed Canadian aquaculture (other than for oysters) would never generate much revenue. Fiscal support remained scanty. Historians of science will appreciate Cook's account of early failures to overwinter penned Atlantic salmon in the Bay of Fundy – despite following the best practices of pioneering Scandinavian scientists – until SABS biologists were able to travel to Norway to get first-hand experience.

Cook writes of political and practical alliances formed to advance Bay of Fundy aquaculture: for example, John Anderson, station director from 1965 to 1971, became the Aquaculture Association of Canada's first president in 1984. Scientists at the Huntsman Marine Science Centre helped to train industry personnel, an obvious necessity if new techniques were to succeed. Cook helped found the Salmonid Demonstration and Development Farm to teach aquaculturists the best practices and to supply them with advice. Cook also recognized the importance of marketing both aquaculture and its products, and describes these efforts. Indeed, many bodies had to be formed to represent the new interests that emerged. The Southern New Brunswick Aquaculture Development Committee organized and set protocols to respond to or preferably avoid aquaculture pitfalls (such as transmitting sea lice and infectious disease due to poor aquaculture practices). Cook himself, trained in agricultural science, introduced "crop rotation"-like practices to salmon aquaculture, to "cleanse" salmon pens between "crops."

While critics might argue that Dr Cook does not dwell on the environmental problems surrounding aquaculture, he does acknowledge them, and his history shows that St Andrews Biological Station scientists have actively researched both initial problems and any shortcomings in remediation measures. Dr Fred Page, for example, developed comprehensive research programs to monitor sea lice and the effects of chemical measures used to counteract them. Efforts to improve aquaculture practice reflect dual concerns for profitability and environmental conservation. One impressive advance was the introduction in 2001 of Multi-Trophic Aquaculture by Dr Shawn Robinson, which attempts to recreate an entire ecosystem near a salmon cage, including seaweeds, mussels, and other species within the cages so that the species' life habits can interact beneficially: filter-feeders such as mussels, for example, can filter salmon debris and thus help clear the water. Research on this practice is ongoing. Cook's descriptions of the many individuals and industry-government partnerships involved in developing Maritime aquaculture projects, provides an excellent history of an endeavour that has not received enough dispassionate attention.

The twelve chapters in *A Century of Maritime Science* cover only a sample of the important programs and contributions made by scientists at the St Andrews Biological Station in its first one hundred years. Not every contributor to St Andrews's "Workshop on the Evolution of Marine Science in Canada" wished to write up their talks, and there were several lacunae in describing the station's research programs. Contributors who generously gave of their time have, however, provided a range of vital articles that indicate how a location can "make" science by influencing research choices, and how the different research programs within an institution intersect as experts' interests converge and diverge. These chapters illuminate not only the importance of a location, but also the importance of local science, which may serve to advance general theory and understanding even as it addresses local problems and issues. Very often, to echo a point made many years ago by Edward Layton,[37] it is the science that directly addresses direct issues such as pollution problems, sustainability, and health concerns that produces applicable results. The science that addresses such issues, arguably, is of equal importance to much of the universal, theoretical science beloved of science historians, but it is often difficult for interested lay readers to see through the turgid titles to understand how each publication fits into a grander picture. This volume will not only assist lay understanding, but will also provide current and future fisheries and aquatic and environmental scientists with an understanding of how their various disciplines were shaped by historical circumstances and how they evolved into now-familiar programs.

NOTES

1 See, for example, Bruce Doern and Jeff Kinder, *Strategic Science in the Public Interest: Canada's Government Laboratories and Science-Based Agencies* (Toronto: University of Toronto Press, 2007); Kenneth Johnstone, *The Aquatic Explorers: A History of the Fisheries Research Board of Canada* (Toronto, University of Toronto Press, 1977).

2 Tim D. Smith, *Scaling Fisheries: The Science of Measuring the Effects of Fishing, 1855–1955* (New York: Cambridge University Press, 1994); Eric Mills, *Biological Oceanography: An Early History, 1870–1960* (London: Cornell

University Press, 1989); Eric Mills, *The Fluid Envelope of Our Planet: How the Study of Ocean Currents Became a Science* (Toronto: University of Toronto Press, 2009); Helen Rozwadowski, *The Sea Knows No Boundaries: A Century of Marine Science under ICES* (London: ICES and University of Washington Press, 2002); Jennifer Hubbard, *A Science on the Scales: The Rise of Canadian Atlantic Fisheries Biology, 1898–1939* (Toronto: University of Toronto Press, 2006); and Carmel Finley, *All the Fish in the Sea: Fish, Fisheries Science, and Foreign Policy, 1930–1960* (Chicago: University of Chicago Press, 2011).

3 Joseph Gough, *Managing Canada's Fisheries: From Early Days to the Year 2000* (Sillery,Quebec: Septentrion and Fisheries and Oceans Canada, 2006); and George A. Rose, *Cod: The Ecological History of the North Atlantic Fisheries* (St Johns, NL: Breakwater Press, 2007).

4 Hubbard, *A Science on the Scales*, 239.

5 Ibid., 10.

6 The name of the station changed in 1949, when Newfoundland joined Confederation and the St Andrews station lost its status as the only Canadian marine biological station on the Canadian Atlantic coast. Both names are used in this volume, in accordance with how the station was named in the period under discussion; in places the acronym SABS is also used.

7 Joseph Gascoigne, *Joseph Banks and the English Enlightenment: Useful Knowledge and Polite Culture* (Cambridge: Cambridge University Press, 1994), 185–6.

8 Ibid., 8–9.

9 Patrick O'Brien, *Joseph Banks: A Life* (Chicago: University of Chicago Press, 1987), 47.

10 His collaborator and friend, Thomas Pennant, formally published these together with Banks's descriptions made during a later voyage to Iceland, in *Arctic Zoology* (1784). See Gascoigne, *Joseph Banks*, 91.

11 Ibid., 14.

12 Joseph Gascoigne, *Science in the Service of Empire: Joseph Banks, the British State and the Uses of Science in the Age of Revolutions* (Cambridge: Cambridge University Press, 1998), 11–15.

13 Gascoigne, *Joseph Banks*, 188–96.

14 Letter from H.B. Hachey to A.G. Huntsman, 2 January 1952. University of Toronto Archives, Huntsman Collection, B1979–0048, file 001.

15 Kathleen Broome Williams, "From Civilian Planktonologist to Navy Oceanographer: Mary Sears in World War II," in *The Machine in Neptune's Garden: Historical Perspectives on Technology and the Marine Environment*, ed. Helen M. Rozwadowski and David K. van Keuren (Sagamore Beach, MA: Science History Publications, 2004), 243–72.

16 M.G. Ainley, ed., *Despite the Odds: Essays on Canadian Women and Science* (Montreal: Véhicule Press, 1990); M.G. Ainley, "Last in the Field? Canadian Women Natural Scientists, 1815–1965," in *Despite the Odds: Essays on Canadian Women and Science*, ed. M.G. Ainley (Montreal: Véhicule Press, 1990), 25–62; M.G. Ainley, "Gendered Careers: Women Science Educators at Anglo-Canadian Universities, 1920–1980," in *Historical Identities: The Professoriate in Canada*, ed. P. Stortz and E.L. Panayotidis (Toronto: University of Toronto Press, 2005), 248–70; M.G. Ainley, "A Family of Women Scientists," *Le Bulletin Newsletter* (Simone de Beauvoir Institute, Concordia University) 7:1 (1986): 5–11; M. Gillett, *We Walked Very Warily: A History of Women at McGill* (Montreal: Eden Press Women's Publications, 1981); M.R.S. Creese, *Ladies in the Laboratory? American and British Women in Science, 1800–1900* (London: Scarecrow Press, 1998; J. Fingard, "Gender and Inequality at Dalhousie: Faculty Women before 1950," *Dalhousie Review*, Winter 1984–5: 687–703; B. Lightman, "'The Voices of Nature': Popularizing Victorian Science," in *Victorian Science in Context*, ed. B. Lightman (Chicago: University of Chicago Press, 1997), 187–211; M.W. Rossiter, "'Women's Work' in Science, 1880–1910," in *History of Women in the Sciences: Readings from Isis*, ed. S.G. Kohlstedt (Chicago: University of Chicago Press, 1982), 287–304; A.B. Shteir, "Elegant Recreations? Configuring Science Writing for Women," in *Victorian Science in Context*, ed. B. Lightman (Chicago: University of Chicago Press, 1997), 236–55.

17 For information on Huntman, the reader is referred to *A Science on the Scales*, and to a film by Rod Langley, *Huntsman: The Fisherman's Friend* (2007), available online at http://www.science.gc.ca/default.asp?lang=en&n=84E110F4-1.

18 Samuel P. Hays, *Conservation and the Gospel of Efficiency: The Progressive Conservation Movement 1890–1920* (Boston: Harvard University Press, 1959).

19 Doern and Kinder, *Strategic Science in the Public Interest*, 4.

20 See Hubbard, *A Science on the Scales*, chap. 5, 120–48; and Jennifer Hubbard "Changing Regimes: Governments, Scientists and Fishermen and the Construction of Fisheries Policies in the North Atlantic 1850–2010," in *A History of the North Atlantic Fisheries*, vol. 2, *From the Mid-Nineteenth Century to the Present*, ed. David J. Starkey, Jon Th. Thór, and Ingo Heidbrink (Bremerhaven: Deutsches Schiffahrtsmuseum, 2012), 129–76.

21 Doern and Kinder, *Strategic Science in the Public Interest*, 4.

22 Ibid., 188.

23 Brian Wilks, *Browsing Science Research at the Federal Level in Canada* (Toronto: University of Toronto Press, 2004), 200, 315; and Morris Zaslow,

*Reading the Rocks: The Story of the Geological Survey of Canada* (Toronto: Macmillan, 1975), 411.

24 Doern and Kinder discuss some of these issues in *Strategic Science in the Public Interest*.

25 Helen Rozwadowski and David K. Van Keuren, eds., *The Machine in Neptune's Garden: Historical Perspectives on Technology and the Marine Environment* (Sagamore Beach, MA: Science History Publications USA, 2004).

26 Michael S. Reidy, "Gauging Science and Technology in the Early Victorian Era"; Christine Keiner, "Modeling Neptune's Garden: The Chesapeake Bay Hydraulic Model, 1965–1984," in *The Machine in Neptune's Garden*, ed. Rozwadowski and Van Keuren, 1–38, 273–314.

27 Stefan Helmreich, *Alien Ocean: Anthropological Voyages in Microbial Seas* (Berkeley: University of California Press, 2009).

28 Mills, *The Fluid Envelope of Our Planet*, 111–36.

29 Ibid., 239–44.

30 John Caddy, personal note. See J.F. Caddy and J.A. Gulland, "Historical Patterns of Fish Stocks," *Marine Policy* 7:4 (1983): 267–78.

31 L.E. Gales and J.F.Caddy, "YRAREA, a Program to Demonstrate Effects of Exploitation on a Contagiously Distributed Shellfish Population," Technical report no. 582 (Canada: Fisheries and Marine Service, Research and Development Directorate), 1–79; J.-C. Seijo, J.F. Caddy, and J. Euan, "SPATIAL: Space-time Dynamics in Marine Fisheries: A Bioeconomic Software Package for Sedentary Species," FAO Computerized Information Series (Fisheries), 6 (1994): 1–116 (plus discs).

32 John Caddy, personal communication, 25 January 2011.

33 In Stanley Lowther, *The Relativity of Journeys* (Victoria: Trafford Publishing, 2001), 24–5. Note that the volume author is John Caddy: but it is dedicated to his grandfather, whose name is indicated as the author. The pilot was an employee of the company that leased the submersible to the St Andrews Biological Station, and his last name cannot at this time be supplied.

34 Scallop fishing boundary negotiations between the United States and Canada were the subject of another poem which reflects a saltier era in science. While the first meeting was held in a hotel opposite the Watergate, in 1974 ("here we bribed the doorman for ashtrays to commemorate the recent scandal"), a later negotiation meeting had a stranger venue: "We meet this time / Straight from the airport / at a small restaurant / which covers as a mid-day / strip-club: sandwiches on maps / with swinging melons and beer, / hot discussions on national interests / took place in enemy territory; high heels / straddling the zone of contention." John Caddy, "Watergate's Washington, October 1974," in Lowther, *The Relativity*

*of Journeys*, 22. The Hague Line was finally established as the US-Canadian boundary in 1984.

35 Richard Peterson, earlier version of his introduction to his chapter in this book; these also reflect comments he made in his presentation at the 2008 "Workshop."

36 The department overseeing the federal fisheries from 1969 to 1971 was the Department of Fisheries and Forestry; in the years 1971–6 the fisheries were subsumed under the Department of the Environment, renamed the Department of Fisheries and the Environment in 1976. From 1979 the fisheries came under the purview of the Department of Fisheries and Oceans.

37 See Edwin Layton, "Mirror Image Twins: The Communities of Science and Technology in 19th Century America," *Technology and Culture* 12: 4 (1971): 562–80.

# 1 Science in Canada: The Context of the Biological Board of Canada's St Andrews Biological Station

ERIC L. MILLS

The Atlantic Biological Station at St Andrews was one of two marine biological stations founded in 1908 in Canada. Its origins are found in the scientific and political environments of post-Confederation Canada, rather than in earlier British exploratory scientific endeavours: the scientific investigations of the Arctic in the early nineteenth century; the magnetic observatory established in Toronto in 1839; and the influential Geological Survey of Canada that originated in 1842. Its origins were also separate from the ubiquitous amateur science and scientific societies throughout nineteenth-century Canada. Although hydrographic and tidal surveys were established in the 1880s, Canadian marine science in the late nineteenth century was rudimentary. Canadian universities were small, scientific research in them or elsewhere was rare, and governments were interested more in scientific inventory than in broader aspects of science. It was not until 1898 that the need for information on fisheries, increasing self-confidence by academics interested in research, and concern about the intellectual influence of the United States led to the first government support establishing a floating marine laboratory, and then in 1908 two land-based laboratories under the direction of what became the Biological Board of Canada. The early history of the St Andrews Biological Station seems cut and dried at first glance. In short, the story begins with the appointment in 1892 of Edward E. Prince, English in origin, although recruited from Scotland, as Dominion Commissioner of Fisheries. Then followed lobbying of the Laurier government for support of a biological station by Prince and others. In 1898, a board of management of a biological station for Canada was established (renamed the Biological Board of Canada in 1912), followed by more lobbying, this time for money to support a biological station. When money was granted, the first biological station was a floating one that was moved from place to place and project to project from 1899 until 1907, when it was badly damaged. This was the precipitating event that led to the establishment in 1908 of two fixed and permanent Canadian marine biological stations, one at Nanaimo, British Columbia, the other at St Andrews, New Brunswick. But marine science in Canada, like Canadian science itself, did not begin with these circumstances, which were the outcome of a particularly Canadian and very loosely linked chain of events going back into the mid decades of the nineteenth century.

## Science in Canada in the Mid and Late Nineteenth Century

There were no organized marine sciences in pre-Confederation Canada, which until 1867 was made up of small colonies dependent on agriculture and forestry in the hands of immigrants mainly from Britain and, late in the eighteenth century, from the newly independent United States. What science existed was the avocation of the small group of relatively leisured upper-middle-class people, including merchants and soldiers.[1] There were also scientific efforts in the north,[2] mainly (but not entirely)[3] by Royal Navy personnel searching for a Northwest Passage, especially in the first four decades of the nineteenth century. These began with John Ross's exploration of Baffin Bay in 1818, followed by W.E. Parry's voyages into the eastern and central Arctic in 1819–1825, and another expedition by Ross (with his nephew James Clark Ross) in 1829–33. Scientific observations in the north were part of the program of the ill-fated Franklin Expedition of 1845, and had its naturalist Harry Goodsir survived, knowledge of Arctic marine animals would certainly have increased earlier than it did. The many expeditions in search of Franklin did their bit, but in a scattered way that was secondary to their aim to discover the fate of Franklin and his men.[4]

Although there was an increase in interest in science in the early decades of the nineteenth century due to the continued increase in the numbers of United Empire Loyalists, wealthy and well-educated, bringing an interest in good schooling to Canada (i.e., Ontario and Quebec), Nova Scotia, and New Brunswick, science in Canada up to the 1850s was done mainly in an ad hoc way by amateurs and military men. The military in particular, mainly from the Royal Engineers and the Royal Artillery, played an important part in increasing knowledge of natural history, geology, hydrography, and astronomy.[5] And it was their interest in contributing to magnetic studies of the Earth in efforts such as the "Magnetic Crusade"[6] that led to the first permanent scientific establishment in the Canadas.

The Toronto Magnetic Observatory was founded in 1839 as part of a worldwide magnetic observing network and was staffed by Royal Artillery officers.[7] Of these, the best known and most influential was J.H. Lefroy, who was in Canada between 1842 and 1853, and whose most ambitious project (although virtually unsung) was a search for the north magnetic pole that took him from Toronto as far as the Mackenzie River and Fort Simpson between May 1843 and November 1844.[8] By mid-century interest in magnetics was waning and that in meteorology was increasing; the Magnetic Observatory, linked to North American telegraph networks, became the nucleus of the Canadian Meteorological Service.[9] Although it played no role in marine research, the Magnetic Observatory did provide an early example of the professional, as opposed to the amateur, pursuit of science.

A good case can be made that the first truly Canadian professional and pre-eminent scientific organization was the Geological Survey of Canada (GSC), established under William E. Logan in 1842 to search for economically valuable minerals (notably coal) in the Province of Canada (now Ontario and Quebec), then expanding in time to the Maritimes, and after Confederation from coast to coast.[10] The GSC was very small at first and based initially in Montreal (it moved to Ottawa in 1881) under the influence of the Montreal Natural History Society. Its extensive surveys extended first from Lake of the Woods to southern Labrador, then shortly after Confederation to Nova Scotia and New Brunswick. It made a major move westward when Manitoba and British Columbia joined Confederation in 1870 and 1871. Throughout, most of the scientific officers of the GSC made general collections in addition to geological ones, including in natural history and ethnology, providing a major scientific survey of the expanding country that included marine biology. As one historian has commented, "Early geologists were also explorers, geographers, botanists, zoologists and anthropologists who played a large role in opening up the West and later the Arctic."[11] One example of this, indicating the broad compass of GSC scientists, is George M. Dawson's (figure 1.1) investigation of the Queen Charlotte Islands in 1878, which included a thorough geological survey and mapping, along with detailed information on the Haida Nation, including their language (by Dawson himself), and lists of the invertebrates and plants collected under the names of a number of specialists.[12]

1.1 George M. Dawson, about 1885

In fact, the GSC had become the Geological and Natural History Survey in 1877, and its collections, including those of Dawson and of the first botanist of the survey, John Macoun (appointed in 1882),[13] formed the basis of the Victoria Memorial Museum, which opened in Ottawa in 1910 and is the ancestor of the modern Canadian Museum of Nature.[14] Macoun, who saw no limits to himself in natural history, encouraged collecting of all kinds, including marine collections on the west coast of Vancouver Island, where he joined his collectors William Spreadborough and C.H. Young at Ucluelet in 1909 (figure 1.2).

But Dawson and Macoun, Spreadborough and Young, were the exception, not the rule until well after 1867. One example of how science was done is the early work of J.F. Whiteaves, a young Englishman who came to Montreal in 1862 as an employee of the Natural History Society of Montreal[15] and who dredged in the Gulf of St Lawrence on his own time from 1863 to 1875, before joining the GSC as a zoologist (later a distinguished paleontologist).[16] His catalogue of the marine invertebrates of eastern Canada,[17] summarizing much early investigation along with his own, is an important milestone in Canadian marine science, showing the benefits of both amateur and professional science in the newly united country.

## Societies, Universities, and the Teaching of Science

Whiteaves's career highlights the fact that in mid- to late-nineteenth-century Canada, most natural history and scientific work in general was done by amateurs, not by professionals such as those in the GSC. Local societies abounded, including the Montreal Natural History Society (founded 1827), the Canadian Institute (1849, later the Royal Canadian Institute), the Hamilton Association for the Cultivation of Literature, Science and Art (1857), the Botanical Society of Canada (1860), the Nova Scotian Institute of Science (1862), the Entomological Society of Canada (1863), the New Brunswick Natural History Society (1863), and the Ottawa Field-Naturalists' Club (1872). Most published their own journals, some of which survive to this day, and the societies and their journals were the main means of passing information around in Canadian scientific circles until the end of the nineteenth century, when amateur science gave way to government agencies and the universities.[18]

The main national learned society, the Royal Society of Canada (RSC), was established in 1882,[19] based not on the purely scientific Royal Society of London but instead on the broader Royal Irish Academy,

1.2 John Macoun (1831–1920), standing centre, with C.H. Young and William Spreadborough sorting marine specimens

taking in all the branches of scholarly knowledge.[20] As the brain-child of the governor general from 1878–83, John Campbell, the Marquess of Lorne, it was intended to counter influence from the United States and stimulate Canadian learning. Ironically, Canadian scientists such as J.W. Dawson and Daniel Wilson were sceptical initially that the RSC had a role and could survive. But survive it did, although never becoming an important patron and accrediting body like the Royal Society of London, probably because it was seen as non-utilitarian or even anti-utilitarian by governments, especially those of John A. Macdonald (1867–73, 1878–91). Nonetheless, the RSC did lobby government for tidal surveys, fisheries research (especially the foundation of a marine station – see later), a dominion observatory, and much later for federal government support of science, leading after the First World War to the foundation of the National Research Council of Canada.[21]

What then of the universities and other institutions of higher learning in Canada during the nineteenth century? The first "university" (little more than a high school in modern terms) was King's College, an Anglican foundation for the children of the establishment, beginning in Windsor, Nova Scotia, in 1789. Thereafter followed Saint Mary's in Halifax (1802, chartered 1852, Roman Catholic); Dalhousie in Halifax (1818, nominally non-denominational, in fact Presbyterian) (figure 1.3); McGill in Montreal (1821, Presbyterian); King's College, Toronto (1827, Anglican; it formed the basis of the secular University of Toronto in 1850); King's College, Fredericton (1828, Anglican); Acadia in Wolfville, Nova Scotia (as Queen's College 1838, Baptist); Mount Allison in Sackville, New Brunswick (1839, Methodist); Victoria College in Coburg, Ontario (later in Toronto) (1841, Methodist); Queen's in Kingston, Ontario (1841, Presbyterian); Bishop's in Lennoxville, Quebec (1843, Anglican); Laval in Quebec City (1852, progeny of the Séminaire de Québec, Roman Catholic); and St Francis Xavier in Antigonish, Nova Scotia (1866, Roman Catholic).[22]

By Confederation in 1867, there were seventeen degree-granting colleges or universities in Canada, most of them with 100 or fewer students.[23] The emphasis, especially in the Anglican institutions, was on the classics or on a broad, general education, but where Scottish influence was strong, some science was taught as part of this, including astronomy, mathematics, physics (as natural philosophy initially), chemistry, botany, zoology, palaeontology, geology, and geography.[24] There were no permanent laboratories for teaching or research. As Jarrell summarizes the situation in nineteenth-century Canada:

1.3 Dalhousie College in 1875

> Science teaching in Canadian universities was a mixture of various elements, with the Scottish and American predominating. From the founding until after the turn of the 20th century, most of the small liberal arts colleges offered general science as part of a general education. For most of the [19th] century, 2 reasons for teaching science were commonly given: science aided the student in learning to think logically, and it exhibited to the student the wonders of God's creation. Little thought was given to preparing future scientists, and those Canadians who became professionals had either to resort to schools overseas (usually German or American) or to virtually train themselves with the help of sympathetic professors ... Typically, 2 professors, one for natural history and geology, the other for physics, chemistry and perhaps astronomy, covered the whole range of science. Laboratory practice was unknown until the present [20th] century.[25]

Under these circumstances, it is not surprising that the marine sciences, like other branches of science, had a tenuous existence in nineteenth-century Canada at least until graduate training became a routine part of the larger university curriculum around 1900.[26]

Even when advanced training and research began in Canadian universities very late in the nineteenth century, the direction of scientific activity was constrained by its distinctive meaning. Governments and the populace at large regarded science as information – for example, on agricultural output, on rainfall, on pest insects, on public health, and the like. Fact-gathering sciences were favoured, especially beginning in the 1880s, notably geology, botany, entomology, and with an emphasis on statistics (e.g., mining information). There was often political pressure (on the GSC, for example) for exactly this, that is, for what has been called "inventory science,"[27] the cataloguing of resources. It was in this utilitarian context that the marine sciences began to expand at the end of the nineteenth century.

## The Beginnings of Canadian Marine Science

It is instructive to compare the development of scientific institutions in the United States with those in Canada. A US Coast Survey was established in 1807. The US Fish Commission dated from 1871. In 1888 the Marine Biological Laboratory opened in Woods Hole, Massachusetts. American scientists united in the American Association for the Advancement of Science in 1848, and established the prestigious National Academy of Sciences in 1863. Advanced degrees on the German plan, like the Ph.D., were being offered by East Coast universities (beginning with graduate degrees at Johns Hopkins in 1876) during the 1880s. By comparison, up to the 1880s Canada could boast only its Geological Survey and Royal Society.

Confederation, coming nearly a century after the independence of the United States, made a difference in this regard, even though Canada had only a tenth of the population of the United States, had achieved political independence much later, was made up of a huge and nearly ungovernable land area, and was dependent on Britain for a good deal of its scientific and technical horsepower. But an even greater influence than Confederation on the spread of learning was the completion of the Canadian Pacific Railroad from central Canada to Vancouver in 1885 (it was preceded by the completion of the Intercolonial Railway in the Maritimes and eastern Quebec, 1872–6). The railway network expanded inexorably, and in 1903 the railroad reached St Andrews, New Brunswick, making that summer destination easy to get to from the inland cities (a factor in the account that follows). Access to both coasts was now straightforward, opening up the continental interior and the coasts to commerce and scientific investigation.

The railroad links were attended by union of most of the north with Canada (the North-West Territories in 1870 and the Arctic islands in 1880), and the entry of Alberta and Saskatchewan into Confederation in 1905, giving Canada pretty largely the geographical form it has now. But even more important was the increase in Canadian population, for example from 5,371,315 to slightly over 7,200,000 between the censuses of 1901 and 1911 and to over 8,000,000 soon after, as well as the nature of its employment. At first, agriculture attracted most of the immigrants, but by the beginning of the First World War they were arriving to take up manufacturing and other urban occupations, centred on the rapidly growing cities of Montreal and Toronto.[28]

In this dynamic context, very rapid changes in Canadian science, including the marine sciences, began in the 1890s and increased after 1900, coinciding with Wilfrid Laurier's Liberal government from 1896 to 1911. They were based on the development of a Canadian scientific profession in government and later the universities in the 1880s and after (in which the GSC played a pioneering role), beginning in government surveys and later university teaching. But there had been earlier needs to investigate the waterways of Canada.

In the late eighteenth and into the nineteenth centuries, most of the charting (hydrography) of Canadian waters had been done by British surveyors, one notable example being the surveys conducted by H.W. Bayfield from Lake Superior to Newfoundland between 1816 and 1856.[29] Inland, much valuable hydrographic surveying during the early nineteenth century was carried out by surveyors of the Board of Works (established in 1841) in preparation for canal building. After Confederation it became necessary for Canada to do more surveying and hydrography, and to this end the Department of Marine and Fisheries asked the British Admiralty to help establish a Canadian hydrographic survey. Commander J.G. Boulton arrived in Canada in 1883 as head of the Georgian Bay Survey – the direct result of the steamship *Asia*'s sinking with loss of life during a fall gale in September 1882.[30] Most of the early surveys were confined to the Great Lakes, but gradually work on the coasts increased.[31] The first Canadian survey of the sea was of Burrard Inlet by W.J. Stewart, an engineering graduate of the Royal Military College. Surveys of the important shipping lanes in the lower St Lawrence River followed in 1905; this and later surveys were more in Canadian hands because British ships were being withdrawn from Canada to meet the increasing naval threat from Germany perceived in Britain.[32]

The Hydrographic Survey of Canada (renamed the Canadian Hydrographic Service in 1928) evolved from the Georgian Bay Survey after 1904–5, first devoting itself to inshore charting with small vessels, although the Royal Navy still did major work on the West Coast until 1910 and on the East Coast until 1913.[33] Canadian hydrography was put on a new basis and given modern equipment when the CSS *Acadia* was commissioned for the Hydrographic Service in 1913, beginning its work in Hudson Bay that very year and playing an important part in the Canadian Fisheries Expedition of 1915.[34]

Hydrography in Canada has been largely independent of other branches of marine research since its inception – as has the study of tides, which, counter-intuitively, began independently of hydrography. Despite the importance of tides to navigation, and the entreaties of merchants bringing cargoes through the shoals of the tidal St Lawrence, John A. Macdonald's governments resisted many requests to begin systematic tidal surveys.[35] By 1884, the only major systematic tidal information from Canadian waters (apart, possibly, from naval dockyards) came from Cumberland Sound, Baffin Island, during the first International Polar Year, 1882–3).[36] That year the British Association for the Advancement of Science met in Montreal and formed two committees to deal with tides. The first, including the great geophysicist G.H.

Darwin and the director of the Toronto Observatory, Charles Carpmael, was concerned with reducing tidal observations and producing tide tables worldwide, concerns central to nineteenth-century tidal science.[37] The second, under Alexander H. Johnson, professor of astronomy at McGill, was set up to ask for government funds for Canadian tidal observation stations. There was no response to its entreaties year after year until finally in December 1889 the committee, bolstered now by prominent businessmen from the Montreal shipping community, again petitioned Minister of Marine and Fisheries Charles Hibbert Tupper for a tidal survey. They got a small parliamentary appropriation in 1890.

The first Canadian tidal observation stations were established at Pointe-au-Père and Anticosti Island, in the Magdalen Islands (Îles de la Madeleine), at St Paul Island (all three on the way into the Gulf of St Lawrence), and at Saint John, New Brunswick. In 1893, the Canadian Tidal and Current Survey was established under W. Bell Dawson (the son of the principal of McGill, J.W. Dawson, and brother of G.M. Dawson), who remained its superintendent until 1924, when it was united with the Hydrographic Service (it became part of the Canadian Hydrographic Service in 1928, uniting charting and tidal surveys). At first, the Tidal Survey was poorly funded, but with time it expanded and tide tables for all major Canadian ports were produced.[38]

But it was the fisheries that were most closely related to the beginnings of the mainline marine sciences in Canada. With Confederation, fisheries became (as they still are) a federal responsibility, at the beginning under the wing of the Department of Marine and Fisheries, established in 1868. Initially there was no regulation of the main fishery, in the Atlantic, and resources were heavily used by Canadian and American fishermen. It was widely believed that finfish, lobsters, and oysters were decreasing, but there was nothing but hearsay to judge one way or another.[39] Pleas for fishery regulation and study of the fisheries, combined with support of pure science, began in the 1880s, one example of which is the Guelph biologist J.P. McMurrich's 1884 article "Science in Canada," calling for fisheries stations and steamers to "adopt measures for the increase of our fisheries, by informing us of the real extent, of which we are comparatively ignorant, and by preventing their wanton destruction."[40] But there was little or no home-grown talent to apply to these problems, and no immediate response to McMurrich's entreaty.

Under these circumstances – the need for advice on fisheries problems and the small, relatively undeveloped science of late-nineteenth-century Canada – it is no surprise that the Canadian government looked once again to Britain for help. In 1892, upon the advice of an eminent Scots fisheries biologist, W.C. M'Intosh of the University of St Andrews, it appointed Edward E. Prince dominion commissioner of fisheries (figure 1.4). Prince's pedigree was excellent, and his political savoir faire proved to be outstanding. "The genial Professor Prince"[41] began early to make a mark in Canada and was an important presence from his arrival in 1893 until his retirement in 1924.[42]

Within the first year, Prince suggested a marine biological station for Canada and a "complete biological survey of coastal waters." As Jennifer Hubbard has shown in her detailed analysis of Prince's influence, there were a number of factors at work, among them certainly the need for more information on Canadian fisheries, the prevailing interest in Europe and North America in marine stations as locations for advanced research in marine biology, the increasing interest in Canadian universities in having laboratory facilities on lakes and the oceans, and an underlying sentiment that Canada should be matching the United States in its research capabilities. The biggest problem, at least in the short term, was that no one in the Department of Marine and Fisheries showed the slightest interest in marine laboratories.

The Royal Society of Canada endorsed the plan for a marine station in 1896, but it was given real impetus when the British Association for the Advancement of Science met in Toronto to great fanfare in 1897. A British Association committee for a Canadian biological station was set up, and a part of the committee, bolstered by Canadian academics from Toronto, Laval, and McGill, along with members of the Royal Society, the Nova Scotian Institute of Science, and the Natural History Society of Montreal, met with the minister of marine and fisheries in April 1898 to ask for support. As Hubbard has written:

> The delegates' arguments were of a kind long used by scientists to get government funding. These "classical" arguments began with an appeal to national pride: "Canada is the only civilized country where no Marine Biological Station has been established." But this argument only works if very tangible benefits are promised. They argued that the fishing industry would generate greater revenues through improved practices. Furthermore, a Canadian station would provide precise scientific observations relating to the conditions of fish stocks, their size and locations, how they could be increased, how harvesting methods could be improved, and so on. But this would require "exact scientific investigation into such questions."[43]

1.4 Edward E. Prince at St Andrews, NB, right, with A.G. Huntsman, left, and an unidentified young researcher in between

Moreover, "university researchers would aid the Dominion government in solving fisheries problems, and the government would have first claim on the information they generated."[44] This, combined with the argument that such a station would keep Canadian researchers at home rather than losing them to the United States, was persuasive – $15,000 was provided to build a marine station and maintain it for five years under the direction of a board of management (named the Biological Board of Canada in 1912) chaired by Prince.

When the first marine station came into use in 1899, it was a floating one (figure 1.5), its design based on a precursor in the United States, but probably inspired by John Murray's *Ark* in the Firth of Forth, which later became the nucleus of the Scottish station at Millport, a vessel and events with which Prince was certainly familiar. The floating station was towed from location to location, beginning its scientific life in St Andrews, New Brunswick, for the first two years, then in Canso, Nova Scotia, in 1901–2, Malpeque Bay, Prince Edward Island, in 1903–4, and Gaspé, Quebec, in 1905–6. In each location research centred on the problems of the area – including, but not exclusive to, "sardines" in New Brunswick, groundfish in Nova Scotia, oysters in Prince Edward Island, and so on. For the first time, university researchers could readily find laboratory facilities on the Atlantic coast and spend their summers as unpaid "volunteers" in marine research that benefited them and the country.

When the floating station was damaged en route to the north shore of the Gulf of St Lawrence and had to be abandoned in 1907, it seemed a disastrous end to such promising investigations. But the speed with which a replacement for the floating station was provided suggests that something more satisfactory than the floating station had been planned for some time. With careful consideration of all the important factors, including the local fisheries, the richness of the marine biota, water quality, the availability of land, access to supplies, social factors, and ease of access from central Canada (the railroad link that opened in 1903 proved a great convenience if not a determining factor), the choice was St Andrews, New Brunswick.[45] The Atlantic Biological Station opened there in 1908, and there it has remained ever since.

## The St Andrews Biological Station in Context

The two marine biological stations that opened in Canada in 1908, the Atlantic Biological Station at St Andrews and the Pacific Biological Station at Nanaimo, British Columbia, institutionalized marine biology

1.5 Canada's First Marine Biological Station

in a literal and figurative way upon their opening. It may seem surprising that their history was so short – a mere decade from initial funding of a movable station, to permanent establishments on both coasts. But such was the case. While the marine environment had been studied episodically in various ways from the French regime in the seventeenth century until permanent stations were founded in the early twentieth century, there is no common narrative, no historical imperative, that unites the varied charting, explorations, tidal surveys, and marine collecting that took place in Canada until 1898. What gave rise to the first government funding of marine biological research in Canada between 1898 and 1908 was partly the need for information about the fisheries. But it was more than "inventory science" – it was the need to develop a Canadian scientific identity uniting utility with scientific understanding that lay behind the origin of Canada's first marine stations. That a small group of Canadian university-based scientists could carry this off in the face of government indifference makes this a particularly interesting case study to historians, and possibly also to today's marine scientists, who should find it easy to identify with the problems that their intellectual ancestors faced in our own time of governmental indifference to science in Canada.

NOTES

1 T.H. Levere and R.A. Jarrell, "General Introduction," in *A Curious Field-Book: Science and Society in Canadian History*, ed. T.H. Levere and R.A. Jarrell (Toronto: Oxford University Press, 1974), 1–24; C. Berger, *Science, God, and Nature in Victorian Canada* (Toronto: University of Toronto Press, 1983); S. Zeller, *Land of Promise, Promised Land: The Culture of Victorian Science in Canada* (Ottawa: Canadian Historical Association, 1996), Historical booklet no. 56; and S. Zeller, "Nature's Gullivers and Crusoes: The Scientific Exploration of North America, 1800–1870," in *North American Exploration*, vol. 3, *A Continent Comprehended*, ed. J.L. Allen (Lincoln: University of Nebraska Press, 1997), 190–243, 564–77.

2 T.H. Levere, "Science and the Canadian Arctic, 1818–76, from Sir John Ross to Sir George Strong Nares," *Arctic* 41 (1988): 127–37; T.H. Levere, *Science and the Canadian Arctic: A Century of Exploration, 1818–1918* (Cambridge: Cambridge University Press, 1993); and Zeller, "Nature's Gullivers."

3 See D. Lindsay, "The Historical Context: Science in Rupert's Land before 1859," in *The Modern Beginnings of Subarctic Ornithology: Correspondence to the Smithsonian Institution, 1856–1868*, ed. D. Lindsay (Winnipeg: Manitoba Record Society Publications 10, 1991), ix–xxx; and S. Houston, T. Ball, and M. Houston, *Eighteenth-Century Naturalists of Hudson Bay* (Montreal and Kingston: McGill-Queen's University Press, 2003).

4 Levere, "Science and the Canadian Arctic," 44–97, 190–238.

5 Berger, *Science, God, and Nature in Victorian Canada*, 3–27; D.W. Haslam, "The British Contribution to the Hydrography of Canada," in *From Leadline to Laser: Centennial Conference of the Canadian Hydrographic Service*, Ottawa, in *Canada: Special Publication of Fisheries and Aquatic Science* 67 (1983): 20–35; and D.W. Haslam, "The British Contribution to the Hydrography of Canada," *International Hydrographic Review* 61 (1984): 17–42.

6 See J. Cawood, "The Magnetic Crusade: Science and Politics in Early Victorian England," *Isis* 70 (1979): 493–518.

7 A.D. Thiessen, "The Founding of the Toronto Magnetic Observatory and the Canadian Meteorological Service," *Journal of the Royal Astronomical Society of Canada* 34 (1940): 308–48; G.A. Good, "Toronto Magnetic Observatory and International Science ca. 1850," *Vistas in Astronomy* 28 (1985): 387–90; G.A. Good, "Between Two Empires: The Toronto Magnetic Observatory and American Science before Confederation," *Scientia Canadensis* 10 (1986): 34–52; M. Thomas, *The Beginnings of Canadian Meteorology* (Toronto: ECW Press, 1991), 43–116; Zeller, "Nature's Gullivers."

8 C.M. Whitfield and R.A. Jarrell, "Lefroy, Sir John Henry," in *Dictionary of Canadian Biography* 11 (2000): 508–10.

9 Thiessen, "The Founding of the Toronto Magnetic Observatory"; Thomas, *The Beginnings of Canadian Meteorology*; and Zeller, "Nature's Gullivers."

10 L. Chartrand, R. Duchesne, and Y. Gingras, *Histoire des sciences au Québec* (Montreal: Éditions du Boréal, 1987), 127–57; W.A. Waiser, "The Government Explorer in Canada, 1870–1914," in *North American Exploration*, vol. 3, *A Continent Comprehended*, ed. J.L. Allen (Lincoln: University of Nebraska Press, 1997), 412–60, 593–9; M. Zaslow, *Reading the Rocks. The Story of the Geological Survey of Canada 1842–1972* (Toronto: Macmillan of Canada, 1975); and S. Zeller, *Inventing Canada: Early Victorian Science and the Idea of a Transcontinental Nation* (Toronto: University of Toronto Press, 1987).

11 R.G. Blackadar, "Geological Survey of Canada," in *The Canadian Encyclopedia*, 2nd ed., vol. 2 (Edmonton: Hurtig Publishers, 1988), 889–90; see also Waiser, "The Government Explorer in Canada."

12 G.M. Dawson, "Report on the Queen Charlotte Islands, by Mr. G.M. Dawson, with Appendices A. to G," In *Geological Survey of Canada. Report of Progress for 1878-79* (Montreal: Dawson Brothers, 1880), 1–239; D. Cole and B. Lochner, eds., *To the Charlottes: George Dawson's Survey 1878 of the Queen Charlotte Islands* (Vancouver: UBC Press, 1993); and Waiser, "The Government Explorer in Canada."

13 See W.A. Waiser, *The Field-Naturalist: John Macoun, the Geological Survey, and Natural Science* (Toronto: University of Toronto Press, 1989); and Waiser, "The Government Explorer in Canada."

14 W.A. Waiser, "Canada on Display: Towards a National Museum, 1881–1911," in *Critical Issues in the History of Canadian Science, Technology and Medicine*, ed. R.A. Jarrell and A.E. Roos (Thornhill, ON, and Ottawa: HSTC Press, 1983), 167–77; and Waiser, "The Government Explorer in Canada."

15 S.B. Frost, "Science Education in the 19th Century: The Natural History Society of Montreal, 1827–1925," *McGill Journal of Education* 17 (1982): 31–48.

16 Anonymous, "Eminent Living Geologists: Joseph Frederick Whiteaves, LL.D., F.G.S., F.R.S. (Canada)," *Geological Magazine*, n.s., decade 5, 3 (1906): 433–42; and W.J. Sollas, "John Joseph Frederick Whiteaves, LL.D (McGill), F.R.S. Canada (1835–1909)," *Quarterly Journal of the Geological Society of London* 66 (1910): xlix–l.

17 J.F. Whiteaves, *Catalogue of the Marine Invertebrata of Eastern Canada* (Ottawa: Geological Survey of Canada, 1901), 271

18 P.J. Bowler, "The Early Development of Scientific Societies in Canada," in *The Pursuit of Knowledge in the Early American Republic*, ed. A. Oleson, and S.C. Brown (Baltimore: Johns Hopkins University Press, 1976), 326–39; Berger, *Science, God, and Nature in Victorian Canada*, 3–27; R.A. Jarrell, "The Social Functions of the Scientific Society in 19th-Century Canada," in *Critical Issues in the History of Canadian Science, Technology and Medicine*,

ed. R.A. Jarrell and A.E. Roos (Thornhill and Ottawa: HSTC Press, 1983), 31–44; Chartrand, Duchesne, and Gingras, *Histoire des sciences au Québec*; and Zeller, *Land of Promise*.

19 C. Berger, *Honour and the Search for Influence: A History of the Royal Society of Canada* (Toronto: University of Toronto Press, 1996); and T.H. Levere, "The Most Select and the Most Democratic: A Century of Science in the Royal Society of Canada," *Scientia Canadensis* 20 (1998): 3–99.

20 R.A. Jarrell, "The Influence of Irish Institutions upon the Organization and Diffusion of Science in Victorian Canada," *Scientia Canadensis* 9 (1985): 150–64.

21 W. Eggleston, *National Research in Canada: The NRC 1916–1966* (Toronto: Clarke, Irwin, 1978); and R.A. Jarrell and Y. Gingras, eds, *Building Canadian Science: The Role of the National Research Council* (Thornhill, ON: Scientia Press / *Scientia Canadensis* 15:2, 1992).

22 R.S. Harris, *A History of Higher Education in Canada 1663–1960* (Toronto: University of Toronto Press, 1976), 27–36.

23 P. Anisef and J. Lennards, "University," in *The Canadian Encyclopedia* (Edmonton: Hurtig, 1985), 3: 1872–3.

24 R.A. Jarrell, "Science Education at the University of New Brunswick in the Nineteenth Century," *Acadiensis* 2 (1973): 55–79; and Harris, *Higher Education*, 27–36.

25 R.A. Jarrell, "Science," in *The Canadian Encyclopedia* (Edmonton: Hurtig, 1985), 3: 1654.

26 See the discussion in Y. Gingras, *Physics and the Rise of Scientific Research in Canada* (Montreal and Kingston: McGill-Queen's University Press,1991), 11–35.

27 Zeller, *Inventing Canada*.

28 R. Cook, "The Triumph and Trials of Materialism," in *The Illustrated History of Canada*, ed. C. Brown, rev. ed. (Toronto: Key Porter Books, 2002), 377–472.

29 R. McKenzie, *Admiral Bayfield: Pioneer Nautical Surveyor* (Canada: Fisheries and Marine Service, 1976), Special publication 32.

30 S. Fillmore and R.W. Sandilands, *The Chartmakers: The History of Nautical Surveying in Canada* (Toronto: NC Press, 1983), 57–63, 86–9.

31 R.W. Sandilands, "Hydrographic Charting and Oceanography on the West Coast of Canada from the Eighteenth Century to the Present Day," *Proceedings of the Royal Society of Edinburgh* 73 (1972): 75–83; Fillmore and Sandilands, *The Chartmakers*; Anonymous, *From Leadline to Laser*, Centennial Conference of the Canadian Hydrographic Service, *Canadian Special Publication of Fisheries and Aquatic Science* 67 (Ottawa: Canadian Hydrographic Service / Canadian Hydrographic Association, 1983):

1–223; O.M. Meehan (edited by William Glover with David Gray), "The Canadian Hydrographic Service from the Time of Its Inception in 1883 to the End of the Second World War. Part 1," *Northern Mariner* 14:1 (2004): i–x, 1–158; 14:2: Part 2 ix–xiv, 159–297; O.M. Meehan (edited by William Glover with David Gray), "The Canadian Hydrographic Service from the Time of Its Inception in 1883 to the End of the Second World War," *Northern Mariner*, Special edition (2006), 46, 78–9.

32 Fillmore and Sandilands, *The Chartmakers*; Meehan, "The Canadian Hydrographic Service" (2006).

33 V. DeVecchi, "The Pilgrim's Progress, the BAAS, and Research in Canada: From Montreal to Toronto," *Transactions of the Royal Society of Canada*, series 4, 20 (1982): 519–32; Meehan, "The Canadian Hydrographic Service" (2006).

34 Hubbard, *A Science on the Scales*, 67–89; Meehan, "The Canadian Hydrographic Service" (2006).

35 DeVecchi, "The Pilgrim's Progress"; and V. DeVecchi, "The Dawning of a National Scientific Community in Canada, 1878–1896," *Scientia Canadensis* 18 (1984): 32–58.

36 Levere, *Science and the Canadian Arctic*.

37 M.S. Reidy, *Tides of History: Ocean Science and Her Majesty's Navy* (Chicago: University of Chicago Press, 2008).

38 Fillmore and Sandilands, *The Chartmakers*.

39 K. Johnstone, *The Aquatic Explorers: A History of the Fisheries Research Board of Canada* (Toronto: University of Toronto Press, 1977).

40 J.P. McMurrich, "Science in Canada," *The Week* 1:49 (1884): 777.

41 J. Hubbard, "An Independent Progress: The Development of Marine Biology on the Atlantic Coast of Canada 1898–1939," Ph.D. thesis, University of Toronto, 1996, 33.

42 Anonymous, "Prince, Edward Ernest," in *The Canadian Men and Women of the Times: A Handbook of Canadian Biography of Living Characters*, ed. H.J. Morgan (Toronto: William Briggs, 1912), 918–19; Anonymous, "Edward Ernest Prince," *Proceedings of the Transactions of the Royal Society of Canada*, 3rd ser., 31 (1937): xx–xxiii; A.G. Huntsman, "Edward Ernest Prince," *Canadian Field-Naturalist* 59 (1945): 1–3; Johnstone, *The Aquatic Explorers*; A.W.H. Needler, "Prince, Edward Ernest," in *The Canadian Encyclopedia* (Edmonton: Hurtig, 1985), 3: 1474; Hubbard, *A Science on the Scales*.

43 Hubbard, *A Science on the Scales*, 29.

44 Ibid., 30.

45 Ibid.

# 2 Some Contributions of Women to the Early Study of the Marine Biology of Canadian Waters

MARY NEEDLER ARAI

During the nineteenth century, the cultural role of women in North America and Europe expanded from strictly the household sphere to a more broadly educated one. In this century, "Natural History" became a socially approved interest for both men and women. This has been better documented in the United States than in Canada. As women's education expanded into schools and then colleges, the opportunities for women to teach also expanded. Although women were not involved in field explorations of eastern Canadian waters, they did work in museums on the resulting specimen collections. They attended classes and did research at marine stations such as the Marine Biological Laboratory at Woods Hole. As male scientists' education became more professional, a few pioneering women also obtained higher degrees and taught in universities. Some taught in household science or similar university faculties, primarily educating women. Others, such as Carrie Derick of McGill, taught in biology departments, and between 1901 and 1907 Josephine Tilden operated the Minnesota Seaside Station on southern Vancouver Island. After the First World War, summer work at several US marine laboratories such as that at Woods Hole, and at the St Andrews and Nanaimo marine laboratories, allowed some women to carry out research and obtain graduate degrees. Unfortunately, the Depression and restrictions on hiring married women or women for fieldwork led to difficulties for most of these women in obtaining employment or even access to research facilities. Opportunities again expanded briefly during the Second World War, and more permanently from the 1960s on.

There has been little written about women's contributions to Canadian marine biology. A landmark 1990 publication on Canadian women in science was edited by Marianne Ainley.[1] Her own article in that volume, and a more recent update, emphasized contributions by women at Anglo-Canadian universities in the first half of the twentieth century.[2] She described the three generations of women scientists in my own family as examples of the factors affecting each generation of marine biologists:[3] my grandmother Edith Berkeley, who taught at the University of British Columbia during the First World War, and then was one of the first and probably the longest-lasting of the volunteer scientists of the Fisheries Research Board; my mother, Alfreda Needler, who was one of Huntsman's doctoral students in the Great Depression and then also a volunteer researcher; and myself. I have had an academic career and continue as a volunteer researcher in retirement. This chapter argues that none of us, even Edith Berkeley, my grandmother, were complete pioneers. Most of the educational, career, and social factors affecting women between the two world wars were actually already in place by the end of the First World War, in Canada as in the United States and Great Britain.

This chapter describes Canadian women who pioneered women's involvement in marine biology; it also examines the development of opportunities for women's training and research inside and outside academia before 1918. The training and subsequent careers of a number of women in Canada and the United States in the period 1918–40 will also be discussed. The limited employment opportunities, especially during the Great Depression, the restricted employment of married women, and the obstructions to women doing fieldwork, led to success for only a few, and truncated or totally prevented most careers. Nevertheless, some of these women succeeded in continuing their research and publishing, even if they had to work out of their home laboratories. Very few succeeded in careers in academic institutions. There is insufficient information to generalize about these women, but a selection of careers to illustrate their range before the First World War will be presented.

## The Increasing Participation of Women up to the First World War

During the eighteenth and nineteenth centuries, "Natural History" became a socially approved interest for both men and women. Making collections and paintings, especially of plants and shells, became widely popular. The first Canadian natural history society was the Natural History Society of Montreal, organized in 1827.[4] By 1883, at least another eight societies large enough to publish journals for their mostly amateur membership were organized in various parts of Canada.[5]

The guide books for amateur naturalists, male, female, and child alike, were often written by women. Women continued to write guide books in Great Britain in the later nineteenth century when many men specializing in biology were turning to more professional and technical subjects.[6] In North America, marine biology manuals did not appear until around 1900. Augusta Foote Arnold of Washington wrote the *Sea-beach at Ebb-tide*, an identification manual for the seaweeds and more common fauna, published in 1901.[7] The emphasis was on forms from the east coast of North America, as far north as Newfoundland, although some Pacific forms were also included. Similarly, the 1908 popular conchology manual, *The Shell Book*, by Julia Ellen Rogers, included not just shells from both coasts of North America, but also some which might have been brought commercially from farther away.[8]

As far as women were concerned, marine biology differed from other natural history fields – in restricting their full participation – in at least two ways. One was that women were not allowed on high-seas research ships until late in the twentieth century. Their work in many countries was confined to what flora and fauna they could collect inshore, or examine when it was brought inshore by others. The other restriction, which was by no means trivial, was that women were expected to wear long skirts and often even corsets while collecting in the nineteenth century.[9] A study of tide pools published in 1906 includes a photograph taken in 1903 near the Minnesota seaside station at the southern end of Vancouver Island (figure 2.1).[10] Note that by 1903 it was allowable for women to shorten the skirts slightly and wear boots. It was possible to wear looser gymnastic suits with voluminous knee-length bloomers, but only when men were not present. Around 1910 it became acceptable for women to wear gym suits or bathing suits (also with bloomers). By the 1920s and 1930s women could wear trousers (first knickers and then slacks) for leisure activities such as collecting, boating, or hiking.

Far longer-lasting was the ban of women from high-seas research ships. Probably the most striking example of this restriction is provided by the eminent oceanographer Mary Sears, who was never allowed a berth on the research vessels of the Woods Hole Oceanographic Institution (WHOI) in the United States.[11] She had been a staff member at WHOI since 1940. During the Second World War she headed a naval oceanographic unit for long-range strategic forecasting, attaining the final rank of Lt. Commander, and then returned to WHOI until

2.1 Attire for collecting

her retirement in 1970. After 1963, when WHOI got its first research vessel with facilities for women, her career had been deflected from research on plankton such as siphonophores to editing oceanographic literature, and so she never had the opportunity to participate in an oceanic expedition.

Shipboard restrictions were directed against including women in an official capacity as scientists or ship's officers. It should be noted that wives of officers and warrant officers had commonly gone to sea, at least from Tudor times to the end of the era of sailing vessels.[12] Similarly, wives often accompanied the captains of whaling and commercial trading vessels.[13] However, wives were also largely barred from fuel-driven steamers (although women were hired as stewardesses to assist female passengers on transoceanic steam ships in the late nineteenth century). As to officers, Molly Kool, a native of Alma, New Brunswick, was the first woman in North America to hold captain's papers. She earned a coastal master's certificate from the Merchant Marine Institute in Yarmouth, Nova Scotia, in 1939, but was considered an oddity.[14]

Through the nineteenth century, general education for girls moved from home tutoring or dame's schools to college degrees. This was a hard-fought and irregularly completed progression, so that all stages might be present in the same city simultaneously. As schools for women developed, a parallel demand for teachers also developed. These teachers might be educated in dedicated normal schools or in colleges. It is outside the scope of this chapter to describe these changes except as they affected the teaching of marine biology itself.

The first Canadian professor of natural history, William Hincks, was appointed in the University of Toronto in 1853.[15] In 1858, natural history teaching was initiated at McGill University and at Queen's for male students. Canadian colleges began to open their doors to allow women to attend co-educational courses during the 1860s.[16] In 1875, Grace Annie Lockhart was granted a B.Sc. in science and English by Mount Allison, Sackville, New Brunswick. It was only a three-year program; nevertheless, Lockhart was the first woman to earn a degree in the British Empire. In 1882 and 1883 Mount Allison also conferred the first four-year B.A. and the first M.A. in Canada. Within the decade, undergraduate co-educational degrees were available to women at most English-speaking Canadian universities.

Canadian women did not, apparently, become involved in professional scientific work in the late nineteenth century. In the last half of the nineteenth century a number of surveys were made of the fauna and flora in various waters off eastern Canada, a listing of which and a catalogue of marine invertebrates identified before 1901 was published by J.F. Whiteaves of the Geological Survey of Canada.[17] No women were involved in the actual dredging or survey work. Whiteaves also does not mention any women working in the museum of the Geological Survey, which had been founded in Montreal but transferred to Ottawa in 1881, and which formed the nucleus of the National Museum. The Redpath Museum of McGill, which opened in 1882, presented lectures sponsored by the Ladies' Educational Association on botany and zoology in its first decade, but again there is no mention of women working on the specimens.[18] Rather, much of the earlier collecting was done by American scientists and those specimens were usually retained in US collections. A number of US scientists made summer visits to Grand Manan, New Brunswick, from 1852 to 1872. Addison E. Verrill of Yale University and co-workers made a number of collections, particularly in the Halifax region, under the auspices of the US Fish Commission, which in 1881 established a permanent research laboratory at Woods Hole. The commission also collected in the Canadian Pacific. In the summer of 1899, the privately funded Harriman Alaska Expedition passed up the inside passage towards Alaska and Siberia. However, although Mrs Harriman, family, and friends were on board the steamer, no women were listed in the extensive scientific party.[19] H.B. Bigelow continued collecting for the Fish Commission into the next century, in the Gulf of Maine and as far north as Newfoundland and Labrador. These US surveys resulted in specimens being deposited in Harvard University's Museum of Comparative Zoology at Cambridge, Massachusetts, in the Peabody Museum, Yale University, at New Haven, Connecticut, and in the National Museum of the Smithsonian Institution in Washington (e.g., see the coelenterate collections).[20]

While women were not involved in surveys or dredging for specimens, several women did work on Canadian specimens in foreign museums. Katherine J. Bush (1855–1937), who had no university training, was hired in 1879 as assistant to A.E. Verrill at the Peabody Institute.[21] She took courses as a special student in the Sheffield Scientific School of Yale. Although working at the Peabody, she was supported, at least in part, by the US Fish Commission.[22] Her first paper was a general catalogue of molluscs and echinoderms collected on the coast of Labrador by W.A. Stearns;[23] her subsequent papers were largely on annelids and molluscs. As well as four co-authored papers, she was sole author of at least fourteen papers.[24] In 1901 she was the first woman to earn a doctorate in zoology at Yale.

The best known of the US women who worked on museum collections was Mary Jane Rathbun (1860–1943).[25] She began her career in 1881 as a summer volunteer and then a paid clerk for the Fish Commission at Woods Hole. In 1886, she followed her brother to the US National Museum, where she was employed in the Division of Marine Invertebrates, which her brother headed in the 1890s. Rathbun organized and catalogued museum collections, but also became an expert on decapod crustacea. She had a regular government salary until 1914, when she gave up her salary so an assistant could be hired. She then continued working full time as an honorary associate in biology until 1939. Although she had no undergraduate degree, in 1916 she received an honorary M.A. from the University of Pittsburgh. Then in 1917, when she was fifty-seven, she earned a Ph.D. from George Washington University. Rathbun was well recognized for her extensive corpus of

166 published articles.[26] She is best known for her four monographs on the crabs of America.[27] Among other papers were a report on the decapod crustacea from the Canadian Arctic Expedition of 1913–18 and a paper on decapods from the Canadian Atlantic Fauna series.[28]

A second carcinologist, who also worked primarily at the US National Museum Division of Marine Invertebrates, was Harriet Richardson (1874–1958), who graduated from Vassar with a B.A. in 1896 and an A.M. in 1901; she earned a Ph.D. from Columbia University (now George Washington University) in 1903, and produced eighty publications.[29] The best known is her monograph on the isopods of North America, published in 1905, which included Canadian records.[30] Few publications, however, followed after her 1913 marriage to William Searle, a Washington lawyer, and the birth of a handicapped son.

Although some Canadian specimens had reached Europe earlier, the *Challenger* expedition initiated an era in which a number of specimens were taken to European museums. The *Challenger* visited Halifax from 9 to 19 May 1873 in the early stages of her voyage.[31] The *Michael Sars*, with naturalists Johan Hjort and Sir John Murray aboard, loaded coal at St John's, Newfoundland, from 3 to 8 July 1910 and collected briefly on the Grand Banks during the North Atlantic Deep-Sea Expedition.[32] Several expeditions, primarily exploring western Greenland, also visited the eastern Canadian arctic and deposited specimens in Scandinavian museums.[33] The *Pteropoda, Heteropoda,* and nudibranchs from the *Michael Sars* voyage were described by Kristine Bonnevie (1872–1948),[34] Norway's first woman professor.[35]

Unlike the European – and also Canadian – research stations when they were developed, most US marine stations not only included research facilities, but also emphasized the importance of teaching marine biology courses, primarily to secondary school teachers.[36] At the Woods Hole Marine Biological Laboratory the four main six-week courses taught over the years were invertebrate zoology, a course established in 1888; marine botany, established in 1890; general physiology, established in 1892; and embryology, established in 1893.[37] A photograph from 1895 illustrates the high proportion of women students in an invertebrate zoology course; the class is seen setting out for a collecting expedition on the schooner *Vigilant*.[38] Although it is known that Canadian women enrolled in these courses and also used the research facilities, it is not known how many did so.

Marine biological research gradually became more professionalized for men. As graduate degrees became more usual, women also attempted to obtain them. One outstanding early Canadian female university faculty member was Carrie M. Derick (1862–1941).[39] After teaching at a private school for girls, she earned her B.A. at McGill in 1890 with the highest marks in the graduating class. In 1892, while still teaching at the school, she also was appointed as a part-time demonstrator in botany, the first woman on McGill's instructional staff. After earning her M.A. in 1896, she was appointed to a full-time position as lecturer in botany and demonstrator in McGill's Botanical Laboratory and in 1904 was promoted to assistant professor. She continued her studies, spending summers at Harvard, the Royal College of Science, London, and the Woods Hole Biological Station. She also spent eighteen months at the University of Bonn in Germany, doing research for a higher degree, but did not receive it because, at that time, Bonn did not yet award the Ph.D. to women. After being acting chair of the McGill Department of Botany 1909–12, she was passed over for appointment as the professor of botany and, instead, was given the "courtesy title" of professor of morphological botany (in 1928 changed to professor of comparative morphology and genetics).[40] In 1929, she became McGill's first professor emerita.

Carrie Derick published widely on the genetics, development, and morphology of a variety of plants, but is included here for her early papers on marine algae – examined during seven summers at Woods Hole.[41] It is not clear whether she ever worked on algae in Canada, although she is listed as a summer visitor at St Andrews in 1923 by Rigby and Huntsman.[42] All four of the marine stations built on Canadian coasts between 1898 and 1908 provided opportunities for use by women. For example, the floating laboratory, built and operated on the east coast by the board of management from 1899 to 1907,[43] spent the summer of 1900 at St Andrews, New Brunswick, where Miss Susan Ganong of the staff of the Halifax Ladies College (later the Ambrae Academy) collected marine specimens for her institution.[44]

Often ignored by historians of Canadian marine science is the Minnesota Seaside Station, operated from 1901 to 1907 by Josephine E. Tilden (1869–1957) on the southwestern coast of Vancouver Island. Tilden (figure 2.2) had received her B.S. degree in 1895 and her M.S. in 1896 from the University of Minnesota, where she became the first woman faculty member.[45] In addition to teaching, she travelled widely and collected plants, especially algae. Since it was considered improper for a woman to travel alone, she was almost always accompanied by her mother, Elizabeth. In the summer of 1898, they discovered the excellent

2.2 Josephine Tilden

possibilities for collecting algae at a potholed beach site she named Botanical Beach near Port Renfrew, BC.[46] A local homesteader donated four acres and Tilden convinced Professor Conway Macmillan to act as director of a station with herself as sub-director and chief organizer. For five years after 1901, the station supported up to sixty students and five or six faculty members each summer.[47] Many of its courses were oriented to teachers, and separate dormitory accommodation allowed a number of women to attend. Research and graduate work also flourished, and two volumes of the yearbook *Postelsia* were published. It had been expected that the University of Minnesota would then begin to support the facility.[48] Instead, in 1906, MacMillan was accused of stealing university microscopes to take to a foreign country and misusing funds for the station, and had to resign. Tilden financed the 1907 courses at a loss and then had to abandon the station. In the years that followed Tilden was promoted to full professor.[49] She continued to travel extensively with students after her mother's death, especially in the South Pacific. She retired in 1937 after further financial difficulties, having by then produced over fifty scientific articles and three books, and mentored many students including six Ph.D.s.

The Pacific Biological Station, first opened in 1908, was to become, unlike its predecessor, a permanent and successful institution. Within its register a record of doubtful date indicates that Miss E. McClughan of McGill University collected *Polyzoa*, possibly in 1910.[50] Of more interest is the June–September 1911 visit of Miss Helen L.M. Pixell, a Reid Fellow from Bedford College, University of London, the oldest women's college in England. Miss Pixell studied polychaete and phoronid worms and collected protozoa. Two papers on her Canadian work described the worms,[51] but Miss Pixell became primarily an authority on protozoa after returning to England. In 1913, she married Professor E.S. Goodrich, a well-known comparative anatomist and fellow of Merton College, Oxford. There were no children. She visited the Atlantic Biological Station in 1924 with her husband.[52] She assisted her husband with some teaching and collaborated on two joint papers,[53] but largely published independently. Until at least 1950, four years after his death, she was working at, and publishing from, the Department of Zoology and Comparative Anatomy in Oxford.[54]

In the summers of 1917 and 1918, Edith Berkeley (1875–1963) (figure 2.3) came to Nanaimo to collect specimens for the University of British Columbia.[55] She had read zoology at University College, London, taught briefly at a girl's school in London, married Cyril Berkeley in 1902, and then spent ten years in India.[56] During the First World War, because so many young men were away at war, it was possible for a woman like Berkeley, even without any graduate school training, to get a position as a laboratory assistant. She helped A.H. Hutchinson to start courses in zoology at a pay rate of thirty dollars per month, becoming the first woman employed at the University of British Columbia in a scientific position. The annual Edith Berkeley Memorial Lecture was inaugurated by the university in 1969.

Women did not participate in summer research at the Atlantic Biological Station from 1908 to 1913.[57] After 1914, however, the situation became quite different. In 1914, Dorothy Duff of McGill University studied the growth of haddock.[58] Marie Currie of McGill University studied the post-larval development of the copepod *Calanus finmarchicus* in 1916. Clara Fritz, also of McGill, studied diatoms in the summers of 1916–18. Bessie Mossop of the University of Toronto studied starfish fertilization with Robert Chambers of Cornell University in 1917. She also began a three-year study of the growth of mussels at St Andrews and Digby.[59]

In 1915, Clara Cynthia Benson (figure 2.4) began summer studies at St Andrews on the chemistry of fish, lobster, and other marine food, also taking samples back to the University of Toronto for further analysis

2.3 Edith Berkeley

in the winter. She had received her Ph.D. in 1903 from the University of Toronto; she and one other woman (Emma Baker, who did her dissertation in philosophy) were the first two women to do so there.[60] Trained as a physical chemist, she became an associate professor of physiological chemistry when the faculty of household science was established in 1906.[61] American universities typically staffed household science faculties with a high proportion of women chemists doing nutrition research.[62] This represented a chance for women to attain promotions to higher academic ranks at a time when science departments remained prejudiced against them, but it also tended to isolate them from traditional science. Benson's own work continued until at least 1924. In addition, she was the convener of a group of co-workers from the University of Toronto, Macdonald College, McGill University, the Macdonalds Institute, Guelph, and Queen's University working to improve methods for preparing fish for food at the behest of the Ministry of Marine and Fisheries, which was attempting to increase consumer demand for fish.[63]

2.4 Clara Benson

**Participation of Women between the Wars**

By the end of the First World War, women in Canada, as well as those in the United States and Great Britain, had already established their right and ability to take training in marine zoology. Nevertheless, their ability to work was severely restricted. They made collections inshore and examined specimens brought from offshore vessels, but they were not allowed to work on those vessels. When hired, they were often paid less than men with comparable training, and were often hired to teach rather than to do research leading to publications. If they married, they often lost their jobs, especially in tough times like the Depression. Women married to other scientists might become one member of a two-person, single-career family. If they continued their own research and publications, they might be able to obtain research space and equipment through their husbands, do unpaid volunteer work at universities or federal laboratories, or simply work at home. The rest of this section will illustrate these different situations using Canadian examples.

One example of a woman scientist who married a scientist and became a member of a team was Lucy Wright Smith (figure 2.5). Smith graduated from Mount Holyoke College in 1912, and taught there from 1912 to 1918. She earned an M.A. in 1911 and a Ph.D. in limnology in

2.5 Lucy (Wright) Smith Clemens

1914 from Cornell University.[64] She married Wilbert Clemens of the University of Toronto in June 1918 and they spent the summer at the Atlantic Biological Station, working together on the mutton fish *Zoarces anguillaris*.[65] They had two children. After 1921, they spent summers at the Ontario Fisheries Research Laboratory on Lake Nipigon.[66] When Clemens became director of the Pacific Biological Station in 1924, his family accompanied him. The Biological Board had agreed to pay the travelling expenses, board, and lodging for approved voluntary investigators from the universities.[67] However, these people were accommodated in tents on the station grounds. When the residence building was completed in 1929, Lucy Clemens planned and supervised its furnishing and then took responsibility for its operation.[68] During the winters, she worked with Clemens on annual sockeye salmon reports for the provincial fisheries department until her early death in 1937.

Another example of a woman scientist who became a member of a scientific team through marriage was Edith Berkeley, who actually became the scientific authority in her field, and whose husband – the salaried scientist of the couple – joined her research program. In the early years, research at the Atlantic and Pacific Biological Stations was mostly done in the summertime by volunteer investigators. Edith Berkeley left her University of British Columbia instructorship and became a volunteer investigator on a year-round basis.[69] After camping on Pacific Biological Station land in the summer of 1919, she and Cyril Berkeley bought a house across the road which had belonged to the first curator, the Rev. G.W. Taylor. She became an authority on polychaete taxonomy and was joined in her work by Cyril after 1930.[70] Although she was already forty-four when she started research, her career was one of the longest among the volunteer investigators. By the time she died in 1963, she was internationally known, with an extensive list of publications.[71] In 1962, she was named an honorary member of the newly formed Canadian Society of Zoologists, and in 1971 an issue of the *Journal of the Fisheries Research of Canada* was dedicated to her and her husband.

Women who married non-scientists might not have such a successful career. Canadian Irene Mounce (1894–1987) made collections of diatoms at the Pacific Biological Station during the summers of 1919–20 for Dr A.H. Hutchinson of the University of British Columbia. She investigated various physical factors that might influence diatom populations.[72] In 1921 and 1922, working with A.T. Cameron, she was based at the University of Manitoba while continuing to research the diatoms of Departure Bay.[73] During 1920–1, she was one of only six women graduate students who received a fellowship or bursary from the National Research Council.[74] She had a successful career from 1924 to 1945 in the mycology division of the Department of Agriculture in Ottawa, reaching the level of agricultural scientist grade three, and doing fieldwork as well as investigating fungal sexuality.[75] Nevertheless, she was required to resign when she married Gordon M. Stewart in 1945. Another Canadian scientist, Gertrude Smith, assisted C.M. Fraser at the Pacific Biological Station in a series of studies of the ecology of commercial clams.[76] She earned her doctorate from the University of California, Berkeley, in 1934[77] and was promoted to assistant professor at the University of British Columbia, but following her marriage, this position was terminated in 1940.

Unlike Irene Mounce, Constance Ida McFarlane (1904–2000), a marine botanist,[78] never married and was able to continue her career. McFarlane graduated in 1929 with Dalhousie University's first honour's degree in biology and won the Governor General's medal as top student in her year. In 1931, she completed a master's degree. Part of her fieldwork had been carried out at the Marine Biological Laboratory at Woods Hole, Massachusetts. Unfortunately, she was unable to find funding support to complete a Ph.D. after the stock market crash of 1929. For most of the 1930s, she taught science and acted as an administrator in high schools, but from 1946 to 1949, she taught botany and was the dean of women at the University of Alberta. Her main body of research on seaweeds such as Irish moss was developed while she served as director of the Seaweeds Division of the Nova Scotia Research Foundation from 1949 to 1970. During this period she also taught part-time at Dalhousie and Acadia universities.

Helen Battle (1903–94) (figure 2.6) provides another example of a woman who never married and was able to pursue a full scientific career. Battle was the most successful of several University of Toronto women students supervised by A.G. Huntsman. Her first experience of marine research was assisting A.D. Robertson in experiments on factors affecting oyster growth at Prince Edward Island 1921.[79] She earned her bachelor's and master's degrees at the University of Western Ontario, spending the summers of 1923 and 1924 at the Atlantic Biological Station working on the effects of extreme physical conditions on the development of fish larvae.[80] After carrying out further research at the Atlantic Biological Station, she received her Ph.D. in 1928. She was hired by the University of Western Ontario, returning there shortly before the Depression. Battle did not marry, and advanced steadily through the academic ranks, being appointed assistant professor in 1929, associate professor in 1934, and full professor in 1949.[81] She also supervised several graduate students herself, including Nancy Frost and Nancy Henderson, and wrote at least thirty-seven articles. She was the second president of the Canadian Society of Zoologists in 1962–3 and received the society's Fry medal.

During the Atlantic Biological Station's first fifty years (1908–56), over 200 students and scientists carried out research there, and of those, approximately 18 per cent were women.[82] It is difficult to identify those women who were working on marine rather than freshwater fauna or flora, or to otherwise generalize about the type of research they did. The percentage of women was apparently high in the 1920s, when there was still a high percentage of volunteer workers at the biological station, but decreased in the 1930s. During the Depression, support for the transportation, board, and lodging of volunteer workers was cut off due to the Biological Board's straitened financial circumstances,

2.6 Helen Battle

although some of the facilities remained available to those scientists and students able to pay their own expenses.[83] In addition, the opportunities for work by women also decreased because women were very rarely hired for fieldwork.

One of the few exceptions was Nancy Frost Button (1909–98) (figure 2.7), the first Newfoundland woman trained and employed in marine biology before Confederation with Canada in 1949. She entered the first class when Memorial University College opened in St John's in 1925, earned a scholarship to Acadia University and graduated with a B.A. in 1929. She went on to spend a summer at Woods Hole and complete an M.A. from the University of Western Ontario in 1931, supervised by Helen Battle. When Newfoundland's Bay Bulls Laboratory began operations the same year, she was appointed as a biologist. There she worked on fish eggs and larvae and other plankton. She received some specialized training at the Marine Biological Station in Plymouth, England, from which she produced at least three papers. However, as with the other women scientists discussed above who married non-scientists, she stopped doing research when she married Carman Button in 1940 and had three children. When she returned to work in 1958, she taught first at the elementary-school level and then in high school.[84]

2.7 Nancy Frost Button

Some women were able to continue as volunteer researchers even during and after the Depression. Three are discussed here. The first, Alfreda Berkeley Needler (1903–51) (figure 2.8) was Edith and Cyril Berkeley's daughter, and hence as a young scientist already had strong connections with the Biological Board. She had been born in India and sent to live with her grandmother for education in England. Alfreda rejoined her parents when they immigrated to Canada in 1914. She obtained a teaching certificate from the Provincial Normal School of British Columbia and taught for one year, before graduating from the University of British Columbia in 1926. She carried out summer research at the Pacific Biological Station, primarily on the systematics, anatomy, and reproduction of shrimp. Her most significant discovery was sex reversal in these animals.[85] She earned a Ph.D. at the University of Toronto under the supervision of A.G. Huntsman in 1930. She also married Biological Board scientist Alfred Needler in 1930 and had three children. Alfred Needler was director of the Atlantic Biological Station's small substation at Ellerslie, Prince Edward Island, from 1930 to 1941,

2.8 Alfreda Berkeley Needler

and Needler was therefore able to continue her research. She did not hold any further paid position, but she was able to use laboratory facilities and publish.[86] At Ellerslie she discovered that sex reversal also occurs in oysters. After Alfred Needler became director of the Atlantic Biological Station in 1941, she continued her volunteer work there on pycnogonids, paralytic shellfish poisoning, and other subjects until her early death in 1951[87] – some of this research is discussed elsewhere in this volume by Jennifer Martin.

Another woman supervised by A.G. Huntsman who received a Ph.D. in zoology in the Great Depression was Viola Davidson (figure 2.9). She had worked on diatoms and fish as a student at the Atlantic Biological Station as early as 1924.[88] After receiving her doctorate in 1933, she continued to teach at the High School of Commerce, Toronto, but she did do further summer work at the Atlantic Biological Station, like Alfreda Needler, as a volunteer researcher, for example in 1949, publishing an account of her research on salmon and eel movements in constant currents.[89]

The final volunteer researcher to be discussed here is Josephine F.L. Hart (1909–93) (figure 2.10), a volunteer researcher at the Pacific Biological Station from 1929 until 1937. She had received a master's

2.9 Viola Davidson

degree from University of British Columbia in 1931, and completed a Ph.D. degree under Dr Huntsman at Toronto in 1937. After her marriage to G.C. Carl in 1938, she worked primarily out of a small basement laboratory in her home in Victoria. Without even a compound microscope, she amazingly was able to rear and document the complete life cycles of at least thirty species of arthropods. In 1960–1, she worked as a laboratory assistant at Victoria College and received a small grant from the US National Science Foundation that allowed her to buy a microscope and other equipment. She was able to produce twenty-four publications despite minimal support for her science and her lack of equipment.[90]

The one woman scientist associated with marine biology in this period who managed to both marry and hold on to her professional career was Dixie Pelluet, a cytologist who earned her M.A. in botany at the University of Toronto in 1920 and held a Canadian Federation of

2.10 Josephine F.I. Hart

University Women Fellowship at University College, London, in 1922–3. She earned her doctorate at Bryn Mawr in zoology in 1929. After serving as warden of a residence at Dalhousie University, she moved to a small American liberal arts college, where she taught in 1930–1. In 1931 she became a lecturer at Dalhousie and then was hired in 1932 as an assistant professor of zoology. In 1934 she married Ronald Hayes, the biology department's other biologist, and, with no children, she continued active teaching and research on the embryonic development of slugs and salmon. As with most of her colleagues, her salary was frozen between 1932 and 1947, but she seemed to have escaped any ill effects from her married status. However, when a new salary scale was initiated in 1949, she was designated a "special case" and held below the floor level of associate professor despite having been at that rank since 1941.[91]

**Conclusion**

As this chapter has shown, women trained as marine biologists before 1940 experienced a number of factors that curtailed their subsequent employment and research. Although there were a few additional academic appointments available during the Second World War, they were temporary and ended when the men returned.[92] Advancement remained slow for most women during the 1940s and 1950s.

In the 1960s, opportunities for women improved. The expansion in size and number of the Canadian universities provided a number of academic positions for both sexes. Museums and government laboratories also expanded employment opportunities. In addition, barriers to the employment of married women were slowly removed.[93] Two changes in the 1950s and 1960s particularly improved the ability of women to do marine research: first, a few women were allowed to work on offshore marine vessels; and second, university marine science centres built on the Pacific and Atlantic coasts of Canada provided broader career opportunities.

The first Canadian woman scientist to participate in a research cruise was probably Moira Dundar (1918–99), a geographer of the Defence Research Board who researched sea ice. In 1955, the deputy minister's office gave her permission to participate in research on a Department of Transport icebreaker in the Arctic.[94] In the 1960s, Betty Bunce (1915–2003) of the US Geological Survey, Woods Hole, became the first North American woman to serve as a chief scientist on oceanographic expeditions.[95]

University expansion in the 1960s was followed by the creation of marine science centres. The Huntsman Marine Science Centre was established in 1969, initially by a consortium of twenty universities, in co-operation with the Atlantic Biological Station of Fisheries and Oceans, St Andrews, New Brunswick. The Bamfield Marine Sciences Centre was established in 1972 on the Pacific by a consortium of five western universities, the Western Canadian Universities Marine Sciences Society. These centres allowed access for university personnel of both sexes, and were especially valuable to women who had been largely denied access to Canadian marine laboratory facilities since the closure of most of the volunteer program of the Fisheries Research Board during the Great Depression.[96]

## NOTES

This chapter could not have been written without the historical knowledge and technical assistance of Gordon Miller, Librarian, Pacific Biological Station.

Charlotte McAdam, a librarian at St Andrews Biological Station, provided photographs of workers at that station. Many people have added to my historical information and records over the years, of whom I would particularly like to thank Marianne Ainley, Helen Battle, Cyril Berkeley, Edith Berkeley, Ed Bousefield, Dale Calder, Charlotte Haywood, Bill Hoar, Alfred Needler, Alfreda Berkeley Needler, Nina Needler, and "Cas" Stevenson.

1 M.G. Ainley, ed., *Despite the Odds: Essays on Canadian Women and Science* (Montreal: Véhicule Press 1990).
2 M.G. Ainley, "Last in the Field? Canadian Women Natural Scientists, 1815–1965," in *Despite the Odds*, 25–62; M.G. Ainley, "Gendered Careers: Women Science Educators at Anglo-Canadian Universities, 1920–1980," in *Historical Identities: The Professoriate in Canada*, ed. P. Stortz and E.L. Panayotidis (Toronto: University of Toronto Press, 2005), 248–70.
3 M.G. Ainley, "A Family of Women Scientists," *Le Bulletin Newsletter* (Simone de Beauvoir Institute, Concordia University) 7:1 (1986): 5–11.
4 M.J. Dymond, "Zoology in Canada," in *A History of Science in Canada*, ed. H.M. Tory (Toronto: Ryerson Press, 1939), 41–57.
5 Ibid.; J.R. Dymond, "One Hundred Years of Science in Canada. 10. Zoology," in *The Royal Canadian Institute Centennial Volume 1849–1949*, ed. W.S. Wallace (Toronto: Royal Canadian Institute, 1949), 108–20.
6 B. Lightman, "'The Voices of Nature': Popularizing Victorian Science," in *Victorian Science in Context*, ed. B. Lightman (Chicago: University of Chicago Press, 1997), 187–211; A.B. Shteir, "Elegant Recreations? Configuring Science Writing for Women," in *Victorian Science in Context*, ed. B. Lightman (Chicago: University of Chicago Press, 1997), 236–55.
7 A.F. Arnold, *The Sea-beach at Ebb-tide: A Guide to the Study of the Seaweeds and the Lower Animal Life Found between Tide-Marks* (New York: The Century Co., 1901; and New York: Dover Publications, 1968).
8 J.E. Rogers, *The Shell Book: A Popular Guide to a Knowledge of the Families of Living Mollusks, and an Aid to the Identification of Shells Native and Foreign* (Boston: Charles T. Brantford Co., 1908).
9 P.C. Warner and M.S. Ewing, "Wading in the Water: Women Aquatic Biologists Coping with Clothing, 1877–1945," *BioScience* 97 (2002): 98–104.
10 I. Henckel, "A Study of the Tide-Pools on the West Coast of Vancouver Island," in *Postelsia: The Year Book of the Minnesota Seaside Station* (St Paul, MN: The Pioneer Press, 1906), 275–304.
11 K.B. Williams, "From Civilian Planktonologist to Navy Oceanographer: Mary Sears in World War II," in *The Machine in Neptune's Garden: Historical Perspectives on Technology and the Marine Environment*, ed. H.M.

Rozwadowski and D.K. van Keuren (Sagamore Beach, MA: Watson Publishing International, 2004), 243–72.

12 S.J. Stark, *Female Tars: Women aboard Ship in the Age of Sail* (Annapolis: Naval Institute Press, 1996).

13 J. Druett, *Hen Frigates: Wives of Merchant Captains under Sail* (New York: Simon and Schuster, 1998); I. Norling, *Captain Ahab Had a Wife: New England Women and the Whale Fishery, 1720–1870* (Chapel Hill: University of North Carolina Press, 2000).

14 M. Fox, "Molly Kool, 93, a Pioneer of the Coastal Waters, Dies," *New York Times*, 3 March 2009, A25.

15 Dymond, "Zoology in Canada"; Dymond, "One Hundred Years"; E.H. Craigie, *A History of the Department of Zoology of the University of Toronto up to 1962* (Toronto: University of Toronto Press, 1966).

16 M. Gillett, *We Walked Very Warily: A History of Women at McGill* (Montreal: Eden Press Women's Publications, 1981).

17 J.F. Whiteaves, "Catalogue of the Marine Invertebrata of Eastern Canada," *Geological Survey of Canada Special Report* 722 (1901): 1–272.

18 S. Sheets-Pyenson, "Stones and Bones and Skeletons: The Origins and Early Development of the Peter Redpath Museum (1882–1912)," *McGill Journal of Education* 17:1 (1982): 49–62.

19 A.A. Lindsey, "The Harriman Alaska Expedition of 1899 Including the Identities of Those in the Staff Picture," *BioScience* 28 (1978): 383–6; R.M. Peck, "A Cruise for Rest and Recreation," *Audubon* 84:5 (1982): 86–99.

20 M.N. Arai, "Research on Coelenterate Biology in Canada through the Early Twentieth Century," *Archives of Natural History* 19:1 (1992): 55–68.

21 J.E. Remington, "Katherine Jeanette Bush: Peabody's Mysterious Zoologist," *Discovery* 12:3 (1977): 3–8.

22 M.W. Rossiter, "'Women's Work' in Science, 1880–1910," in *History of Women in the Sciences: Readings from* Isis, ed. S.G. Kohlstedt (Chicago: University of Chicago Press, 1982), 287–304; and M.R.S. Creese, *Ladies in the Laboratory? American and British Women in Science, 1800–1900* (London: Scarecrow Press: London, 1998).

23 K.J. Bush, "Catalogue of Mollusca and Echinodermata Dredged on the Coast of Labrador by the Expedition under the Direction of Mr. W.A. Stearns in 1882," *Proceedings of the US National Museum* 6 (1884): 236–47, pl. 9.

24 Remington, "Katherine Jeanette Bush"; Rossiter, "'Women's Work' in Science."

25 Rossiter, "'Women's Work' in Science"; Creese, *Ladies in the Laboratory.*

26 See bibliography in W.L. Schmitt, "Mary J. Rathbun 1860–1943," *Crustaceana* 24 (1973): 283–97.

27 M.J. Rathbun, "The Grapsoid Crabs of America," *U.S. National Museum Bulletin* 97 (1918): 1–461; M.J. Rathbun, "The Spider Crabs of America," *U.S. National Museum Bulletin* 129 (1925): 1–598; M.J. Rathbun, "The Cancroid Crabs of America of the Families *Euryalidae, Portunidae, Atelecyclidae, Cancridae*, and *Xanthidae*," *U.S. National Museum Bulletin* 152 (1930): 1–593; M.J. Rathbun, "Oxystomatous and Allied Crabs of America," *U.S. National Museum Bulletin* 166 (1937): 1–278; F.A. Chase, "Mary J. Rathbun (1860–1943)," *Journal of Crustacean Biology* 10 (1990): 165–7; and P.A. McLaughlin and S. Gilchrist, "Women's Contributions to Carcinology," in *History of Carcinology*, ed. F.Truesdale (Rotterdam: A.A. Balkema, 1993), 165–206.

28 M.J. Rathbun, "The Decapod Crustaceans of the Canadian Arctic Expedition 1913–18," in *Report of the Canadian Arctic Expedition 1913–18.* Vol. 7: *Crustacea.* (A) (Ottawa: F.A. Acland, 1919), 1–14a; M.J. Rathbun, "Decapoda." *Canadian Atlantic Fauna* 10 (1929): 1–38.

29 McLaughlin and Gilchrist, "Women's Contributions to Carcinology"; D.M. Damkaer, "Harriet Richardson (1874–1958): First Lady of Isopods," *Journal of Crustacean Biology* 20 (2000): 803–11; Creese, *Ladies in the Laboratory.* A bibliography of her publications is appended to Damkaer.

30 H. Richardson, *A Monograph on the Isopods of North America, Bulletin of the U.S. National Museum* 54 (1905): 1–727 [repr. 1972, Lochem, Netherlands].

31 T.H. Tizard, H.N. Moseley, J.Y. Buchanan, and J. Murray, J. *Narrative of the Cruise of H.M.S. Challenger with a General Account of the Scientific Results of the Expedition. Report on the Scientific Results of the Voyage of H.M.S. Challenger*, in *Narrative* 1:1 (1885): 1–509.

32 J. Murray and J. Hjort, *The Depths of the Ocean: A General Account of the Modern Science of Oceanography Based Largely on the Scientific Researches of the Norwegian Steamer* Michael Sars *in the North Atlantic* (London: Macmillan and Co., 1912).

33 M.J. Dunbar, "Eastern Arctic Waters," *Bulletin of the Fisheries Research Board of Canada* 88 (1951): 1–131.

34 K. Bonnevie, "*Pteropoda* from the *Michael Sars* North Atlantic Deep-sea Expedition, 1910," *Report of the Scientific Results of the* Michael Sars *Expedition 1910*, vol. 3:1 (1913): 1–69, pl. I–IX; K. Bonnevie, "*Heteropoda* Collected during the *Michael Sars* North Atlantic Deep-sea Expedition 1910," *Report of the Scientific Results of the* Michael Sars *Expedition 1910*, vol. 3:2 (1920): 1–16, pl. I–V; K. Bonnevie, "Pelagic Nudibranchs from the *Michael*

*Sars* North Atlantic Deep-sea Expedition 1910," *Report of the Scientific Results of the* Michael Sars *Expedition 1910*, vol. 5:2 (1929): 1–9, pl. I–IV.

35 Creese, *Ladies in the Laboratory*.

36 J. Maienschein, "Agassiz, Hyatt, Whitman, and the Birth of the Marine Biological Laboratory," *Biological Bulletin* 168, suppl. (1985): 26–34; K.R. Benson, "Why American Marine Stations? The Teaching Argument," *American Zoologist* 28:1 (1988): 7–14.

37 F.R. Lillie, "The Woods Hole Marine Biological Laboratory," *Biological Bulletin* 174:1, suppl. (1944): 1–284.

38 Ibid.

39 Gillett, *We Walked Very Warily*; and M. Gillett, "Carrie Derick (1862–1941) and the Chair of Botany at McGill," in *Despite the Odds: Essays on Canadian Women and Science*, ed. M.G. Ainley (Montreal: Véhicule Press, 1990), 74–87.

40 Gillett, "Carrie Derick."

41 C.M. Derick, "Notes on the Development of the Holdfasts of Certain *Florideae*," *Botanical Gazette* 28 (1899): 246–63.

42 M.S. Rigby and A.G. Huntsman, "Materials Relating to the History of the Fisheries Research Board of Canada (Formerly the Biological Board of Canada) for the Period 1898–1924," Fisheries Research Board of Canada Manuscript Report 660 (1958): 1–272.

43 F.S. Jackson, "The Canadian Marine Biological Station," *Canadian Record of Science* 8:5 (1901), 308–14; Rigby and Huntsman, "History of the Fisheries Research Board of Canada"; and K. Johnstone, *The Aquatic Explorers: A History of the Fisheries Research Board of Canada* (Toronto: University of Toronto Press, 1977).

44 Rigby and Huntsman, "History of the Fisheries Research Board of Canada."

45 G.I. Hansen, "Josephine Elizabeth Tilden (1869–1957)," in *Prominent Phycologists of the 20th Century*, ed. D. Garbary and M. Wynne (Hantsport, NS: Lancelot Press Ltd., 1996), 185–95; Creese, *Ladies in the Laboratory*.

46 R.B. Scott, *People of the Southwest Coast of Vancouver Island: A History of the Southwest Coast* (Victoria: Morriss Printing Co., 1974); Hansen, "Josephine Elizabeth Tilden"; and T. Brady, "The Algae of Acrimony," (January–February 2008) available from http://www.minnesotaalumni.org/s/1118/content.aspx?pgid=1077.

47 See photographs in Scott, *People of the Southwest Coast of Vancouver Island*.

48 Ibid.

49 Hansen, "Josephine Elizabeth Tilden"; Creese, *Ladies in the Laboratory*; Brady, "Of Algae and Acrimony."

50 Rigby and Huntsman, "History of the Fisheries Research Board of Canada."

51 H.M.L. Pixell, "Two New Species of the *Phoronidae* from Vancouver Island," *Quarterly Journal of Microbial Science* 58 (1912): 257–84; and H.M.L. Pixell, "*Polychaeta* from the Pacific Coast of North America. Part. I. *Serpulidae*, with a Revised Table of Classification of the Genus *Spirobis*," *Proceedings of the Zoological Society of London* 1912: 784–805.

52 Rigby and Huntsman, "History of the Fisheries Research Board of Canada."

53 G.R. de Beer, "Edwin Stephen Goodrich, 1868-1946," *Obituary Notices of Fellows of the Royal Society* 5:15 (1947): 477–90.

54 H. Pixell Goodrich, "*Aggregata leandri*," *Quartery Journal of Microbiological Science* 91 (1950): 465–467; and H. Pixell Goodrich, "Sporozoa of *Sipunculus*," *Quarterly Journal of Microbial Science* 91 (1950): 469–76.

55 Rigby and Huntsman, "History of the Fisheries Research Board of Canada."

56 Ainley, "A Family of Women Scientists"; and M.G. Ainley, "Marriage and Scientific Work in Twentieth-Century Canada: The Berkeleys in Marine Biology and the Hoggs in Astronomy," in *Creative Couples in the Sciences*, ed. H.M. Pycior, N.G. Slaack, and P.G. Abir-Am (New Brunswick, NJ: Rutgers University Press, 1996), 143–55.

57 Rigby and Huntsman, "History of the Fisheries Research Board of Canada."

58 D. Duff, "Investigation of the Haddock Fishery, with Special Reference to the Growth and Maturity of the Haddock," *Contributions to Canadian Biology 1914–15* (1916): 95–102.

59 M.E. Currie, "Exuviation and Variation of Planktonic Copepods with Special Reference to *Calanus finmarchicus*," *Transactions of the Royal Society of Canada Sect. IV* 12 (1918): 207–33; C. Fritz, "Plankton Diatoms, Their Distribution and Bathymetric Range in St Andrews Waters," *Contributions to Canadian Biology 1918–20* (1921): 49–62; C. Fritz, "Experimental Cultures of Diatoms Occurring near St Andrews, N.B.," *Contributions to Canadian Biology 1918–20* (1921): 63–8; R. Chambers and B. Mossop, "A Report on Cross-fertilization Experiments (*Astrias* X *Solaster*)," *Transactions of the Royal Society of Canada Sect. IV* (1918): 145–7; B.K.E. Mossop, "A Study of the Sea Mussel (*Mytilus edulis Linn.*)," *Contributions to Canadian Biology 1921:2* (1922): 17–48, chart I-III; and B.K.E. Mossop, "The Rate of Growth of the Sea Mussel (*Mytilus edulis L.*) at St Andrews, New Brunswick; Digby, Nova Scotia; and in Hudson Bay," *Transactions of the Royal Canadian Institution* 14 (1922): 3–22.

60 M.L. Friedland, *The University of Toronto: A History* (Toronto University of Toronto Press, 2002); Ralph A. Bradshaw, "Clara Cynthia Benson,"

*American Society for Biochemistry and Molecular Biology Today* (March, 2006), available from http://www.asbmb.org/uploadedFiles/AboutUs/ASBMB_History/200603Clara%20Bensen%20-%20March%202006.pdf.

61 A.R. Ford, *A Path Not Strewn with Roses: One Hundred Years of Women at the University of Toronto 1884–1984* (Toronto: University of Toronto Press, 1985).

62 Rossiter, "'Women's Work' in Science"; and Rossiter, *Women Scientists in America*.

63 Rigby and Huntsman, "History of the Fisheries Research Board of Canada."

64 P.S. Brown, "Early Women Ichthyologists," *Environmental Biology of Fishes* 41 (1994): 9–30.

65 W.A. Clemens and L.S. Clemens, "A Contribution to the Biology of the Muttonfish (*Zoarces anguillaris*)," *Contributions to Canadian Biology 1918–20* (1921): 69–83.

66 E.H. Craigie, *A History of the Department of Zoology of the University of Toronto up to 1962* (Toronto: University of Toronto Press, 1966).

67 W.A. Clemens, "Education and Fish: An Autobiography," Fisheries Research Board of Canada Manuscript Report 974 (1968): 1–102.

68 W.A. Clemens, "Reminiscences of a Director," *Journal of the Fisheries Research Board of Canada* 15:5 (1958): 779–96; and Clemens, "Education and Fish."

69 Ainley, "A Family of Women Scientists"; Ainley, "Canadian Women Natural Scientists"; and Ainley, "Marriage and Scientific Work in Twentieth-Century Canada."

70 J.C. Stevenson, "Edith and Cyril Berkeley – An Appreciation," *Journal of the Fisheries Research Board of Canada* 28 (1971): 1360–4.

71 M.N. Arai, "Publications of Edith and/or Cyril Berkeley," *Journal of the Fisheries Research Board of Canada* 28 (1971): 1365–72.

72 Rigby and Huntsman, "Fisheries Research Board of Canada."

73 A.T. Cameron and I. Mounce, "Some Physical and Chemical Factors Influencing the Distribution of Marine Flora and Fauna in the Strait of Georgia and Adjacent Waters," *Contributions to Canadian Biology and Fisheries (New Series)* 1:4 (1922): 39–72; and I. Mounce, "Effect of Marked Changes in Specific Gravity upon the Amount of Phytoplankton in Departure Bay Waters," *Contributions to Canadian Biology and Fisheries (New Series)* 1 (1922): 81–94.

74 M.G. Ainley and C. Millar, "A Select Few: Women and the National Research Council of Canada, 1916–1991," *Scientific Canadian* 15:2 (1991): 105–16.

75 Ainley, "Canadian Women Natural Scientists."

76 M.N. Arai, "Charles McLean Fraser (1872–1946): His Contributions to Hydroid Research and to the Development of Fisheries Biology and Academia in British Columbia," *Hydrobiologia* 530/531 (2004): 3–11.

77 Ainley, "Canadian Women Natural Scientists."
78 Ibid.
79 Rigby and Huntsman, "History of the Fisheries Research Board of Canada."
80 Ibid.; and Johnstone, *The Aquatic Explorers.*
81 Ainley, "Canadian Women Natural Scientists"; and Brown "Early Women Ichthyologists."
82 J.L. Hart, "Fisheries Research Board of Canada Biological Station, St Andrews, N.B., 1908–1958: Fifty Years of Research in Aquatic Biology," *Journal of the Fisheries Research Board of Canada* 15 (1958): 1127–61.
83 Johnstone, *Aquatic Explorers;* and Hubbard, *A Science on the Scales.*
84 J.R. Smallwood, ed., *Jr. Encyclopedia of Newfoundland and Labrador,* vol. 2 (St John's: Newfoundland Book Publishers Ltd., 1981), 428–9.
85 A.A. Berkeley, "Sex Reversal in *Pandalus danae,*" *American Naturalist* 63 (1929): 1–3.
86 Ainley "A Family of Women Scientists"; and McLaughlin and Gilchrist, "Women's Contributions to Carcinology."
87 A.B. Needler, "Paralytic Shellfish Poisoning and *Gonyaulax tamarensis,*" *Journal of the Fisheries Research Board of Canada* 7 (1949): 490–504.
88 Rigby and Huntsman, "Fisheries Research Board of Canada"; and Hubbard, *A Science on the Scales.*
89 Viola Davidson, "Salmon and Salmon Eel Movement in Constant Circular Current," *Journal of the Fisheries Research Board of Canada* 7:7 (1949): 432–48.
90 P.A. McLaughlin and P.L. Illg, "Josephine F.L. Hart (1909–1993)," *Journal of Crustacean Biology* 14 (1994): 396–8. Hart's twenty-four publications are listed in this article.
91 J. Fingard, "Gender and Inequality at Dalhousie: Faculty Women before 1950," *Dalhousie Review* 64 (Winter 1984–5): 687–703; and Ainley, "Canadian Women Natural Scientists."
92 Ainley, "Gendered Careers."
93 Ibid.
94 C. Thomas, "Moira Dunbar: Woman Scientist the Navy Refused to Take on Board," *The Guardian,* Wednesday 12 January 2000.
95 D. Lavoie and D. Hutchinson, "The U.S. Geological Survey: Sea-going Women," *Oceanography* 18 (2005): 39–46.
96 Fuller documentation of women's advances in the 1960s and later must await future publications. A preliminary discussion of academic women in this period is included in "Gendered Careers," a 2005 paper by the late Marianne G. Ainley that encompasses developments to 1980.

# 3 The Gospel of Efficiency and the Origins of Maximum Sustained Yield (MSY): Scientific and Social Influences on Johan Hjort's and A.G. Huntsman's Contributions to Fisheries Science

JENNIFER HUBBARD

Scientific institutions such as the St Andrews Biological Station develop specific cultures shaped equally by the personal interests and characters of those who interact there, and the scientific and social beliefs and values shared by the larger culture in which they are imbedded. The resulting institutional contributions in turn reinforce these ideals. Two individuals who did the most to imprint a certain style of fisheries science on the station's early research program were first director Archibald Gowanlock Huntsman and Norwegian fisheries biology pioneer Johan Hjort. However, this article will argue that they both shared common ideas shaped by the larger culture: the ethic of service, a feature of nineteenth-century British science; and the conservation ideals then in vogue, which amounted to wise-use conservation. The service ethic, embodied in Joseph Banks, the English botanist who smuggled science policy into British government, enabled the creation of government-funded but independent science bodies, such as the ones that created the Atlantic and Pacific Biological Stations. The service ethic also had implications for the kinds of scientific problems addressed. "Gospel of Efficiency" conservation, by contrast, was the popular version of resource and wildlife conservation that emerged around the beginning of the twentieth century. Powerfully influenced by German scientific forestry, it was the wellspring of the maximum sustained yield ideal in both forestry and fisheries.

Nineteenth-century German scientific forestry, with its focus on "optimal" and "sustained" yield, shaped the discourse of early-twentieth-century fisheries scientists like Hjort and Huntsman through shaping the era's conservation zeitgeist. This influence exceeded that of fish farming, which has been argued to be the inspiration for maximum sustained yield in American fisheries science. Forestry science also directly inspired Huntsman to make direct analogies between trees and fish for productivity management. Huntsman was informed by the writings of Bernhard Fernow, chief of the US Department of Agriculture Division of Forestry (1886–98), who spent his later career (1907–17) teaching at the University of Toronto. Scientific service and sustained yield ideals strongly influenced fisheries biologists to ask questions of nature that had practical and utilitarian implications. Conservation as understood today did not figure in fisheries science until the late twentieth century.

The Atlantic Biological Station and its slightly younger sister station on the Pacific were founded due to a convergence of interests between Canadian university biologists and the Department of Marine and Fisheries. The scientists wanted facilities to pursue the new, trendy marine biology, while the department recognized that fisheries management required greater scientific understanding. Eric Mills, in the first chapter of this volume, notes the seemingly anomalous nature of a scientific organization, established by government funding, yet run by independent scientists with minimal government oversight. Such an institution was made possible because of the service ethic that characterized much of the practice of nineteenth-century science. Of particular importance is the tradition of independent but government-funded scientific institutions to serve the British government, the legacy of Joseph Banks, whose work and powerful influence are described in this volume's introduction. As president of the Royal Society from his election at the age of thirty-five, in 1778, until his retirement forty-one years later, he championed applied and public science. His position put him in the centre of powerful British interests, enabling him to smuggle science into English government; he acted as a scientific adviser and co-founded several bodies that operated as quasi-government departments, albeit run by volunteers steeped in the gentlemanly tradition of service. These individuals saw that they could improve their own economic positions while simultaneously helping their agricultural or other dependents and assisting Britain's economic development. For example, the Royal Institution, co-founded by Banks in 1799, served as a research and development body to solve problems of interest to the government. The Board of Agriculture, founded by Banks and the Scottish agricultural reformer Sir John Sinclair in 1793, was a public-private hybrid body created to improve British agriculture, government-funded but not answerable to the Treasury.[1] Such bodies slowly evolved into government ministry departments, with important ramifications for the kinds of scientific questions emphasized in these institutions.

The Board of Agriculture was similar in funding and functions to the early Biological Board of Canada. It existed because the government trusted gentlemen volunteers to run it altruistically, using annual endowments only for supplies, equipment, and labour. The British political tradition, then, developed a hands-off approach to scientific institutions; government funding was procured, but scientists worked as volunteers. As experts they set the research agenda, but received no remuneration, except for expenses derived from their volunteer activities. This became the model, as argued in the introduction, for Canada's first marine biological station, and the board of management of the Marine Biological Station. It inspired the creation of the permanent Atlantic Biological Station in Saint Andrews, New Brunswick, its sister station in Nanaimo, British Columbia, and all later fisheries research and experimental stations in Canada, run by the Biological Board (later the Fisheries Research Board of Canada). Participating university scientists, in the gentlemanly tradition, used their expertise to serve government and nation, and were trusted by Members of Parliament (if not always by government bureaucrats) to act on behalf of the dominion government.

This scientific tradition had practical consequences. No centralized research program was dictated by the Department of Marine, through which government funding came. Generally, Canada's government and its bureaucrats had only marginal interest in fisheries matters; at the turn of the century, no civil servants, aside from Commissioner of Fisheries Edward Ernest Prince (1858–1936), understood how science could improve Canadian fisheries. This hands-off model of government funding would appear to guarantee complete freedom in research. Prince could not impose any particular program: he relied on scientists' goodwill to investigate the effects of dynamite fishing, study problems of oyster cultivation in Prince Edward Island, and so on. However, ethical constraints within the gentlemanly tradition governed the choice of research and even dictated approaches to be followed in studying fish and fisheries. So strongly was the service ideal shared by university professors, moulded by the gentlemanly virtues of nineteenth-century gentlemen naturalists, that Prince had no trouble finding scientists willing to address fisheries problems. In return, they gained marine facilities to further their own research interests. The resulting fisheries biology was geared to help fishermen improve fishing efficiency, find more fish, and improve fish processing techniques.[2] Conservation or other environmental goals were secondary except where justified by recognized depletion, as with the work of A.P. Knight (1849–1935) on lobster conservation.[3]

The Canadian government, and therefore Canada's citizens, benefited by obtaining, virtually free of charge, the services of skilled Canadian scientists, some with stellar reputations as experts in their fields. However, scientists' desire to aid fishermen and fishing communities also shaped the questions asked, and conservation issues as we would understand them were not explored. For decades, Canadian scientists focused mainly on improving fishing efficiency, finding more fishable stocks, and working to help fish processors improve their products – canned, salted, dried, or frozen – so as to get better prices. Furthermore, this service ideal reinforced the "inexhaustible fisheries" bias Prince introduced to Canadian scientists. Prince had served as scientific assistant to William Carmichael McIntosh, scientific expert to the 1883 Commission on Trawl Nets and Trawl Fishing, headed by Thomas Henry Huxley. The commission concluded that since trawl nets swept the sea floor, while groundfish eggs and spawn float, trawling could not possibly harm the fisheries. Prince absorbed and defended Huxley's belief that no amount of fishing (given the technologies then current) would ever deplete the sea fisheries, since fish reproduce so prolifically. Prince was among a large coterie of European and North American scientists who dismissed fishermen's and some scientists' concerns about overfishing and diminishing catches.[4]

## Johan Hjort and the Introduction of Fisheries Biology to Canada

While Prince's science was somewhat blinkered, he did recognize Johan Hjort's significant contributions. Hjort (1869–1948), the director of the Fisheries Directorate headquartered in Bergen, Norway, worked in conjunction with other researchers under the auspices of the International Council for the Exploration of the Sea (ICES), to pioneer fish population demographic studies, laying the foundation for fish-stock dynamics and fishing-theory studies. Prince and the Biological Board's secretary, Dr A.B. Macallum invited Hjort (figure 3.1) to come to Canada in 1914 to head up the Canadian Fisheries Expedition of 1914–15. They wanted Hjort to introduce Canadian biologists to European oceanography, and to train them in the methods that were defining and creating fisheries biology. Ostensibly they invited him to help locate new concentrations of herring for Canadian fishermen, fulfilling the Biological Board's mandate to aid the government and fishermen. Sadly, the "concrete" results did not impress the Department of Marine and Fisheries – Canadian fishermen then had no need of underused fish stocks. Hjort's visit to Canada, however, achieved just what the Biological Board was really angling at – a newly invigorated scientific enterprise in Canada.[5] Archibald Gowanlock Huntsman (figure 3.2), Hjort's Canadian assistant, later recalled:

> His tremendous vitality and clear thinking made a great impression on us all ... His main contribution was to apply to our Atlantic waters the methods of investigation which had been developed in Europe and for which he had been to a considerable extent responsible ... He ... opened up for us in marvellous fashion the fishery-biological problem throughout these waters.[6]

Hjort's example instigated full-blown fisheries biology in Canada. Huntsman (1883–1973), Canada's first professional fisheries biologist, was hired full-time by the Biological Board within a year of the Canadian Fisheries Expedition.

After Hjort introduced new methods of fish population analysis developed by ICES and himself, Canadian biologists demanded improved fisheries statistics from the Department of Marine and Fisheries; records collected after the First World War improved markedly. Indeed, when the North American Council on Fisheries Investigations (NACFI) was created in 1921, in imitation of ICES, Canadian fisheries statistics were superior to US statistics. Hjort's most important convert was Huntsman, Canada's most influential fisheries biologist well into the 1930s. Curator to the Atlantic Biological Station since 1911, Huntsman spent his winters at the University of Toronto, where he became a full professor. When Hjort and his collection of herring samples arrived at Toronto in December 1914, Huntsman was appointed to assist him. With no background in fisheries biology, Huntsman soon began to criticize Hjort's method of scale analysis, which proposed that herring scales grew "in direct proportion to the length of the fish." Huntsman was supposed to analyse the age profile of western Atlantic herring stocks, but soon pointed out that "the scales could not be growing proportionately to the whole fish, since in the small herring, of which [he] ... had one preserved in a bottle, the scales did not touch each other, while in a large herring, they overlapped greatly."[7] Hjort, considerably annoyed, declared

3.1 Johan Hjort, left with Captain Robson ca. 1915

3.2 Archibald Gowalock Huntsman

Huntsman unfit to work with scales, and switched him to counting vertebrae and studying plankton. Despite this disagreement, Hjort and Huntsman respected each other, and Huntsman assisted all of Hjort's other expeditionary work. Thus, Hjort himself introduced Huntsman to the fundamentals of European fish population analysis, plankton studies, and hydrography, the main components of ICES research.

Inspired by Hjort, Huntsman, upon being hired full-time in 1916 by the Biological Board, dove whole-heartedly into fisheries biology.

(Before the expedition, his scientific focus was ascidian taxonomy.) Within two years he published articles in hydrography,[8] fisheries biology methodologies,[9] an analysis of the fish scale method – with proposed improvements[10] – and the world's earliest mathematical model of fish population dynamics and a fishery's effects on the fished population's age structure (see figure 3.3).[11] His graphical representation of the effects of fishing showed the diminishing numbers in older year-classes of fished populations.[12] Tim D. Smith argues that Huntsman's mathematical work did not receive the attention it deserved, appearing as it did in the Biological Board's first *Bulletin*, directed at fishermen. However, it is nevertheless believed to have influenced W.F. Thompson's work for the Pacific Halibut Commission in the 1920s and 1930s,[13] and Huntsman himself returned to it in the 1940s and developed further elaborations, which will be explored later.

Even before Hjort's arrival, summer researchers at the Atlantic Biological Station in St Andrews were studying how to improve canned and frozen fish products. Hjort's example vastly increased Biological Board research into fish-processing methods. Hjort was, to put it mildly, appalled by what Maritimers did with their weir-caught herring. Their cured and pickled herring was notoriously inferior; their poor practices showed little concern for quality. Also they ignored the growing North American preference for fresh fish, or fresh-tasting frozen fish. Hjort was eager to teach Canadian fishermen a new Norwegian fast-freezing process, in addition to the scientific work that justified his leave-of-absence from Norway's Directorate of Fisheries. The Canadian government, however, forbade Hjort's fish-processing demonstrations. Nevertheless, Hjort inspired later Biological Board fish-processing research by Huntsman and others. A.P. Knight crusaded for Canadian lobster cannery improvements, and Huntsman in the 1920s developed the world's first commercial-scale system for fast-freezing fish. While Huntsman's "Ice Fillet" scheme was sunk by the government, and American Clarence Birdseye went on to frozen food fame, Huntsman did give us the lasting and now standard technology of his "jacketed cold storage" – the use of plate coverings over refrigerant pipes in freezers to reduce condensation and loss of moisture in stored products.[14] By improving processed fish, Huntsman hoped to increase efficiency, conserving fish by reducing losses due to wastage.

Hjort thus further strengthened the service ideal and commitments of scientists at the Atlantic Biological Station. On the other hand, his example also weakened concerns about overfishing. As Norwegian

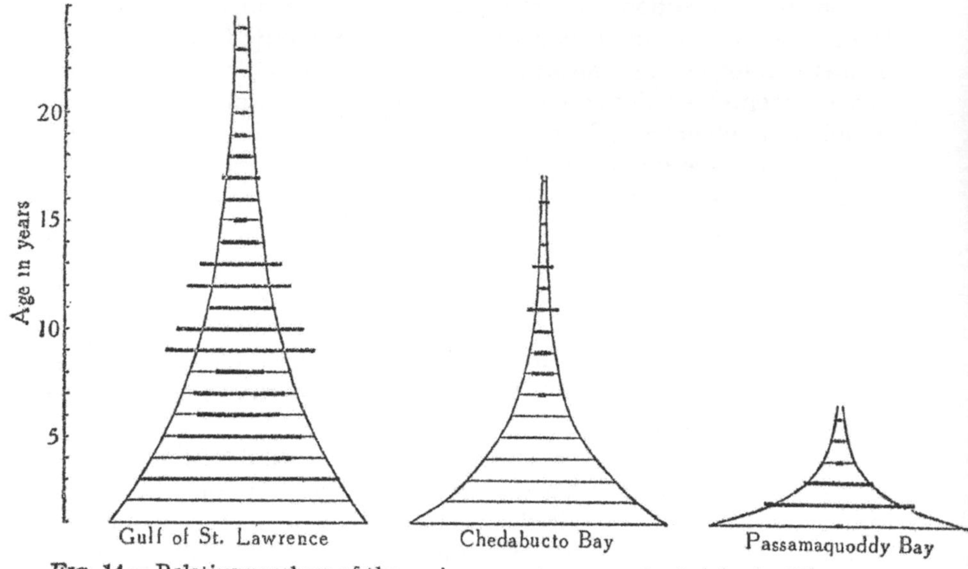


3.3 Huntsman's pioneering graphical representations of Maritime plaice stocks

fisheries historian Vera Schwach states, while British and German scientists were preoccupied with the overfishing problem, Hjort focused on natural variations in the catches. Scandinavians, in particular Norwegians, saw variations in catches as natural fluctuations, not symptoms of overfishing. Leading Norwegian scientists were mostly uninterested in overfishing until about 1957 for cod and the late 1960s for herring (following stock crashes), in part because of the low technological level of the fishing fleets compared to England's fishing industry.[15] Hjort strongly believed that the state should reform and modernize the economy: fisheries scientists' research should help modernize the fishing industry. Hjort promoted motorized boats, expanding fishing capacity, and exploiting more varied species of fish such as eel, halibut, and plaice and shellfish like shrimps. "There was no reason to limit fishing; on the contrary 'fish more' was the motto."[16]

Hjort was a scientist of international stature, so it is not surprising that Canadian fisheries biologists followed his example. The Biological Board's subsequent activities emphasized studies of fish behaviour, growth, and ecology, to enable a "rational" fishery and exploit "underutilized" or unused fish species and stocks. But it is important to recognize that Hjort's and the Biological Board's agendas also reflected larger societal trends that dominated the Progressive Era, shaping and forming their ideals and thus influencing later fisheries science.

## The Progressive Era and the Philosophy of Conservation: The Gospel of Efficiency

The year 1908 was significant in the history of conservation. Not only did the permanent Atlantic Biological Station in St Andrews open, but US president Theodore Roosevelt (1858–1919) appointed the National Conservation Commission. Conservation fever was at its height, as various interest groups came together to fight the disappearance of wilderness and industrial encroachment on the American wilderness. The National Conservation Commission succeeded in popularizing conservation, and inspired the foundation of the Canadian Commission of Conservation in 1909. In creating the American commission, with its four sections on water, forests, lands, and minerals, Roosevelt for the first time officially extended the word "conservation" beyond its original meaning – that is, saving and preserving water resources.[17] But in 1908, this more inclusive conservation had a very different meaning from its current connotations, as Samuel P. Hays made clear in his seminal 1959 work *Conservation and the Gospel of Efficiency: The Progressive Conservation Movement 1890–1920*. Conservation in 1908 meant maintaining or even improving nature for human use. This meshed with the Progressive Era's improvement ideals; reformers sought to improve workers', women's, and immigrants' rights, promote education and health, and so on.

The two men most influential in shaping the conservation agenda, aside from Roosevelt himself, were both trained in German scientific forestry. These were Gifford Pinchot (1865–1946), who served as the chief of the Division of Forestry in the Department of Agriculture, and who wrongly gets credited for the National Parks movement in the United States; and Bernhard Fernow (1851–1923, figure 3.4), who introduced German scientific forestry to the United States and Canada. Fernow, Pinchot's predecessor as chief of the Division of Forestry (1886–98), successfully campaigned for national parks to be created for public recreation and as resources for careful and efficient exploitation by the timber industry – both goals later adopted by Pinchot. As chief forester, Fernow sought to establish a national forest system and

3.4 Bernhard Fernow

introduce German scientific forest management to America, and laid most of the groundwork for creating national forests to protect watersheds, work for which Pinchot usually gets the credit.[18]

Fernow pioneered scientific forestry in both the United States and Canada. During the optimistic Progressive Era (1890–1920) reformers and activists successfully changed laws and institutions to abolish child labour, improve women's wages, educate immigrants and the poor, and eliminate political corruption at the city level. The American conservation movement fitted in with this desire for improvement. A few, preservationists like John Muir and his Sierra Club, wanted huge nature reserves free of human exploitation. When the first American national parks were being legislated into existence, they hoped these would remain untouched lands, unsullied by human exploitation – even limiting human visitors who wanted to enjoy nature. In contrast, the prevailing conservation ideal can be described as thoughtful exploitation of nature. The "Gospel of Efficiency" conservationists, led by US president Theodore Roosevelt during his presidency (1901–9), and Gifford Pinchot, chief of the Division of Forestry (1898–1910), successfully campaigned for national forests to be used as national parks for recreation and as reserves for careful and efficient exploitation by the timber industry. Water, forest, wildlife, and mineral resources were to be conserved by the "rational" exploitation of nature, in which scientists would direct natural-resource harvesting and use, eliminating any waste, "for the direct benefit of mankind."[19]

Forest reserves created through the new national parks system resonated more with the American general public than any other measure: Americans' self-identity was built on the American wilderness, differentiated from the built-up or intensively farmed and domesticated European landscapes. But conservationists not only lobbied for government-run national parks, but also popularized Gospel of Efficiency conservation ideals among the receptive general populace.

Conservation, then, had a very different meaning in 1908 than it does now. It meant maintaining nature for human use, and fitted in with earlier utilitarian programs to improve or increase resources. The ideals of rational exploitation and wise use of resources became common currency among scientists – who would serve as centralized authorities and decision makers in resource management – although they were not as popular with all the different end-users, such as woodsmen, stockmen and grazers, boatmen, fishermen, and so on. But the idea that scientific and engineering experts should be the ultimate authorities over resource use, because scientists were carefully studying resources, their exploitation, and renewability, resonated with government agencies and scientists alike.[20] Fisheries science was among these resource sciences influenced by Gospel of Efficiency conservation ideals.

While Samuel Hays vividly highlighted the conservation movement's goals, political adversities, and effects, he did not explain the intellectual origins of the Gospel of Efficiency. These origins lay in German scientific forestry, which he hints at by mentioning the German forestry training of both Pinchot and his predecessor, Fernow. It was not in the resource-abundant United States, where measurable exploitation of resources dates only from the 1820s onward, but rather in Germany, with its intensively used lands and deforestation that dates from the late medieval period, that the ideals of rational and scientific exploitation of resources originated.

In Europe, forested lands were valuable and coveted. Growing populations, cleared farmland, and wood shortages led to laws that mandated building in stone and restricted the use of timber as fuel. By the 1700s, timber and wood use were centrally controlled and rationed, and foresters began to practise silviculture – the science of fostering the reproduction of trees. Silviculturists learned to rotate timber harvests, preserving some stands while others were felled, with rotations of up to eighty or even one hundred years.[21] German silviculturists introduced science to forest management in the mid-1700s, using surveys, compasses, and plane tables. They measured tree circumferences and heights to estimate the volume of wood in a given area, and so determine when to harvest. Enormous worries around 1800 that Germany would suffer from a timber famine due to the enormous demand for timber generated by the Napoleonic Wars led to the creation of national forests. Owned collectively by the state instead of belonging to a specific prince or ruler, these were brought under controlled scientific management.[22] German scientists elaborated the characteristic features of scientific forestry management in the nineteenth century, some of which will be detailed later. Forestry science shared with industrialism the ideal that "maximizing economic efficiency in the short term" was also compatible with the social goal of getting the greatest level of social and economic benefits in the long term.[23] Both German forestry science and the ideal of service, then, were compatible, and guided the leader of the Gospel of Efficiency conservation movement.

The Polish-born Fernow, who studied at the University of Königsberg and the Forest Academy at Münden, first introduced German scientific forestry to the United States. As chief forester, his main goals were establishing a national forest system and introducing German scientific forest management. He laid the groundwork for the United States Forest Service, founded in 1905.[24] In 1898, he resigned, frustrated by politics, and founded the New York State College of Forestry – the first forestry school with a four-year degree in the United States – at Cornell University.

Fernow's successor as chief of the Bureau of Forestry had similar but less impressive training. Pinchot, came, like Banks, from a wealthy, landowning family. His parents demanded that their privileged children offer service to the world. Pinchot's mother introduced her son to the Social Gospel teachings of the 1880s. The Social Gospel's powerful influence redirected religious campaigns away from improving individual moral values "towards measures designed to improve man's environment."[25] Pinchot ardently followed these teachings, which in America were linked with the quest for maximum productivity and efficiency, since such conditions would benefit workers and consumers alike. Pinchot combined his love of the outdoors with his desire to be socially useful, and turned to forestry science, intending to help preserve America's endangered forests and natural heritage.[26] But with the Social Gospel also came the belief that it was not enough to preserve nature – nature must also be allowed to be useful.

After graduating from Yale in 1889, Pinchot went to London, England, to study under the leading forestry authority, German-born and trained Sir Dietrich Brandis, who taught a six-year program. But Pinchot's studies only lasted one year. He reasoned that as America had no professional forestry schools, his scanty training would still give him the status of a forestry expert. Also, since the United States had many seemingly untouched forests, and very different tree species, German training devised for Europe's extensively managed forests was not entirely relevant. But Pinchot absorbed the German forestry scientist's conviction that forests were a crop that could be nurtured like any other crop. The goal was to create a regular annual yield, while enhancing the condition of the whole forest. The result would be a sustained yield at optimal levels.

To demonstrate the financial value of systematic forest management, so the private sector would accept state forestry, he convinced Cornelius Vanderbilt to hire him to cut and manage his private forest in North Carolina. While Pinchot's experiment was a failure, he publicized his ideas at the World's Columbian Exposition in Chicago in 1893, using evocative photographs of managed forests in Europe. According to historian Brian Ballogh, this was the first extensive use of photographs rather than samples of material at an exhibition, and the public loved it. The utilitarian rationale for forestry science was conveyed via a romantic medium: pictures of lush and healthy forests. The public thus learned that nature and its technological exploitation were compatible; the wilderness, through systematic forestry, could be converted to economic use.[27] As chief of the Division of Forestry from 1898 to 1910, Pinchot's vision of efficient and productive use of natural resources shaped US federal and state policy for managing state-owned forests and national parks. Because he promised Roosevelt (a personal friend), as well as Congress and the US administration, that he could make the national forest reserves pay, he gained control over these reserves, and pushed for more national parks.[28]

Gospel of Efficiency conservationists like Pinchot, who dominated mainstream conservation measures until at least the 1960s, promoted the efficient and rational use of resources to ensure sustainable yields – goals taken directly from German forestry science. These goals, and the language used to express them, permeated the discourse of Canadian conservation also; the successful campaigns in the early twentieth century to conserve Canadian wildlife were affected by those who wished to use that wildlife for recreational purposes, such as hunting. Ducks

Unlimited, for example, achieved more to conserve the wetlands and spaces required for migrating birds than any preservationist group,[29] while the Hudson's Bay Company was instrumental in wildlife conservation to help conserve its fur trade.[30]

Nor was this approach limited to North America. While the progressive reform movement is best documented in its American manifestation, it was a North Atlantic movement: American reformers were inspired by progressive reformers in Germany and often were educated in German universities; in addition, there were contacts with Fabian socialists in England and social reformers in France. Reformers from Canada, the United States, and European countries corresponded and visited one another to investigate ongoing reforms in different locations.[31] The English and German conservation movement originated around the desire to conserve "national monuments" and places of national historical interest, but in the twentieth century grew to include conserving beautiful natural spaces for city-dwellers' recreation and enjoyment, and in Germany for logging or other economic uses as well. In Europe as in America, the Gospel of Efficiency also prevailed.[32]

**Forestry Influences on Fisheries Biology? Importing MSY into Fisheries Science**

"Gospel of Efficiency" ideals shaped natural-resource conservation policies, including those related to forests and fish stocks, through the 1970s and even beyond. Since this approach was strongly influenced by German forestry practices it can be argued that, at the very least, German forestry indirectly influenced fisheries biologists' promotion of the "rational use" of fisheries. The argument here is not that the mathematical models used to achieve the economic goal of Maximum Sustainable Yield (MSY) were influenced by forestry science: they were not. The history of population analysis from Verhulst through to Beverton and Holt, Ricker, Schaefer, and beyond is well documented.[33] Rather, the argument here regards the importation of a philosophy of management.

Forestry science–inspired conservation for efficiency can clearly be seen in the language of fisheries science; in 1923, for example, when Henry G. Maurice, the assistant secretary of Great Britain's Ministry of Agriculture and Fisheries, and president of ICES (1920–38) addressed the Challenger Society's "Conference of Associations and Institutions Conducting Marine Research," he told the gathered assembly that ICES "exists to promote 'the rational exploitation of the Fisheries.'" He continued: "To promote the rational exploitation of the fisheries must, in my view, be the chief function of a Fisheries Department," requiring knowledge of the life history of fishes and thus scientific investigation. The goal was to take "full advantage of the supplies Nature provides" while avoiding "the gradual exhaustion of the supply."[34] Here is "Gospel of Efficiency" conservation in a nutshell.

It is easy to see how German forestry science, by shaping the American conservation movement and the Gospel of Efficiency, also influenced fisheries biology. Of greater interest here, however, is an argument that the fisheries management ideal of MSY grew out of the Gospel of Efficiency. This seems to be apparent in terms of the Gospel of Efficiency's agenda of using resources wisely and efficiently, to prevent waste and promote conservation. However, the interdisciplinary origins of MSY in fisheries management have not received much attention, and no link between the Gospel of Efficiency and MSY has been documented before now.

In fact, a completely different theory has been put forward elsewhere. American fisheries biologist Daniel L. Bottom argues that American fisheries conservation ideas, including MSY, emerged out of the American mid-nineteenth-century tradition of trying to increase fish stocks in rivers, lakes, and in the case of anadromous fish, the sea, through propagating fish in fish hatcheries, and planting eggs and fry in streams and lakes. The methods and goals of the US fish culture movement were essentially agricultural. Bottom argues that the agricultural goals of US fisheries science shaped the questions asked and that this "may explain why fisheries science developed its own ideas and theories distinct from those of community or systems ecology." Bottom points out: "From the economic perspective of commercial fishermen, harvest restrictions to protect fish were simply unwarranted." Early advocates of fish culture believed hatchery eggs had a high survival rate that would result in rivers bursting with fish. Spencer Fullerton Baird (1823–87), the US Commission of Fish and Fisheries' founding commissioner, was a strong advocate. Artificial hatching was seen as a way to enhance nature, and even improve existing stocks by selecting seemingly tamer and more domesticated fish.[35] No one had concerns about contaminating or threatening native populations; attempts by the Biological Board in 1912 and 1913 to introduce Atlantic salmon to British Columbia water were enthusiastically promoted. Farmers supplemented their incomes with artificially stocked fish ponds, and discussed the productivity of an acre of water; American farmers had thus already equated fish conservation and production with agriculture.

The literature and correspondence of the American Fisheries Society, of which A.G. Huntsman was a member from 1931, and president from 1934–7, is replete with explicit references to fish as crops, and confident predictions that fish could be managed as easily as agricultural crops (agricultural output had been increasing until checked by the Great Depression and the Great Plains droughts). However, time and experience slowly offered disillusionment. Canadian scientists learned to suspect fish culturists' claims of success in restocking rivers and lakes by fish hatcheries, especially following Biological Board experiments with lobster hatcheries in the early 1910s, and R.E. Foerster's British Columbia salmon hatcheries investigations in the 1920s. The North American Fish Policy Committee, convened in 1937 to harmonize the American Fisheries Society with Canadian and Mexican policies, stated:

> It has been appreciated by biologists that fishes should be considered as crops, with a definite annual production from each body of water. As in the case of other crops, the annual production is dependent on many factors ... The food supply, temperature and depth of the water, available oxygen, rate of growth of the fish ... The effects of these ... must be amplified by more intensive research.[36]

Huntsman consistently warned the American Fisheries Society about its fish culture policies. Rather than confidently asserting the success of every intervention, he advised the head of the Bureau of Fish Management and Propagation, E.L. Wickliff, that "every fish cultural or fish betterment procedure be designated doubtful that has not been shown to be effective by accurate and adequate experiments."[37] Problems with fish culture were recognized, but in the United States, fishermen and scientists alike preferred it to conservation regulation up to the Second World War.

What one does not find in the committee reports or letters is any mention of sustainable yield or optimal yield or the rational use of resources as a means of conservation. In fact, the language of sustained yield is absent. This language makes no sense in the fish-culture approach to conservation, which promised limitless resources; rather, it is found in scientific papers devoted to restricting or measuring fishing effort. Sustained yield management is justified by the assumption of a scarcity of resources.[38] It is hard to see how American fish propagation could have influenced European and Canadian fisheries scientists in shaping a conservation program for marine fisheries.

Overfishing was a problem in certain ocean fisheries, for which it was difficult to determine species' stock sizes and demographics. Scientists at the 1937 meeting of the North American Council on Fisheries Investigations discussed restricting mesh size for certain fisheries. This organization was purely scientific, unfunded, and with no fisheries management mandate.[39] Conservation was rarely an issue for Canadian Atlantic fisheries biologists, however, except in the case of locally exploited stocks, such as lobster. In Europe, by contrast, fisheries depletion after the introduction of steam trawlers had led to the creation of the International Council for the Exploration of the Seas in 1902. Concern abated just following the First World War, but depletion became an increasingly important issue in Europe in the 1930s. Indeed, it is then, as Tim D. Smith has documented, that Johan Hjort introduced the language of optimal yield to fisheries biology in his paper "The Optimum Catch," which he published with his students and colleagues Guar Jahn and Per Ottestad in 1933. The question is, was there any link between the optimum yield concept in fisheries biology and its older iteration for managing Germany's limited forest resources for over a century?

While this paper did not call the desired goal of fishing intensity "maximum sustained yield," Hjort described the optimum catch as "a definite level to which the stock can be reduced, while preserving the maximum capacity of regeneration and at the same time securing an optimum catch."[40] Hjort's study, motivated by his concern for the future of the lucrative Antarctic whale fishery, followed his voyage to the Antarctic to observe the whaling factory ships. Hjort, according to Smith, had many cross-disciplinary influences: his self-imposed exile in Cambridge, England, from 1916 to 1921 (due to his opposition to Norway's wartime fisheries policies) would have exposed Hjort– who studied physiology and zoology at the university– to the nascent animal population studies of Charles Elton, then a student of Julian Huxley's at Oxford. Smith also speculates that he would later have indirectly encountered the statistical population modelling of Johns Hopkins professor Raymond Pearl,[41] whose English counterpart, Edward S. Russell, the director of the Fisheries Laboratory at Lowestoft, and a fellow member of ICES, dedicated his 1942 monograph, *The Overfishing Problem,* to Pearl. Certainly in his 1933 paper, the product of a long whaling voyage, Hjort applied the logistic curve and mathematical population modelling methods introduced in the 1920s by Pearl. But while Pearl had dealt with the fluctuations of natural, unharvested populations, Hjort was looking at a whale population under severe hunting pressure. Therefore Hjort and his colleagues,

> instead of concentrating on the maximum population size, as Pearl had ... focussed on the middle part of the growth of populations. Their insight was that if a harvest was to be taken that did not cause the population to continuously decline, and hence, would avoid the usual pattern of population depletion, it could be no larger than the amount by which the population would have increased in the absence of a harvest when the population was growing at its fastest.[42]

It is important to note several things here. The first is that even this step involved the importation of ideas and techniques from animal population studies: besides Pearl's logistic equation, Hjort drew upon the Norwegian bear population studies of Amund Helland, who was concerned about the declining population of hunted bears. But while Helland was simply trying to estimate what proportion of the bear population had been lost since 1848, so as to estimate the "original" population size,[43] Hjort was trying to figure out how to obtain a consistent optimal yield of whales. Although the mathematics comes from population ecology, the intent and language mirror German forestry, which had by this time become well established across Scandinavia.

The pragmatic Hjort, as Smith has observed, "was able to bridge academic disciplines," bringing mathematical population ecology to "real world" fishery problems.[44] It is most likely he was aware of trends in forestry science in Norway, a burgeoning field begun "in 1897 with the opening of the Department of Forestry at the Agricultural University of Norway,"[45] in Oslo. Hjort, in 1921, returned to Norway from his English exile to take up a new post as professor of marine biology at the University of Oslo, where pragmatic agricultural and forestry research was being carried on. But even if he paid no attention to the forestry work there, it was impossible to miss forestry's fruits sprouting up along the western coast of Norway, because the Department of Forestry at the Agricultural University of Norway was just one institution among many carrying out forestry education, research, planting, and propaganda. In 1916, a private forest research institute for western Norway (Vestlandets forstlige forseksstasjon) was established to research afforestation of this region, joined in 1917 by the Norwegian

Forest Research Institute (Det norske Skogforseksvesen), which became a centre for Norwegian forestry research.[46]

Norway was in many ways resource poor, aside from its fisheries and forests. The supply of timber was vital; historically an important export to other nations for shipbuilding and construction, it also provided fuel, building material, and other wood necessities for home consumption. In 1921 one-sixth of Norway's labourers were employed "in the wood-refining industries" and "one-tenth in the paper industry." Under Norway's strict forest conservation policy, all available means were employed to build up a future forest supply.[47] Norway, not a rich nation, in 1921 still had to import considerable quantities of wood from Finland, but was cut off from its former Russian supplies following the Russian Revolution. The main forestry research issues, identified in 1927 by Erling Eide, the director of the Norwegian Forest Research Institute, included "the regeneration of the pine forests in Northern Norway" and "the relation between growth and consumption in Norwegian forestry."[48] The entire program was based on German forestry science. "In 1927, probably more than 98 per cent of research activity was on silviculture and the rest on inventory/mensuration,"[49] financed by the Ministry of Agriculture.[50] Norway generously supported its forest industry, encouraging its development and productivity.

Aside from government efforts, private individuals vigorously participated in the Norwegian Forestry Association, which by 1921 had over 15,000 members, including lawyers and teachers. This indicates a well-coordinated publicity program to draw in the general public. Founded by Axel Heiberg in 1898, this association aimed to restore forests along the whole Norwegian western coast. It educated and enlisted the public and obtained generous public and private financial support. The association also worked to improve and preserve existing forests, through "rational cutting and ... drainage." In the west coast districts, Heiberg observed, "The hill-sides, once bare, are now covered with young trees, firs and pines, all in a healthy state of growth and development. There is great future wealth in store for us."[51] Before the Second World War the primary problem facing Norwegian forestry was "possible over-cutting of Norwegian forests," leading to an emphasis in the 1920s and 1930s – and even after the war – on research "in the fields of yield, inventory and mensuration ... regarding inventories at national, regional, county and forest ownership levels, and growth and yield models."[52]

One does not have to look as far afield as Germany or the United States, then, to see how forestry likely shaped the optimum yield philosophy that Hjort introduced to fisheries science. Although Hjort was worried about whale overfishing, his primary concern was the whaling industry's continuance. Hjort, as has been stated earlier, "fought with those who thought that overfishing was the main problem for the fishing industry."[53] His message regarding the whales was optimistic. With careful attention, an optimum, and by inference, continued yield could be achieved, by tailoring the catch rate to be in equilibrium with the maximum population growth rate. The objective was economic control, not conservation as we would understand it.

**Social Upheaval and the Politics of Sustained Yield**

Sociologist Robert G. Lee, who studied the history of the idea of sustained yield in forestry, observes that it has changed meaning in different periods, just as in fisheries the definition of MSY may have had a precise meaning for scientific experts, but a much looser meaning for policymakers and the general public; MSY also has changed connotations today compared to its understood meaning in 1950. Meaning, Lee observes, "varies in time and place in response to prevailing social, economic, and political conditions."[54] The concept of sustained yield emerged around 1800, a period of social upheaval in the German states; the state grabbed vast areas of common forest land, traditionally collectively owned by peasants with rights to cut portions of the forests. The forests were in terrible shape; Germany was facing a timber famine. The ideal of sustained yield emerged in response to the instability and uncertainty due to war and social and economic change wreaking havoc with traditional communities. Sustained yield also came to the forefront in the United States first during the Progressive Era, when industrialization, population growth, and the closing of the western frontier led to fundamental social and economic change. The language of sustained yield passed from foresters such as Pinchot into the vocabulary of government officials. Sustainable yield was advocated by officials to promote social stability during periods of social unrest, promising future stability.[55] This language emerged again at the end of the First World War, when the government worried about soldiers returning from Europe, job shortages, and problems with the disposition of labour. Sustained yield was once more emphasized during the

Great Depression. Government officials promised the sustained yield of forest resources "as a means of creating stability and prosperity in small rural communities." During the Second World War, the Sustained Yield Forest Management Act of 1944 was "an attempt to answer the question of how small communities could avoid instability in times of rapid change."[56] The sustained yield concept, then, was an instrument for promoting social order, conveying the comforting idea that experts were managing resources on the basis of the best scientific information.[57]

Now consider when the similar language of optimal yield and maximum sustained yield emerged in fisheries science. The fact that Hjort's article on the optimum catch "reoriented the thinking of a generation of fisheries biologists"[58] shows that it reflected the concerns of the era in which it appeared. Any new scientific theory or approach must be appreciated by other scientists if it is to change the science. Hjort's paper, published at the height of the Great Depression, with its optimistic message of an optimum yield, meshed well with the hopes of other fisheries scientists similarly trying to improve the economic conditions of fish harvesters. Michael Graham (1898–1972), a Quaker socialist, later the director of the Fisheries Laboratory in Lowestoft, published a paper in 1935 that incorporated Hjort's optimum catch.[59] Graham advocated an overall reduction in fishing effort to allow the populations of North Sea fish to increase, which he claimed would not harm the yield[60] – an important argument since fish were needed to feed a hungry populace hurting from the economic setbacks of the Depression. He argued that if fishermen's efforts were reduced, their expense, waste of time, and lost profits would go down, since if they fished beyond the point where the overall catch fell, they expended more effort in finding fewer fish. Graham referred to his ideal as a "maximum steady yield," and when he later defended his calculations and ideas from critics like Huntsman, he defended them on economic grounds.[61] As he told Huntsman in 1948, "Some regulations, such as those of minimum mesh, may be justifiable … when it can be shown … that the fishermen would make no important sacrifice by adopting them. But for more far-reaching measures, I personally consider that it is necessary to produce fair estimates of the benefits that may accrue from regulations."[62]

How quickly the idea of an optimum catch or optimum fishing level spread is revealed by a 1937 letter by A.L. Tester from the Pacific Biological Station at Nanaimo to E.L. Wickliff of the US Fish and Game Commission. Tester objected to the commission's goals, stating that fish hatcheries and fish planting were irrelevant to most marine species. For commercial fisheries, the objective of science and fisheries management, he argued, was

> to determine the maximum weight of fish that can be removed from a body of water each year while maintaining the supply at an approximately constant level of abundance. This requires the determination of rate of growth in weight, of natural mortality, and rate of fishing mortality to arrive at the most efficient level of exploitation, and if necessary, the adjustment of this level to take into consideration the rate of replacement of the supply in order *to arrive at the optimum level* of fishing intensity.[63]

Whatever the provenance of Hjort's ideal of an optimal catch, it melded well with the existing Gospel of Efficiency ideal, popularized in the Progressive Era in the United States and in Canada. Trees and wildlife could be harvested at a rate that could guarantee a sustained yield, through the "rational" exploitation of these resources. This ideal was emphasized by US government officials again in the Great Depression. But so widespread was the idea of "rational" exploitation of resources that Henry Maurice, the president of ICES, used it quite unconsciously in his 1923 address to the Challenger Society, as mentioned above. Its provenance in German forestry science would have already been forgotten, if indeed it was ever remarked upon.

**A Direct Influence? Huntsman Draws on Forestry Science Ideas**

A.G. Huntsman himself provides the only clear evidence of forestry science directly influencing ideas in the early debates about MSY. Huntsman was one of those individuals who read widely, and had wider ecological interests than most fisheries scientists; he liked to incorporate ideas from other fields, and this, even if not attributed by him, was the case for forestry science. Huntsman had before him the example of David Pierce Penhallow (1854–1910), who preceded Huntsman as curator of the Atlantic Biological Station and was a founding member of the board of management. Trained at the Massachusetts Agricultural College and the Imperial College of Agriculture in Sapporo, Japan, Penhallow was appointed a professor of botany at McGill, where he became involved in timber and forestry research. While Penhallow was interested in palaeobotany and the taxonomy and evolution of different tree species based on their vascular structures and evidence from fossil trees, he also studied the strength, durability, and decay of different timber trees, perhaps under "pressure at McGill to produce results of use to their patrons in the Montreal business community."[64] He was among a corps of researchers investigating which woods worked best for different building applications. Penhallow also published articles on Canada's timber reserves and forestry "in trade journals and popular magazines such as *Canada Lumberman* and *Popular Science Monthly*."[65]

Huntsman would also have become aware of forestry science directly through Bernhard Fernow. In 1907, Fernow, the former chief of the Division of Forestry, and pioneer of scientific forestry in the United States, became the founding dean of the University of Toronto's Faculty of Forestry, Canada's first university school devoted to forest science. He had departed from Cornell University – after founding the first US four-year program of forestry science there – because his felling of a forest on Cornell lands to scientifically cultivate a forest from scratch ran afoul of Cornell's rich neighbours in the Adirondacks. New York governor Benjamin Barker Odell, Jr, vetoed an appropriation for the college's operating funds in 1903, effectively closing it.[66] Thus Fernow came to Canada, importing German scientific methods with him, and pioneered forestry science in Canada. He had started the *Journal of Forestry* at Cornell in 1902, and served as its editor-in-chief until his death in 1923. He published books that became classics in the history of forestry, including in 1907 *A Brief History of Forestry in Europe, the United States and Other Countries*.[67] Fernow came to the University of Toronto in the same year that the freshly graduated Huntsman was hired as an instructor of zoology. Fernow did not retire until 1917, by which time Huntsman was a professor. Huntsman must have met Fernow on campus, and both may have attended the same talks or listened to each other's papers. Huntsman knew many people in other University of Toronto faculties, including historian Harold Innes, with whom he collaborated.

Huntsman did not recount how well he knew Fernow, but did write about having a discussion with him. In Huntsman's words: "On hearing ... that deforestation was bad for the water supply and had caused our streams to dry up, and realizing the importance for fish of having water in the streams I consulted Dr. Fernow, Dean of the Faculty of Forestry of the University of Toronto, as to the facts of the matter."[68] Huntsman would not have seen any applicability to fisheries of forestry ideas then; it was not relevant to the fisheries biology then extant. Nor did he focus on sustainability prior to other scientists' overfishing research, because in the Maritimes there was no recognized overfishing, except for lobster, with which A.P. Knight had dealt so competently. However, with increasing rhetoric about sustainable yield during the Great Depression, he began to engage with ideas drawn from forestry science. As the chair of the Royal Society of Canada's Committee on Wise Utilization of Resources, he worked with Robert Newton, then acting president of the University of Alberta and a forestry professor who prepared the agriculture and forestry chapter for their wartime pamphlet *The Wise Use of Our Resources*. Huntsman, as chair, felt compelled to produce a résumé – a summation of contributed ideas – in which he briefly discussed forestry management.

He realized that forestry provided a better analogy for the fisheries than the American Fisheries Society's fish farming modelled on agricultural production, which viewed fish as crops. He circulated his "Memorandum on Animal Resources – a personal estimate" advising that

> when we "harvest" long-lived species, such as most fishes are, we are dealing with "trees," rather than "wheat." The virgin "forest," the result of many year's growth, is bound to show a diminishing yield if only a moderate amount is removed annually, unless there is a continual shift to new grounds. This is true even though the removal of the slowly growing old individuals permits more rapid production of new stock by making room for the rapidly growing young.[69]

Huntsman thus drew from forestry science the theory that removing mature, slow-growing trees opens up space and nutrients for faster-growing younger trees, thus increasing productivity. There is no agricultural analogy possible, since crops are planted annually, and any thinning would involve removing excess plants of the same age (which is also essentially how fish stocks were modelled in the theory of fishing of that period). This is apparently the first time anyone proposed that there would be a positive effect on the growth rate of younger fishes by the removal of older ones, and it proved to be influential. O.E. Sette included this fishing effect in his analysis of the California sardine fisheries a year later;[70] as will be seen, it entered the lexicon of ideas about how fish stocks were to be managed.

Several developments in German forestry science must be described here to show the parallels that existed between this science and the later fisheries biology. National forests for forest conservation were introduced around 1800; marine protected areas elsewhere were first created around 1860. Perhaps the most fundamental parallel is the use of growth rings to determine the age of a given specimen. Carl Oettelt (1724–1802) in his *Practical Proof That Mathematics Performs Indispensable Services for Forestry* (1765) introduced the practice of determining a tree's age by counting its rings. Linnaeus learned of this practice, and wrote of tree rings as "the chronicles of winters,"[71] and this practice thus crossed from forestry into common knowledge.[72] C. Hoffbauer, a German scientist, in 1898 validated the aging of fish using the ring dating of fish scales.[73]

Methods for determining "age classes" by the change of trunk diameter were introduced in 1789 by Johan Jacob Trunk (1745–1802), the Oberforstmeister in Austria. Trees were thereafter identified by age class – in decade increments – as a means of forestry management, a method adopted by the United States Forestry Bureau in the early twentieth century. Again, the parallel with fisheries biology, with its reliance on year-classes as a demographic tool in MSY management analysis, is evident. The idea of sustained yield itself was introduced into forestry by university-trained Heinrich von Cotta (1763–1844), director of forest surveys in Saxony. Cotta, the founder of the Academy of Forestry – after 1816 a state institution, and after the First World War affiliated with Dresden University – inaugurated sustained yield forestry using the estimate of wood volume in a set area to secure an approximately uniform felling budget.[74]

In the 1820s, the idea of the "normal forest" – a concept created in 1788 by an anonymous Austrian tax official – was adopted: a normal stock, distributed in normal age classes, was maintained so as to ensure a sustained yield management. Carl Heyer (1779–1856), a professor of silviculture at Darmstadt, developed an equation and several mathematical formulae to determine the felling budget. Heyer's methods involved regulating a forest, assumed to be in an "abnormal" condition, to bring it into a "normal condition." (Huntsman, as will be seen, introduced a version of this idea into fisheries science in 1947.) Trees were identified by age class (in decade increments) to create a regular distribution of trees of different age classes. Once the forest was made "normal," the forester determined the existing stock and the time needed to add necessary growth increments before harvesting. From the total amount of lumber in the forest, the forester deducted the proper normal stock needed for sustained yield management. The balance was available for periodic felling instalments.[75]

Johann Hundeshagan (1783–1834), professor of forestry at Tübingen in 1817 and Giessen in 1825, developed the formula method or "rational method" to regulate felling budgets and in 1826 founded "Forest Statistics," the basis of later scientific forestry management. The rational use of resources dominated forestry thinking until the 1980s. Forestry scientists throughout the nineteenth century debated what should be considered the proper felling age or period of rotation. Was it better to achieve a maximum volume of production, or some other volume of production to achieve the highest profits? Yield tables became a vital tool in German forestry from the 1870s onwards, and the new Forestry Experiment Stations, the first of which were established in Saxony in 1862, produced more reliable and detailed yield tables.[76] Forestry experiment stations, modelled on the agricultural experiment stations (as were the slightly later marine biological stations), also introduced climatic studies of forests, and experimented with methods to increase soil fertility and use of artificial fertilizers, influenced by Justus von Liebig. Despite the strong element of agricultural science in German forestry science, the two remained quite separate, and forestry science remained entirely utilitarian. "To most foresters, trees past their peak of growth are like savings bonds that no longer earn adequate interest; not cashing in and replacing with higher-yielding young stands makes no economic sense."[77]

Gustav Heyer (1826–83), the son of Karl Heyer, and Max Pressler (1815–86) introduced the idea of "soil rent," which calculated the cost of allowing trees to grow and stand as if they were indexed by the cost of the soil on which they were growing. The ideal is to attain the highest rate per cent on the capital invested in forest production. Slowly growing trees nearly at their maximum height and girth are far more expensive to maintain than rapidly growing less mature trees.[78] They are an economic liability, since it makes more economic sense to harvest these trees and plant new ones to replace them, to shorten the time to the next harvestable generation.

What the forestry scientists of the nineteenth and first half of the twentieth century did not notice was that their science was based on many unquestioned assumptions, some shared with later fisheries biology but with at least one major difference. Among these assumptions were economic and social justifications for sustained yield: that the demand for timber would be essentially unchanging and that a continued and stable flow of lumber and timber products would ensure social stability by ensuring available economically essential materials. With the caveat that fisheries catches were increasing in the early twentieth century, many of the basic assumptions were similar. There was no consideration of the effects of technological change or changing tastes.[79] Late-nineteenth-century German forestry science built up static models of production, promoting the continuance of a set level of production for an unchanging number of harvesters and producers.[80] By 1900 forestry scientists' yield-tables "allowed foresters to calculate the usable volume of timber from a standardised tree (*Normalbaum*) of a known species, age, and diameter. Foresters could more accurately predict the value of a forest if the trees growing in it conformed to the standardised, predictable trees in the mathematicians' tables."[81] Therefore foresters turned to planting forests in rows of monoculture. "The ultimate goal of the scientific forester was to know how much timber was in a plantation or forest without leaving the office – the data could be simply read from tables and maps."[82] Fisheries scientists, faced with different historical realities, were forced to take a more sophisticated approach. Hjort, Graham, E.S. Russell, and others knew that the numbers of fish and fishermen are constantly in flux, and beyond that economic factors such as changes in consumer demand and price fluctuations for goods also play an important role in fisheries' economic success. These considerations did not enter forestry science until the Great Depression, when timber prices crashed. Later, the failure of scientific forestry's early promise to greatly increase productivity through managed forest growth also forced a reconsideration of its early assumptions after the Second World War; productivity in one plantation after another dropped markedly after the first rotation.[83]

There are two other obvious differences between forestry and fishery science. First, forestry deals with visible, immobile objects while fisheries scientists must quantify mobile objects veiled by the waters' surface. Also, in a sense, trees define forest space or geography in a way that fish cannot define lake, river, or ocean geography. Nevertheless, these differences should not lead to us ignoring similarities in sustainable yield goals in forestry and fishery science.

Certainly it seems that Huntsman saw some points of comparison, including the longer time it takes fish to mature and trees to grow compared to agricultural crops. Huntsman's 1942 analysis argued that removing more mature fish would facilitate the faster growth of immature fish by removing competition for resources. While he saw that "harvesting" fish resulted in only smaller and younger fish being available, which he warned "may be economically undesirable," he continued to believe that "the cry of 'depletion' of the resources is usually based upon an uneconomic condition, which may not involve any decrease in abundance of the species concerned."[84]

Huntsman's forestry-inspired ideas had coalesced by 1947, when he attended the Symposium of Fish Populations held at the Royal Ontario Museum in Toronto, in January 1947. This symposium was actually convened in response to Huntsman's endless quest to define what is meant by "overfishing." In attendance were W.C. Herrington ("in charge" of US North Atlantic Fisheries Investigations); Daniel Merriman, director of the Bingham Oceanographic Laboratory; future giant in the field W.E. Ricker, then at Indiana University's department of zoology; M.D. Burkenroad, chief biologist with the North Carolina Survey of Marine Fisheries; R.E. Foerster, director of the Pacific Biological Station; J.C. Medcof, oyster specialist at the Atlantic Biological Station; A.W.H. Needler, director of the Atlantic Biological Station; A.L. Tester, a former Fisheries Research Board scientist who was now professor of zoology at the University of Honolulu; John Van Oosten, first director of the US Geological Survey Biological Resource Division's Great Lakes Science Center; and T.H. Langlois, director of the Franz Theodore Stone Laboratory at Put-in-Bay, Ohio. This is the meeting at which Burkenroad fired the first salvo of the Thompson-Burkenroad debate, questioning Thompson's interpretation of fisheries declines as being due only to overfishing.[85]

Huntsman had been reading the publications of Graham, Russell, and Thompson on the theory of fishing, and was quite critical of their ideas; he used the occasion to make a bravura presentation of a range of possibilities for the effects of fishing. Revisiting his pioneering paper of 1918, he reinterpreted its graphical illustrations of the effects of fishing on the Canadian plaice, *Hippoglossoides platessoides*. His discussion ranged widely over the historical effects of fishing on populations of herring – for which he could find no signs of overfishing in over one hundred years of the fishery in the Fundy region – and Atlantic salmon, the focus of most of his research from the 1930s onward, which he acknowledged could be easily overfished, because they are so readily captured during upstream river migration. Fisheries biologists, however, were too ready to assume a reduced catch was symptomatic of overfishing; he described ten conditions, including lower spawning survival, which could also cause that reduced catch.

One of these mirrored a tenet of forestry: a reduced catch, he said, might simply occur by removing from the overall stock a proportion that had taken years to accumulate – the old, large fish. German forestry scientists acknowledged that removing trees in old growth forests, and creating a managed forest, would result in a lower, but sustained, harvest, since it made no economic sense to wait extra decades for trees to reach their full size. Huntsman argued that fishing may result in "a decrease in old large fish, but a greater take of all sizes as has already been described for the Passamaquoddy herring. This may be the result of young fish making better use of food in growing than old fish. The *oversenility* has been remedied by fishing"[86] (emphasis his).

Here we see a version of Surplus Production Theory, although it is not so named, before US fisheries scientist Milner B. Schaefer introduced it as a plank of MSY fisheries management: the idea that there are surplus and unnecessary fish that play no important role in the continuance of the population. Huntsman's idea that removing older fish would remediate the fish population and improve its productivity is clearly one drawn from forestry science practices that sought to create a "normal" forest. This idea quickly gained widespread acceptance: two years later, Merriman observed: "Our Symposium has had considerable influence ... This laboratory has never been burdened with so many reprints as it has since the publication of [the special Symposium on Fish Populations] ... issue of the Bulletin [of the Bingham Oceanographic Collection]. The demand has been tremendous both here and abroad, and requests continue to arrive in almost every mail."[87] While W.C. Herrington was critical of Huntsman's arguments as not being conducive to protecting fish stocks, he later championed the most aggressive interpretation of MSY, using Shaeffer's surplus-production theory, as the goal of fisheries science when he represented the United States at the International Technical Conference on the Conservation of the Living Resources of the Sea in Rome in 1955.[88] Wilbert M. Chapman, the University of Washington ichthyologist who profoundly shaped US high-seas fisheries policy, working as an attaché in the State Department, introduced MSY as the goal of American fisheries policy, and strongly believed that "fishing out" the older fish would increase fish stocks.[89] More strikingly, Erik M. Poulson, the secretary of the International Commission for the Northwest Atlantic Fisheries, argued at this same conference:

> In a dense population the fishes are slow growing and old ... In such a population a thinning out by an intense fishery is a great benefit – from our point of view – causing the old, slow growing individuals to be

> replaced by young, fast-growing individuals, who use their food in adding to their size and weight, thus increasing the yearly crop.[90]

He added, "Many examples of this are known, fully supported by evidence," but in fact did not back this assertion up with any of that so-called evidence. Thus, an idea introduced through forestry entered the mainstream of fisheries biology.

## Conclusion

In fisheries biology, the management goal of MSY coalesced after the Second World War, when fish population demographic studies, used to develop fishing equations for predicting stock abundance, became the mainstay of American fisheries biology. Tellingly, MSY emerged at an unsettled time. Just as sustained yield in forestry was a response to social instability, MSY was invoked as the basic concept in marine fisheries management in the period when Europe was rebuilding after the war; England, Europe, and Japan responded to the need to find cheap food to feed their war-battered populations by encouraging increased fishing. The desire was to attain a sustained fishery with minimal but enforceable regulations, or if such was not possible, to move on to other fishing grounds. America, deeply involved with post-war reconstruction, but also trying to maintain territorial prerogatives with regards to its domestic fisheries, enshrined MSY as a goal of fisheries management in "U.S. Policy on High Seas Fisheries" published in the *Bulletin of the US State Department* in 1949.[91] Michael Graham cited the director of the US Fish and Wildlife Service (the wartime incarnation of the former Bureau of Fish and Fisheries) as stating in 1945 that the US government's natural resource program was geared to "ensuring the maximum sustained yield from the natural resources."[92]

With fisheries policies now being overseen in the United States by the Fish and Wildlife Service, fisheries biologists were brought under the purview of an organization with close contacts with the Forestry service, since the wildlife managed by this new service involved mainly forest species, and fire control was a major issue. The US government had also earlier created the Civilian Conservation Corps (1933–42), a New Deal make-work project which put thousands of unemployed workers into a conservation army dedicated to planting trees and rehabilitating the habitats of fish and wild species.[93] The trans-disciplinary concerns of these organizations facilitated the transfer of the sustained-yield forestry agenda to the fisheries branch of the Fish and Wildlife Service; in fact, Dr Ira N. Gabrielson, the director of the Fish and Wildlife Service, included "fur seals, halibut, blue crab, trout and black bass, Alaskan salmon, deer and elk, and waterfowl" when he stated that "the purpose of the conservation programme of the Government of the United States of America is to insure a maximum sustained yield from our natural resources."[94] When a US State Department reorganization in 1944 placed fisheries under the commodity division of the economic branch,[95] the strengthening of the economic ideal of sustainable yield as a fisheries goal was inevitable.

Wilbert McLeod Chapman, appointed in 1948 the special assistant to the secretary of state for fish, used fisheries science as "a tool of diplomacy within the U.S. State Department" and endorsed the use of MSY as a means of conserving fish stocks within territorial waters; hence, maximum sustainable yield became the official program of the US government, and was enshrined within the High Seas Policy published by the Secretary of State in January 1949, the US-Mexico fisheries treaty of 1949, and the policies of a series of scientific fisheries commissions from 1949 onward of which the United States was a charter member, beginning with the Inter-American Tropical Tuna Commission (IATTC) and the International Commission on Northwest Atlantic Fisheries (ICNAF).[96] The desire to ensure an abundance of protein for First and Third World populations at the outset of the Cold War made MSY appear to be the ideal strategy to ensure the provision of a high and sustainable yield of fish. America, working to "re-arm and build its allies and bolster its own armaments to contain communism" placed great emphasis on keeping a strong economy, with a rising gross domestic product. "Policymakers and influential opinion molders constantly reiterated that America needed to develop its resources at the quickest pace possible for freedom to survive," and fed the American belief that prosperity preserved democracy. "Maximum material production thus became a national duty and a moral imperative."[97] In Europe, too, the 1958 Oceans Convention at Geneva, formulated under United Nations auspices, formally invoked "Optimum Sustainable Yield" as "the basic concept of marine fisheries management," but this concept was invoked within "the narrow goal of securing "a maximum supply of food" from the renewable ocean resources."[98]

What those who chose to uphold MSY – the vast majority of North American scientists, and many elsewhere – did not notice was that it was not a scientific construct: as Carmel Finley expounds, it was an economic and social ideal which melded closely with prevailing political and business mentalities that extolled efficiency and scientific management[99] – even the scientific management of nature. Its German scientific forestry roots were forgotten or unnoticed, since the Gospel of Efficiency conservation movement which influenced early fisheries science never highlighted these origins in the first place. The utilitarian biological, economic, and social assumptions underlying both forestry and fisheries science went largely unquestioned until the late 1970s, when a new understanding of conservation began to permeate the natural sciences. The quest for efficiency and the assumptions that fed MSY were so much a part of the political, economic, and social thinking of the inter-war and post-war eras that they were invisible to the scientists of that time, because they sounded rational, practical, and obvious. This quest for efficiency meant "biologists tended to regard any unused surplus as waste,"[100] and the management regime that developed treated MSY as a goal rather than a maximum, resulting in fishing that often exceeded the maximum.

Not all scientists embraced MSY: Michael Graham, for one, noted that there were a variety of options for fisheries management. At the "International Technical Conference on the Conservation of the Living Resources of the Sea," held in Rome in April and May 1955, he observed that European fisheries authorities "ha[d] not as yet made any explicit choice among the possible qualities of the fishery – annual yield, catch per unit effort and average size of fish – but in the New World, the choice of maximum yield has been explicit in all recent international conventions."[101] However, even without embracing MSY, Graham was a man of his times, and like Hjort, was working for the fishermen in seeking to safeguard an optimum, sustainable yield; Russell and Graham favoured measures like restricting fishing tonnage, thus boosting the efficiency (catch per unit effort) and therefore the profits of remaining fishermen.

The service ethic of nineteenth- and early-twentieth-century scientists like Joseph Banks, Edward E. Prince, and Johan Hjort, highlighted at the beginning of this article gave way to a business-like approach among environmental scientists in the post-war era;[102] the principles of the later professional code, however, simply served to reinforce the utilitarian program advanced by earlier fisheries scientists, which dovetailed with Gospel of Efficiency conservation. The fisheries were to be conserved or even enhanced if possible so as to serve human needs, with little concern for the needs of the fish or their ecosystems; the Gospel of Efficiency understanding of nature did not raise questions concerning ecosystem functions and species interactions. MSY in mid-century fisheries science, then, can be seen to have conformed to a rhetoric that had its ultimate roots in German scientific forestry. Just as German scientific sustained-yield forestry was blind to its economic and social, non-scientific formative assumptions, MSY was not recognized by most fisheries scientists for what it was: a philosophical and political construct. While MSY itself did not necessarily cause poorly managed fish stocks, the goal of maximizing the catch under all the varied conditions of the stocks, and treating MSY as a goal of management rather than a dangerous maximum, fed into poor management practices that were shaped by a conservation ethic that was ultimately a product of the Victorian and Progressive eras: sustained yield forestry and the Gospel of Efficiency.

NOTES

The author wishes to thank Eric Mills and Dave Wildish for their comments and constructive criticism, St Andrews Biological Station librarian Charlotte McAdam for her assistance in obtaining historical images, Rob Stephenson for inviting me to give the paper from which this chapter grew, and Adrian Hubbard for photographing Huntsman's graphs. Special thanks are due to Carmel Finley, whose inspiration, ideas, and conversations about MSY helped shape this chapter.

1 J. Gascoigne, *Joseph Banks and the English Enlightenment: Useful Knowledge and Polite Culture* (Cambridge: Cambridge University Press, 1994), 188–96. David Wildish, in reviewing this chapter, questioned whether the earlier organization of the Fisheries Research Board was indeed anomalous, commenting: "In my personal experience arm's length from central government has been the best and most productive way to run a research institution. When I joined SABS in 1969 there was a culture of service to the public, which was lost after 1973 when FRB status was lost … Today our service is not to the general public but to the political will of the current government."
2 Hubbard, *A Science on the Scales*, 38–66.
3 Ibid., 98–100.
4 See ibid., 149–72.

5 See ibid., 67–89.
6 Letter from A.G. Huntsman to Johan T. Ruud, 12 April 1949. Huntsman Collection, University of Toronto Archives, Accession no. B1978–0010, box 9.
7 A.G. Huntsman, "Methods in Fisheries Research," Canada: Fisheries Research Board, unpublished mimeograph (1951), University of Toronto Archives, Accession no. B1978–0010, 10.
8 A.G. Huntsman, "The Effect of the Tide on the Distribution of the Fishes of the Canadian Atlantic Coast," *Transactions of the Royal Society of Canada* Vol. XII, Series III (1918): 61–7.
9 A.G. Huntsman, "Fisheries Research in the Gulf of St Lawrence in 1917," Canadian *Fisherman* 1918, 740–4.
10 A.G. Huntsman, "The Scale Method of Calculating the Rate of Growth in Fishes," *Transactions of the Royal Society of Canada*, vol. 12, series 3 (1918): 47–52.
11 See V. Schwach and J.M. Hubbard, "Johan Hjort and the Birth of Fisheries Biology: The Construction and Transfer of Knowledge, Approaches and Attitudes, Norway and Canada, 1890–1920," *Studia Atlantica* 13 (2009): 22–41.
12 A.G. Huntsman, "The Canadian Plaice," *Biological Board of Canada, Bulletin No. 1* (1918).
13 T. Smith, *Scaling Fisheries: The Science of Measuring the Effects of Fishing, 1855–1955* (Cambridge, MA: Cambridge University Press, 1994), 205–8.
14 Hubbard, *A Science on the Scales*, 111–18, 134, 141–8.
15 V. Schwach, "An Eye into the Sea: The Early Development of Fisheries Acoustics in Norway, 1935–1969," in *The Machine in Neptune's Garden: Historical Perspectives on Technology and the Marine Environment*, ed. H.M. Rozwadowski and D.K. van Keuren (Sagamore Beach, MA: Science History Publications, 2004), 211–42.
16 Schwach and Hubbard, "Hjort and the Birth of Fisheries Biology," 38.
17 S.P. Hays, *Conservation and the Gospel of Efficiency: The Progressive Conservation Movement 1890–1920* (Toronto: McClelland and Stewart, 1959), 123.
18 A. Chase, *In a Dark Wood: The Fight over Forests and the Myth of Nature* (New Brunswick, NJ: Transaction Publishers, 1995, 2001), 28–9.
19 Hays, *Conservation and the Gospel of Efficiency*, 124.
20 Ibid., 265–6, 271–2.
21 B. Fernow, *A Brief History of Forestry in Europe* (Toronto: University of Toronto Press, 1913), 39–62.
22 Ibid., 86–91.
23 D.A. Perry, "The Scientific Basis of Forestry," *Annual Review of Ecology and Systematics* 29 (1998): 438.

24 Chase, *In a Dark Wood*, 28–9; and Henry Clepper, *Professional Forestry in the United States* (Baltimore: Johns Hopkins University Press, 1971), 20–30.
25 B. Balogh, "Scientific Forestry and the Roots of the Modern American State: Gifford Pinchot's Path to Progressive Reform," *Environmental History* 7:2 (April 2002): 201–2.
26 Ibid., 203.
27 Ibid., 213.
28 Ibid., 215.
29 J. Foster, *Working for Wildlife: The Beginning of Preservation in Canada* (Toronto: University of Toronto Press, 1978), 120–48.
30 T. Loo, *States of Nature: Conserving Canada's Wildlife in the Twentieth Century* (Vancouver: UBC Press, 2006), 93–120.
31 R. Payre, "A European Progressive Era," *Contemporary European History* 11: 3 (2002): 489–97.
32 K. Ditt and J. Rafferty, "Nature Conservation in England and Germany 1900–70: Forerunner of Environmental Protection?" *Contemporary European History* 5: 1 (1996): 1–28.
33 Smith, *Scaling Fisheries*, 194–235.
34 Henry G. Maurice, "Address to the Challenger Society," 21 January 1923. National Archives of Canada, RG 23, vol. 1200, file 726-1-11 [1].
35 D.L. Bottom, "To Till the Water: A History of Ideas in Fisheries Conservation," in *Pacific Salmon and Their Ecosystems: Status and Future Options*, ed. D.J. Stouder, P.A. Bisson, and R.J. Naiman (Detroit: Chapman and Hill, 1997), 573–4.
36 North American Fish Policy Committee of the American Fisheries Society, "A Progress Report," Huntsman Collection, University of Toronto Archives, B1978–0010, box 18, file 11.
37 A.G. Huntsman, March 1937, Letter to E.L. Wickliff. Huntsman Collection, University of Toronto Archives, B1978–0010, box 30, file 10.
38 R.M. Alston, *The Individual vs. The Public Interest: Political Ideology and National Forest Policy* (Boulder, CO: Westview Press, 1983), 16.
39 North American Council on Fishery Investigations, Minutes of the Twenty-Fourth Meeting, 23–5 September 1937. Huntsman Collection, University of Toronto Archives, B1978–0010, box 72, file 3.
40 J. Hjort, G. Jahn, and P. Ottestad, "The Optimum Catch," *Hvalradets Skrifter* 7 (1933): 127.
41 Smith, *Scaling Fisheries*, 216–18.
42 Ibid., 226.
43 Ibid., 221–5.
44 Ibid., 215.

45 B. Solberg and A. Svendsrud, "Development of Forest Research in Norway Since 1927: Some Issues," *Forestry* 70: 4 (1997): 359.
46 Ibid.
47 Canadian Pulp and Paper Association, "Forestry Conditions in Sweden, Norway, Great Britain and France," *Bulletin Number Thirty-four*, article VI, 15 October 1921. Internet archive: http://www.archive.org/stream/forestryconditio00beckrich/forestryconditio00beckrich_djvu.txt.
48 Solberg and Svendsrud, "Forest Research in Norway," 359.
49 Ibid., 362.
50 Ibid., 364.
51 Canadian Pulp and Paper Association, *Forestry Conditions in Sweden, Norway, Great Britain and France.*
52 Solberg and Svendsrud, "Forest Research in Norway," 360, 364.
53 Schwach and Hubbard, "Hjort and the Birth of Fisheries Biology," 38.
54 R.G. Lee, "Sustained Yield and Social Order," in H.K. Steen, ed., *History of Sustained Yield Forestry: A Symposium* (Portland, OR: Forestry History Society, 1984), 93.
55 Ibid., 93. See also R.G. Lee, "The Classical Sustained Yield Concept: Concept and Philosophical Origins," in *Sustained Yield: Proceedings of a Symposium Held April 27 and 28, 1982, Spokane, Washington*, ed. D.C. LeMaster, D.M. Baumgartner, and D. Adams (Pullman: Washington State University, 1982), 1–10; and T.R. Waggoner, "Community Stability as a Forest Management Objective," *Journal of Forestry*, November 1977: 710–14.
56 Lee, "Sustained Yield and Social Order," 90–100.
57 Carmel Finley, personal communication, 2009.
58 R. Angelini and C.L. Malony, "Fisheries Ecology and Modelling: An Historical Perspective," *PanamJAS* 2:2 (2007): 1935.
59 Graham had served as the fishery expert during the International Passamaquoddy Fisheries Commission (1931–3) to determine the probable effects of damming Passamaquoddy Bay for tidal power.
60 Smith, *Scaling Fisheries*, 231–2; Michael Graham, "Modern Theory of Exploiting a Fishery, and Application to North Sea Fishing," *Journal du Conseil* 10 (1935), 264–74.
61 M. Graham, *The Fish Gate* (London: Faber and Faber, 1943), 158–89; Michael Graham, "The Theory of Fishing," manuscript of presentation to the BAAS, Section D, 13 September 1948. Huntsman Collection, University of Toronto Archives, B1978–0010, box 17, file 9.
62 Letter from Michael Graham to A.G. Huntsman, 25 October 1948, ibid. Graham visited Huntsman at St Andrews on several occasions: they

became friends who disagreed on the interpretation of fish population fluctuations reflected in varying annual catches.

63 Letter from A.L. Tester to E.L. Wickliff, 1 April 1937. Huntsman Collection, University of Toronto Archives, B78–0010, box 30, file 10. Italics mine.

64 S. Zeller, "Darwin Meets the Engineers: Scientizing the Forest at McGill University, 1890–1910," *Environmental History* 6:3 (July 2001): 432.

65 Ibid., 439.

66 A.D. Rodgers II, *Bernhard Eduard Fernow: A Story of North American Forestry* (Durham, NC: Forest History Society, 1991), 308–18; and H. Clepper, *Professional Forestry in the United States* (Baltimore: Johns Hopkins University Press, 1971), 20–30.

67 Fernow, *Brief History of Forestry in Europe.*

68 A.G. Huntsman, "Résumé," in *The Wise Use of Our Resources: Papers for the Joint Session of Sections of the Royal Society of Canada, May 21, 1941*, ed. A.G. Huntsman (Ottawa: Royal Society of Canada, 1942), 37–8.

69 A.G. Huntsman, "Memorandum on Animal Resources – A Personal Estimate," March 1942. Huntsman Collection, University of Toronto Archives, B1979–0048, file 008.

70 O.E. Sette, "Studies on the Pacific Pilchard or Sardine (*Sardinops caerula*). I. Structure of a Research Program to Determine How Fishing Determines the Resource," *United States Fish and Wildlife Service, Special Scientific Report*, 1943: 19.

71 R.A. Studhalter, "Early History of Crossdating," *Tree Ring Bulletin* 21 (1956): 33.

72 Fernow, *Brief History of Forestry*, 68–72.

73 J.R. Jackson, "Earliest References to Age Determination of Fishes and Their Early Application to the Study of Fisheries," *Fisheries* 32:7 (2007): 323.

74 Fernow, *Brief History of Forestry*, 121–9.

75 Ibid., 109.

76 Ibid., 94, 126–8.

77 David A. Perry, "The Scientific Basis of Forestry," *Annual Review of Ecology and Systematics* 29 (1998): 451.

78 Ibid.

79 Alston, *Individual vs. Public Interest*, 16–17.

80 M. Clewson and R. Sedjo, "History of Sustained Yield Concept and Its Application to Developing Countries," in H.K. Steen, ed., *History of Sustained Yield Forestry: A Symposium* (Portland, OR: Forestry History Society, 1984), 4.

81 Ibid.

82 Chris Land and Oliver Pye, "Blinded by Science: The Invention of Scientific Forestry and Its Influence in the Mekong Region," *Watershed* 6:2 (2000–1): 27. For a comprehensive overview of the shortcomings of German scientific forestry management see J.C. Scott, *Seeing Like a State: How Certain Schemes to Improve the Human Condition Have Failed* (New Haven: Yale University Press, 1998), 11–22.
83 Land and Pye, "Blinded by Science," 28.
84 Huntsman, "Memorandum on Animal Resources."
85 See B.E. Skud, *Revised Estimates of Halibut Abundance and the Thompson-Burkenroad Debate*, International Halibut Commission (Seattle), Scientific report no. 56 (1975): 1–33.
86 A.G. Huntsman, "Fishing and Assessing Populations," in *A Symposium on Fish Populations Held at the Royal Ontario Museum of Zoology, Toronto, Canada, Jan. 10 and 11, 1947*, ed. D. Merriman, *Bulletin of the Bingham Oceanographic Collection*, vol. 11, Peabody Museum of Natural History, Yale University, p. 17.
87 Letter from Daniel Merriman to A.G. Huntsman, 10 April 1949. Huntsman Collection, University of Toronto Archives, B1978–0010, box 23, file 4.
88 C. Finley, *All the Fish in the Sea: Maximum Sustainable Yield and the Failure of Fisheries Management* (Chicago: University of Chicago Press, 2011), 88; C. Finley, "The Tragedy of Enclosure: Fish, Fisheries Science, and Foreign Policy, 1920–1958," University of California, San Diego, doctoral dissertation, 2007, chap. 12.
89 C. Finley, "The Social Construction of Fishing, 1949," *Ecology and Society* 14:1 (2009): 6, p. 4 [online], http://www.ecologyandsociety.org/vol14/iss1/art6/.
90 E.M. Poulson, "Conservation Problems in the Northwestern Atlantic," in *Papers Presented at the International Technical Conference on the Conservation of the Living Resources of the Sea, April 18 to May 10, 1955, Rome* (New York: United Nations Publications, 1956), 184.
91 Finley, "Social Construction of Fishing, 1949," 3.
92 Michael Graham, "Concepts of Conservation," in *Papers Presented at the International Technical*, 7.
93 J.A. Salmond, *The Civilian Conservation Corps, 1933–42* (Durham, NC: Duke University Press, 1967), 124–5.
94 Graham, "Concepts of Conservation," 7.
95 Finley, "Social Construction of Fishing, 1949," 3.
96 Ibid.
97 P.W. Hirt, *A Conspiracy of Optimism: Management of Forests since World War Two* (Lincoln: University of Nebraska Press, 1994), xxi–xxii.

98 R.H. Stroud, "Introductory Remarks," in *Optimum Sustained Yield as a Concept in Fisheries Management*, ed. P.M. Roedel (Washington: American Fisheries Society, 1975), 1.

99 See Finley, *All the Fish in the Sea*.

100 D.H. Wallace, "Keynote Address," in *Optimum Sustained Yield as a Concept in Fisheries Management*, ed. P.M. Roedel (Washington: American Fisheries Society, 1975), 5.

101 Graham, quoted in Smith, *Scaling Fisheries*, 329.

102 See B. Golley, *A History of the Ecosystem Concept in Ecology: More than the Sum of the Parts* (New Haven: Yale University Press, 1993).

# 4 St Andrews Biological Station: A Case Study in the History of Public Science in Canada

ROBERT L. STEPHENSON

The Atlantic (St Andrews) and Pacific (Nanaimo) Biological Stations were the first laboratories in Canada, and among the first in the world, devoted to the study of the marine and aquatic science. They offer a unique case history in both the development of aquatic science and in the institutional response to the evolving needs of public science since their establishment in 1908. The St Andrews Biological Station's hundredth anniversary in 2008 offered an opportunity to explore how its history exemplifies the evolution of aquatic science, the development of the role of public science, and the changing administration and delivery of public science over the past century.

The history of programs at the St Andrews Biological Station (SABS) demonstrates the development of scientific concepts, advances in scientific methodology, and especially scientific information and tools required for fisheries management, aquaculture, and environmental monitoring. The station's legacy includes major contributions to the understanding of the oceans, the development of marine fishing and aquaculture, the study of human impacts on the aquatic environment, substantial changes to fisheries and oceans management paradigms, and the institution of international committees and marine initiatives. St Andrews Biological Station's research programs have reflected the conceptual evolution of aquatic science and the development of science for an increasingly elaborate management regime that has emerged over the past century.

The history of the St Andrews Biological Station is in essence a history of the development and governance of aquatic science in Canada. It is made up of the histories of individual careers, of research programs, of evolving disciplines and institutions, and of the fisheries and other marine and aquatic activities that have required new scientific methods and approaches.[1]

The St Andrews Biological Station has been, and remains, a prominent laboratory in many aquatic scientists' careers in Canada. For many years it was the major Atlantic laboratory, and although now small in size in comparison with other eastern Canadian laboratories (e.g., the Bedford Institute of Oceanography in Dartmouth, the Northwest Atlantic Fisheries Science Centre in St John's, and the Institut Maurice-Lamontagne in Mont-Joli), it remains an important site for its scientific programs, for its unique seawater and laboratory-culture facilities, and the taxonomic collection of the Atlantic Reference Centre, and as a training centre. There has been a legacy of mentoring and training students at all levels (see for example Arai, this volume). Until recently, a large fraction of aquatic researchers and those administering aquatic research in Canada have had a St Andrews connection.[2]

In addition, the biological station's history demonstrates the changing roles of public science. For most of its history, the station has had a dual role: generating scientific information for developing fisheries and aquaculture, but also for restricting fishing and other activities to achieve sustainability. Its history has been marked by debates over the responsibilities of government in providing science for the "public good," and over the appropriate mixture of "fundamental" versus "applied" science activities, and an ongoing discussion of the interaction between public laboratories and the universities. While the context and some of the driving forces have changed, there has been consistency in much of the underlying responsibility of "public science" to undertake relevant research in support of legislation, to collect and maintain data, and to provide credible information and advice.

The station's history also demonstrates how the administration and delivery of public science has changed to facilitate perceived public needs. The biological station at St Andrews has evolved from being an institute of academic investigation under an independent board of management, through a series of changing arrangements to become a federal government laboratory belonging to Fisheries and Oceans Canada, and from being the sole Canadian Atlantic research station to being one of a network in the east.

This chapter is structured (figure 4.1) around three periods in the biological station's history based on its supervisory organization: (1) the period before 1937, when the station was staffed primarily by academic "volunteers"; (2) the period from 1937 to 1973, when the station was

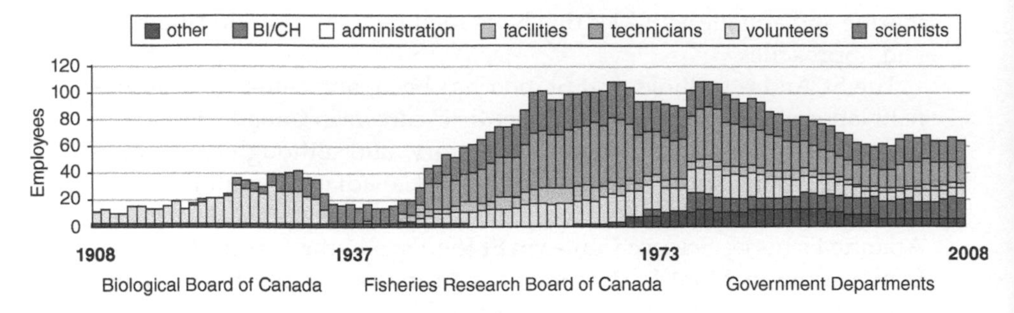


4.1 Employment structure within Canadian scientific departments, 1908–2008

under the jurisdiction of the Fisheries Research Board of Canada; and (3) the era from 1974 to the present, when the station became an entity within a government department. Table 4.1 shows a time line of major events referred to in the text.

Information and ideas for this chapter were drawn from primary and secondary literature, interviews, and discussions at a workshop held to commemorate the one hundredth anniversary of the St Andrews Biological Station.[3] A workshop, "Evolution of Marine Science in Canada," was held 15–16 October 2008, as part of the station's centenary commemoration. This workshop brought together station alumni and current staff, and included a number of historical presentations and case studies. The workshop included a panel of former directors, dating back forty years, which examined some of the significant issues of scientific investigation and management of SABS during the latter half of its history.[4]

## First Period: Volunteer Investigators of the Biological Board of Canada

The rationale for the establishment of the Atlantic Biological Station and its early history have been synthesized in a 1999 article by E.A. Trippel, and in more detail in Jennifer Hubbard's 2006 monograph.[5] Additional context is provided in chapters within this volume by Eric Mills, Jennifer Hubbard, and Mary Needler-Arai. Although the oceans had been significant in the economies of coastal nations for centuries, only at the end of the nineteenth century did several nations, including

Table 4.1 Major events in the history of the St Andrews Biological Station

| Date | Event |
|---|---|
| 1898 | Board of Management of the Marine Biological Station founded |
| 1899 | Portable (floating) Marine Biological Station at St Andrews, NB |
| 1908 | Permanent stations established on Atlantic and Pacific: Atlantic Biological Station at St Andrews, Pacific Biological Station at Nanaimo, BC – both staffed primarily by academic volunteers |
| 1912 | Biological Board of Canada established |
| 1915 | Canadian Fisheries Expedition |
| 1919 | A.G. Huntsman becomes first professional employee of the Biological Board |
| 1920 | North American Council of Fisheries Investigations (NACFI) |
| 1923 | Strait of Belle Isle Expedition carried out under auspices of NACFI |
| 1924 | Experimental stations at Halifax, NS, and Prince Rupert, BC, approved: both opened in 1926 |
| 1928 | NACFI approves investigation into environmental consequences of proposed tidal power dams around Passamaquoddy Bay |
| 1931–3 | International Passamaquoddy Fishery Investigation |
| 1932 | Atlantic Biological Station destroyed by fire; rebuilt |
| 1937 | Biological Board becomes the Fisheries Research Board of Canada (FRB) |
| 1944 | Atlantic Oceanographic Group established, followed by the Joint Committee on Oceanography, linking scientific oceanography with the Royal Canadian Navy |
| 1949 | Following Newfoundland's confederation with Canada, the FRB acquires fisheries research laboratory in St Johns, NL; Atlantic Biological Station is renamed the St Andrews Biological Station (SABS) |
| 1951 | International Commission for the Northwest Atlantic Fisheries headquartered at SABS |
| 1956 | A second tidal power proposal results in an International Joint Commission to determine potential impacts of damming Passamaquoddy Bay |
| 1958 | Post-war expansion of facilities and programs peaks; 242 employees at SABS |
| 1960 | Atlantic Oceanographic Group transferred to Halifax |
| 1962 | Bedford Institute of Oceanography established ; more SABS personnel transferred to Halifax |
| 1970 | Huntsman Marine Laboratory established |
| 1973 | SABS transferred from the FRB to the Department of Fisheries and Marine |
| 1977 | FRB dismantled entirely |
| 1997 | Canada's Oceans Act |
| 2002 | Canada's Species at Risk Act |

Canada, establish marine biological stations. Scientists began to found marine stations in Europe in the mid-nineteenth century (around 1870); the first major marine station was the Naples Zoological Station, which was completed in 1874. The marine biological laboratories at Plymouth, UK, and Woods Hole, US, were established in 1888.

Marine biological studies had been carried out in Canada by Canadian and foreign scientists since the 1830s, but it was not until the 1890s that the need for information on fisheries, increasing self-confidence by academics interested in research, and concern about the intellectual influence of the United States led to the first governmental support for a coastal laboratory, as discussed by Eric Mills in chapter 1. Mills points out that, in addition to the need for information about fisheries, there was the need at that time "to develop a Canadian scientific identity uniting utility with scientific understanding." In 1898, the Parliament of Canada approved the establishment of a portable (floating) marine biological station for a period of five years. The Marine Biological Station was a government-owned and financed facility, provided primarily for the use of university researchers. It was governed by the board of management of the Marine Biological Station, chaired by Dr E.E. Prince, the dominion commissioner of fisheries of the Department of Marine and Fisheries, but otherwise made up of eminent Canadian university scientists, notably Dr D.P. Penhallow (McGill University), Prof. R.R. Wright (University of Toronto), Prof. L.W. Bailey (University of New Brunswick), Prof. A.P. Knight (Queen's University), Dr A.B. Macallum (University of Toronto), Canon V.-A. Huard (Université Laval), Dr A.H. MacKay (Dalhousie University), and Prof. E.W. MacBride (McGill University).

The portable lab was used first in St Andrews in 1899. It was stationed for two seasons each, first in St Andrews, followed by Canso, NS, Malpeque, PEI, and Gaspé, PQ, before developing a leak and being abandoned in 1907. Research conducted at the portable station resulted in fifty-eight published papers and reports by twenty different investigators.

The perceived success of the portable station (and at least in part because it was wrecked in 1907), led the board to establish permanent laboratories on the Atlantic and Pacific coasts. The Atlantic (now St Andrews) Biological Station, first as a temporary lab in 1898, and then as a permanent facility in 1908, made Canada among the first to establish dedicated laboratories for the study of aquatic issues. St Andrews was chosen as the Atlantic Station site with the rationale that it was convenient via rail to Montreal, Saint John, and other major centres, was one of the real fishing centres of eastern Canada, had diverse habitats (including the especially rich feeding grounds of the Quoddy Region), and offered ease of sampling and study, as it had local fish traps and herring weirs, from which samples could be taken.[6] The Atlantic Biological Station and the Pacific Biological Station (Nanaimo, BC) began operations in the spring of 1908.[7]

Former station director and deputy minister of fisheries Dr Alfred Needler pointed out that the government's action in 1898 had been "epoch making in two ways: it marked the beginning of a long and effective support for fisheries (marine) research, and it established a board of management of eminent scientists, which evolved into the Biological Board of Canada (1912) and then the Fisheries Research Board of Canada (1937)."[8]

From the beginning, there were issues of governance and of the applied versus academic nature of science. In launching investigations, the Biological Board felt the need to prove to the government that its work was important, relevant to the industry, and therefore worth funding; and to the scientific community that the station was a viable research institution, so that people would come to do research there. Prince satisfied government requirements by producing frequent departmental reports on the station's activities. Also, many research and published papers were described as being "on matters of practical importance dealing with problems that faced the Canadian fishing industry and which thereby justified the enterprise in the eyes of Parliament and successive administrations."[9] Prince saw to the publication of scientific papers to benefit and connect to the scientific community, and initiated a journal called *Contributions to Canadian Biology* which published reports done by those working at the station. It would later become the *Journal of the Biological Board of Canada*, the *Journal of the Fisheries Research Board of Canada*, and the *Canadian Journal of Fisheries and Aquatic Sciences*.

The annual appropriation for the board was administered by the Department of Marine and Fisheries; this arrangement entailed "restrictions and delays."[10] The resulting conflict between government administrators and the board's scientists contributed to pressure for a more independent status, and under the Biological Board Act of 1912, the board was given authority to manage its own appropriation.[11] While the Biological Board became administratively independent, the Biological Board Act of 1912, and subsequent revisions, refer explicitly to the desire for a station that would undertake applied research, stating, in part, that "the Board ... has the conduct and control of investigations of practical and economic problems connected with the marine and freshwater fisheries, flora, and fauna, and such work as may be assigned to it by the Minister."[12]

Until the mid-1920s, the biological stations were seasonal field stations, generally occupied from May to September. However, in the 1908 report of the first year, Director Penhallow noted "the Board has directed that the St Andrews station should be constructed with a view to winter occupation, in order to preserve perfect continuity of work. Such occupation will be not only desirable, but in all probability a necessity in the near future."[13] It appears that some winter work (collection of hydrographic data and plankton) was conducted from around 1919, but it was not until 1928, when Harry B. Hachey was appointed resident oceanographer, that year-round scientific research really started at the station. Research in the early years was done almost entirely by volunteer investigators – university staff members and students visiting the station during the summers, who received travelling and living expenses but never any more. An exchange program also allowed some American scientists to work at the station.

Dr Archibald G. Huntsman, of the University of Toronto, became the biological station's first scientific staff member, when he was appointed permanent curator in the fall of 1914. As Hubbard elaborates in chapter 3, Huntsman was quickly assigned to work with Norwegian fisheries scientist Johan Hjort in 1914 and 1915 on the Canadian Fisheries Expedition. In the preface to the report of the Canadian Fisheries Expedition, Prince explains that the Government of Canada had been trying to entice Hjort to visit for almost a decade in order that he might lead the type of

> comprehensive fishery investigation in Canadian waters on the lines of researches which, under his direction, had proved so beneficial to the fisheries of Norway. In spite of the fact that the great cod and other fisheries off our Atlantic coast had been carried on for centuries, it was felt that there were doubtless hidden possibilities of development and expansion that awaited only a basis of accurate knowledge to turn them to account.[14]

Prince also commented on the plan for study:

> The plan embraced the necessary physical, hydrographical, chemical, and biological researches, including collection of water samples, etc., and the

> determination of the plankton collections, distribution and varied abundance of the young fry, as well as eggs of the cod, haddock, flat-fishes, and other species, in addition to the thorough study of the varieties, distribution, migrations, and breeding of the herring.[15]

The Canadian Fisheries Expedition (1914–15) was an early example of the concerted effort to develop fisheries, a theme that was to be prominent through much of the century, but it was arguably more important in its development of scientific concepts, in that it was a pivotal early attempt to link biological distribution with physical environment, and in emphasizing the importance of oceanography. It is significant that Hjort was the world's pre-eminent fishery scientist at that time, as is evident from his publication "Fluctuations in the Great Fisheries of Northern Europe."[16] As both Mills and Hubbard point out elsewhere in this volume, the expedition and collaboration with Hjort had a profound influence on Huntsman, whose prime interest turned from ascidian taxonomy towards fisheries biology, or what we would recognize today as fisheries oceanography. Following the Canadian Fisheries Expedition, Huntsman used the expedition approach in subsequent years for multidisciplinary study of other areas. These expeditions included those to Cheticamp (1917, southern Gulf of St Lawrence), Mirimichi (1918), St Mary's Bay district (1919), Cape Sable district (1921), St Margaret's Bay (1922), Strait of Belle Isle (1923), and, in 1930, the Hudson Bay Fisheries expedition.

The work undertaken in the Atlantic Biological Station's early years was diverse and differed from year to year, depending upon the visiting researchers. The report from 1921 demonstrates this diversity:

> The work done at the station during this season included: lobster rearing and lobster canning (Knight); temperature and light in the rearing of lobster larvae (Prince); study of diatoms (Bailey); bacteriology of canned and dried fish (Harrison); anaerobiosis in marine animals (Collip); larvae of decapods (Connolly); study of bioluminescence (Newton Harvey); factors influencing growth and reproduction (Huntsman); the histology of frozen fish muscle (Slater-Jackson); the culture of ostracods and copepods (Klugh); the rearing of lobster larvae (MacKay); effects of varying H-ion concentrations on copepods (Leim); the chemistry of fish muscle stroma (Panton); the chemistry of fish muscle juice (Dempsey); the bacteriology of reddened dried fish (Kennedy); the spore-forming bacteria (Newton); influence of varying H-ion concentrations on marine animals (Taylor);

success in reproduction in the cunner (Reid); effect of light on growth in intertidal animals (Fraser).[17]

Increasing recognition in the 1920s of the potential value of fisheries research, however, led to pressure for greater efforts and more concentration on practical fisheries problems, and as a result, the board started to develop full-time scientific staff at both biological stations.

Dr A.H. Leim was hired as the second full-time scientist in 1924. In that same year, four additional "seasonal" scientific staff members were appointed, including A.W.H. Needler, who many years later would become director. This was the beginning of a shift from university volunteer researchers to full-time scientists working for the board. It was felt that full-time scientific employees were better able to address the issues (primarily fisheries-related) facing the government than were university volunteers. For a while, scientific employees and volunteers worked side by side at the station; however, university volunteers were phased out by 1934, as financial assistance was cut due to the fiscal constraints caused by the Depression. This resulted in a large reduction (but not total elimination) of the number of university scientists conducting research at the biological station.

The debate over governance and the utility of research continued during the 1920s. An attempt in 1919 by the deputy minister of fisheries W.M. Found to create a science division (to include the biological stations) within the Department of Fisheries was passed by the House of Commons, but defeated in the Senate after much lobbying by the Biological Board.[18] A.P. Knight, chairman of the board during the period 1920–5, noted that there was "a definite orientation towards practical studies," but that during that period he had to engineer a mutually acceptable agreement between the board and the department, which had proposed creating a scientific division within the Fisheries Service.[19] Knight encouraged diversifying fishery research activities (towards expansion of the fisheries), and encouraged directors of the Atlantic and Pacific stations to be active in proposing practical research programs.[20] The board was reorganized to include two representatives of the fishing industry, and in 1925 and 1926, new "experimental" stations were established on the two coasts, at Halifax and Prince Rupert, to investigate problems connected with the handling of fish.[21]

In 1920, at the instigation of the Canadian Fisheries Association, Canada initiated the North American Council on Fisheries Investigations (NACFI), formed in cooperation with the United States and

Newfoundland. This permitted Canadian fisheries research on the Atlantic to be coordinated with (and stimulated by) the investigations of the US Bureau of Fisheries and the laboratory at Woods Hole. Fisheries of the Bay of Fundy and Gulf of Maine had close associations with the Northeast US fisheries, and scientists recognized the need for scientific collaboration. Further, as Canada was not yet a member of the International Council for the Exploration of the Sea (ICES), NACFI facilitated the establishment of contacts with Europe, particularly with France, which became a member of NACFI from 1922 because of her fisheries on the Grand Banks.[22]

A proposal to instal tidal dams for developing power at the mouth of Passamaquoddy Bay was brought before the Royal Maritime Fish Commission of Canada (January 1928) and the North American Council on Fisheries Investigations (June 1928). In 1931, an International Passamaquoddy Fisheries Commission was appointed by the governments of Canada and the United States to determine the likely effects. A two-year investigation on herring, zooplankton, phytoplankton, and hydrography was initiated under the International Passamaquoddy Fishery Investigation. The study (the oceanographic aspects of which are described in chapter 6 by Blythe Chang and Fred Page) involved a multinational team including Edmund E. Watson (Queen's University, hydrography), Haaken H. Gran (Oslo, phytoplankton), Michael Graham (Lowestoft, ichthyology), and Charles J. Fish (zooplankton). Huntsman also initiated "The Fundy Survey," an analysis of all species of economic importance within the Bay of Fundy. Reports on various fishes are appended to the biological station's annual report for 1931.[23]

On 9 March 1932, during the Passamaquoddy Fisheries Expedition, the biological station laboratory building was completely destroyed by fire. While other buildings were not affected, the library, records, notes, manuscripts, and extensive study collections of marine material were lost. Immediate offers of assistance came from the Woods Hole Oceanographic Institution for lab space, and from the local sardine producer Connors Bros. Ltd. of Black's Harbour.[24] A new, fireproof lab was built on the same site. The 1932 report states: "As there seemed no possibility of securing any special grant, this reconstruction was proceeded with, owing to the urgency, largely from funds intended for investigations."[25] Huntsman, the director, apparently rebuilt the lab without the board's approval. He found it necessary to cut part-time employment, and reduced spending (including salaries) to have enough money to rebuild.[26] This action, while later seen as important for the station's survival, was in large part responsible for Huntsman's removal from the directorship to the position of "consulting editor."

During the 1930s and 1940s, scientists and fishermen viewed herring and several other finfish resources as being large and under-exploited. Huntsman's report on the herring fishery from the Fundy survey noted that herring were by far the most abundant in the area, and that the fishery was restricted to the mouth of the Bay of Fundy, where it formed the basis for the canned sardine industry. He stated further: "It has been claimed by some that the comparatively small catches elsewhere are due to lack of proper facilities for making use of them and not to any lack of the fish themselves."[27] The biological station researchers experimented with different types of fishing and fish-finding gear, and explored fishing in non-traditional seasons to enable exploiting the herring stock to its full potential. Purse seining had been developed recently and gillnet fishing trials were being encouraged. Towards the end of the 1930s, as the Depression's financial effects eased, the station enjoyed a period of modest growth with an emphasis on practical research in support of fisheries.[28]

## Second Period: Fisheries Research Board of Canada – Expansion

The "Biological Board" became the "Fisheries Research Board of Canada," by act of Parliament in 1937. Kenneth Johnstone noted that "the new Board with its new act was now definitely on the road of applied research."[29] The 1941 annual report described the aim of the Station's work as being "to conduct biological investigations which will provide an adequate basis of knowledge for sound administration and exploitation of our fisheries."[30]

The war had a major impact on activities beginning in 1940, including the loss of a vessel and of personnel (including hydrographer H.B. Hachey) to war service.[31] While the station's output suffered, station director A.W.H. Needler attempted to preserve work that could only be accomplished through long-term investigations.[32] The protection of long-term investigation points to another important feature of the biological station; it was able to initiate and maintain long-term data series. The war also increased the trend towards research directed at increasing fisheries yields, with the objective of providing increasing amounts of protein food.[33] In 1942, Needler reported that this focus was "justified under present conditions by its immediate application to problems connected with the war effort and by its great importance in preparing for reconstruction."[34] In 1958, his successor, J.L. Hart, pointed out that the war effort increased the need for research into, and the promotion of, new types of fishing methods and fisheries.[35]

A.W.H. Needler, director of the biological station from 1941 to 1954, saw the evolving importance of the station's advisory role as being part of a collaboration that would lead to the emergence of modern fisheries management: "Contacts with the administration and the industry, which use the results of the work, have continued to improve. Much effort has been spent on assisting in the proper application of results and planning work to meet real needs. The industry and the various government agencies, Dominion and Provincial, which are concerned in the Station's work, have on their part shown great readiness to ask advice, to use results and to cooperate in the work itself."[36] Indeed, the post-war period was one of great economic expansion. Needler recalled in 1972 that during the period 1945–1960 "research was the magic work in government finance. Research, on the whole, received more assistance than anything else. It was allowed a higher rate of increase."[37] Needler wrote in 1984: "The Station's customers, whether government, fishermen, or the industry started pressing for more research than we could carry out."[38] He noted that the station scientific personnel had become both more closely engaged in discussions with the Department, fishermen, and the fishing industry on matters of domestic fisheries management and development, and more involved as scientific advisers in international matters.

Part of the expansion included increased interest in oceanography and in developing ocean technologies. The Atlantic Oceanographic Group, based at St Andrews, was formed in 1944. Its initial purpose, oceanographic research in aid of submarine detection, is discussed by Chang and Page in this volume. In 1946, the Joint Committee on Oceanography was established as a cooperative initiative involving the Fisheries Research Board, the Naval Service, and the National Research Council. The senior oceanographer at St Andrews, Harry B. Hachey, was highly involved in this initiative and was oceanographer-in-charge of the committee's Atlantic Oceanographic Research Group.[39]

The post-war expansion encouraged further research in support of fisheries development. The 1947 station report optimistically stated: "Some of our potentially valuable species are not used at all, others are poorly exploited, others intensively fished and still others can be increased by artificial means." The station's work included "exploration for new resources, improvement of methods of exploiting others, investigation of intensive fisheries to devise the regulations which will give the best yield and development of cultural methods."[40] In 1949, the minister of fisheries announced an expanded program to develop Canada's fisheries. Confederation brought Newfoundland and its research station under the Canadian government's jurisdiction, but the work of the two stations was seen as easily divided on both a geographical and a subject basis.[41] The biological stations on Canada's Atlantic and Pacific coasts continued to lay the basis for developing fisheries management as well as the oceanographic context for fisheries and management; according to Needler, during that period the credibility of the Fisheries Research Board in the eyes of the department, of the fishing industry, and of the public, improved from the earlier "tolerant skepticism to a positive attitude."[42]

This period was also marked by progress in international research and collaboration. Canadian representatives, including Needler, played an important role at an international conference held in Washington, DC, in February 1949, which resulted in the creation of a convention between the countries participating in the groundfish fisheries of the north-west Atlantic; this provided for the establishment of a commission to coordinate research in this area.[43] The resulting organization, the International Commission for the Northwest Atlantic Fisheries (ICNAF), was originally based in St Andrews. Needler considered ICNAF one of the most valuable examples of an international commission for fisheries regulation.[44] St Andrews provided two of the commission's first executive secretaries: the first was W.R. Martin; in the later 1960s and 1970s L.R. Day filled this role. In 1951, part of one of the cottages was turned into office space for ICNAF.[45] This kind of collaborative co-location has continued from that time up to the present: space has also been allocated, in different periods (some ongoing) for the Joint Committee on Oceanography (Atlantic Oceanographic Group), Atlantic Salmon Federation, Huntsman Marine Laboratory, Fundy Weir Fishermen's Association, the Atlantic Reference Centre, and the Aquaculture Association of Canada.

In defining the scope of work done at the station, Needler explained how the broad research of oceanography allowed the discovery of "principles governing the abundance and movements of the commercial species"; the station's primary concern was with

> the more particular investigations necessary to apply these principles to the problems of the fishery [which] are carried out along four main lines:

> exploration for new resources, discovery of the basis for regulating intensive fisheries to assure the maximum long-term yield, improvement of methods of increasing production by positive cultivation, and development and testing of more effective means of finding and catching fish.[46]

The station saw its scientific challenge as assisting fisheries development in the context of the need to feed a growing world population.

> The general failure of other sources of animal proteins and fats to keep pace with the rapid growth of human populations is producing a worldwide intensification of use of the fishery resource. In the waters off the Atlantic coast of Canada fishing is steadily increasing not only by our own fishermen but by those of other countries, some of them thousands of miles away. There is every reason to believe that this trend will continue both in our own inshore fisheries and in the international offshore fisheries; it will affect species as yet little used as well as those already fished intensively. The problem of obtaining the best use of our resources can be solved intelligently only with the help of information which can be obtained neither easily nor quickly. It is the important and interesting job of biological fisheries investigations, by this and others of the Fisheries Research Board's Stations, to obtain this information and pass it on to the fishing industry and the administration.[47]

Needler's 1951 report alluded to the trade-off between developing and restricting fishing: "On the whole we must regard our fisheries as underdeveloped and, while regulating some and holding a watching brief on others, put emphasis on development rather than restriction."[48] St Andrews Biological Station's fisheries scientists engaged in exploring under-exploited fisheries resources, population studies to avoid overfishing, developing culture methods to enhance fish supply, and studies of predators (discussed by Richard Peterson in chapter 9), diseases, and handling methods.[49]

By the end of Needler's directorship in the early 1950s scientists were recognizing the increasing importance of human fisheries activities and the complexity of their management. Needler's 1952 report, for example, spoke of the need to consider economic and administrative, as well as biological aspects of fisheries management: "The problem itself, however, is complex and has not only scientific but also administrative and economic aspects, all three being inseparable. The station, in its work, is therefore closely associated with various branches of the fisheries administration and of the industry. Its work should be judged only as a part of a joint attack on the fisheries problem as a whole."[50] Needler thus warned of the evolution of a more complex and demanding fisheries management scheme that began to shape the station's research, and continues to shape it to this day. Needler had been director "during the trying war years and the period of rapid expansion which followed. He was adept at perceiving specific economic needs of the Maritimes and the way in which the results of fisheries research could be applied to them. This led to increasing recognition of the Station's usefulness."[51] Needler also raised awareness of the station's contributions by promoting it in the media, and tracking communications such as contributions to the CBC "Fisherman's Broadcasts," and the production of films and filmstrips as well as articles and papers.

Under J.L. Hart's directorship (1954–67), more emphasis was placed on the negative impacts of human activities and the need to restrict some fisheries. Considerable effort at the station was directed towards recognizing pollutants, and at overcoming and dealing with them, efforts that are discussed in chapter 11 by Peter Wells. By 1955, there was also a more urgent attitude to the problem of overfishing. Hart noted that

> fisheries resources of the Maritime Provinces have a wide range of needs in research. Some resources are being fully exploited now. These call for careful biological study of how to harvest the crop so as to make good use of the biological potential. Such biological studies are underway on all the major fisheries of the area. In some cases these studies lead to recommendations to curtail fishing. Restrictions for scientific conservation must be carefully explained so that the regulations may have general acceptance. Effort has, accordingly, been diverted to explaining to interested groups the reasons for regulation and the way in which fishing operations will be affected.[52]

In addition, scientists continued their exploratory efforts, experimenting with improvements in fishing gear, and improving culturing methods for population enhancement.

In 1956, the issue of tidal power in Passamaquoddy Bay was raised again when the Canadian and US governments referred the matter to the International Joint Commission, requesting a study "to determine the effects, beneficial or otherwise, which such a power project might have on the local and national economies in the U.S. and Canada, and, to this end, to study specifically the effects which the construction, maintenance, and operation of the tidal power structures might have upon the fisheries in the area."[53] By the following year a series of highly collaborative efforts were under way:

> Many of the Station's research programs are the result of efficient co-operation with other nations as in the International Commission for the Northwest Atlantic Fisheries, International Passamaquoddy Fisheries Board, and St Croix Engineering Board. Co-operative planning gave good results in such national bodies as the Interdepartmental Shellfish Committee and the Federal-Provincial Co-ordinating Committee on Atlantic Salmon. Problems are shared with the Maritime Provincial and Federal departments of government, as well as Canadian universities. Exchanges of mutual benefit continue to forward the work of investigating species which know no international boundary.[54]

The station, funded by the department's Industrial Development Service, encouraged the application of research results through demonstrations.

The late 1950s saw a major expansion of facilities. During this period "a strong effort was made to increase physical facilities at the stations to match the increase in work load and in personnel ... At St Andrews a new wing was built that more than doubled the space available."[55] New equipment in general was also seen as important, and the station installed such things as mass spectrometers and facilities to support data entry and computing. In the year between 1 April 1957 and 31 March 1958, St Andrews Biological Station employed 242 people. Of these, twenty-nine were research scientists, fifty were scientific support staff, twelve were maintenance workers, and there were eighteen administrative and clerical, twenty-five boat operation, fifty-two seasonal and term, forty-eight casual, and eight part-time employees.[56]

The 1960s marked an end to the rapid growth the station had experienced since 1945. Science administration in Canada was changing, accompanied by the first moves to incorporate the Fisheries Research Board into federal government departments. This agenda was broader than the Fisheries Research Board alone. The Royal Commission on Government Organization, known as the Glassco commission (1960–2) evaluated governmental research and concluded that Canada needed to develop science policies, improve the application of scholarly excellence to Canadian needs, and develop a way to assess the effectiveness and value of research expenditures.[57] It recommended considerable governmental reorganization. For example, the Department of Mines and Technical Surveys was establishing a dedicated multidisciplinary oceanographic research institute in the Halifax and Dartmouth area, and in 1960, the headquarters for the Atlantic Oceanography Group was moved from the biological station in St Andrews to Halifax, and to the Bedford Institute of Oceanography (BIO) when it opened in 1962. Most of the ships (and crews) were transferred by 1968. There is evidence that there were plans to move other units. Hubbard suggests that this transfer was most likely part of an attempt by the Fisheries Research Board chairman in that period, J.L. (Jack) Kask, to close down the St Andrews Station. Hubbard refers to "ambiguity regarding St Andrews that dramatically weakened the support for the former proud pilot station of Canadian marine science." She cites J.C. Stevenson's 1972 interviews with board chair Ronald Hayes (1964–9) and J.L. Hart that indicated "There was poor morale in St Andrews because they didn't know then, and they have never known until very recently whether they were going to be moved to Halifax or what was going to happen." Hart, director of the station from 1954 to 1967, said that St Andrews was "a station that essentially in 1954, and every 2 years thereafter was condemned to death ... I did manage to keep the body breathing, and all I can say is that John Anderson had something to work with when he came."[58]

John Anderson, director from 1967 to 1972, reminisced at the 2008 St Andrews history workshop about the changes that took place. When his directorship began, he and the directors of the other biological stations in the country (there were ten at the time) had "bureaucratic freedom" because "the Fisheries Research Board had control of three important things – the people, the money, and the programs." That changed during the time J.R. Weir was board chair, beginning in 1969, as the Fisheries Research Board became integrated into governmental line management. "The first to go was the people – that went to the Public Service Commission, and later went the programs and the money."[59]

The late 1960s and early 1970s also saw a focused effort on environmental research initiated in response to a growing public awareness of environmental issues and some high-profile problems, including the decline in salmon populations in the Miramichi River system; some of these issues are discussed by Dick Peterson in chapter 9 and Peter Wells in chapter 11.[60] John Anderson, at the St Andrews history workshop, highlighted the advances in underwater observation through the use of SCUBA, research submarines – scientists at the St Andrews Biological

Station used five during his time – and the development of underwater vehicles or platforms, the main focus of Tim Foulkes's chapter in this volume.[61]

Scientists greatly advanced fisheries surveys and resource distribution quantification, including developing in 1970 annual bottom trawl surveys of the entire Scotian Shelf aimed at finfish (groundfish) resources.[62] These were quickly applied using stratified-random survey design to both the Scotian Shelf and Gulf of St Lawrence, beginning a time series that continues to this day, which has formed the basis for assessments and management of most fish stocks.

The St Andrews Biological Station has had a dual role of contributing, on the one hand, to the development of aquatic resources through documenting resources and developing gear and fishing methods, and, on the other hand, to environmental monitoring and research – the latter resulting in restricting or reducing aquatic resource exploitation, and enhanced conservation measures. At its founding, fisheries development was seen as an opportunity for Canada. In the early 1900s, Canada sought expertise internationally, for example, through hiring E.E. Prince as commissioner of fisheries and Johan Hjort as a consultant to develop fisheries. Hubbard, both in her 2006 book and this volume, suggests that the emphasis on fishery development, especially in early years, reflects the influence on Prince (and subsequently on Huntsman) of Thomas Henry Huxley, and William C. M'Intosh, the director of the biological station at St Andrews, Scotland, who were both proponents of unrestricted fisheries on the basis that it was nearly impossible to exterminate sea fish due to their high reproductive capacity. Certainly the pattern of government-supported activities for economic development, followed by the need for restriction and reduction, is a pattern that seems to have been repeated. By the early 1970s several of the fisheries that had experienced expanded fleet size and technological capability were in need of some restraint, so fisheries managers added an increasing set of restrictions (including formal catch quotas) that required increasingly onerous biological evaluations.

The herring fishery is a good example. A major herring fishery conference (under the Federal-Provincial Atlantic Fisheries Committee) held in Fredericton, New Brunswick, in 1966 had been optimistic that herring stocks could support much more intense exploitation.[63] However, the fleet expanded fivefold during the 1960s with the arrival of new vessels and with technological improvements to the existing fleet, and changed from traditional fixed gear using weirs, traps, and gillnets for food and bait, to one dominated by mobile vessels (mostly purse seiners) serving fish-meal and fish-oil reduction plants.[64] Licence limitations had to be introduced in 1970 and quotas in 1972; a herring management committee was created in 1972, the industry was restructured in 1975, and by 1983, an explicit fleet reduction program was introduced.[65] Yet it is interesting to note that the department is still engaged in research supporting developing fisheries, although there is a greater level of attention to sustainability and to a government role of monitoring development, undertaken largely by proponents of new fisheries.

In the meantime, St Andrews Biological Station also branched out to re-integrate academics. Director John Anderson worked with a consortium of eastern Canadian universities to form the Huntsman Marine Science Centre (HMSC). This centre (originally named the Huntsman Marine Laboratory) was opened in 1970 at a site adjacent to the biological station. Named in honour of the Station's former director, A.G. Huntsman, it was formed primarily to provide access to a marine environment for university scientists and students, and to encourage cooperation between government and university researchers. This was an attempt to re-establish university links which had been so important in the early years of the biological stations, but which had been greatly reduced since the 1930s. While no longer the sole eastern research station, the St Andrews Biological Station remained an important "nursery ground" for scientific training.

## Third Period: Government Departments (Fisheries and Oceans) – Evolution of Management and Government Science

At the 2008 history workshop Ralph Brinkhurst, the first director (1972–5) during the period when the government was restructuring the Fisheries Research Board, noted that "the Fisheries Research Board was looking for a new mandate, and we were all busily being absorbed into the public service, and it was obvious that the process was by no means over." The Fisheries Research Board was moved to the Department of Fisheries and Environment, programs were reorganized and the mandate was transferred. Brinkhurst reported that "I had a whole lot of people tied up in bits of programs that had transferred to Halifax there for a while, so we had a lot of fragmentation." He observed:

> One of our biggest problems ... was there was still all this insecurity about the future of the station. Everybody coming down from Ottawa reminded us that we were in such a remote location and nobody gave any reference

to the fact of the bounty of seawater right off the wharf, which was really the reason why we were here in the first place.

The labs "lost their national focus and became part of the regional focus." Brinkhurst lumped projects into two groupings – environmental research and fisheries management-related research – hoping to make better use of resources and increase cohesion among projects, or, as he put it, "trying to make the best use of them by putting them all in a sort of multi-species flock."[66] Major transition issues continued during the directorship of R.H. Cook beginning in 1977: "The St Andrews Biological Station, at the time I was appointed the director, was undergoing a significant change from being a fully integrated, self-contained research laboratory ... to a system of matrix management." Under this arrangement, the director was responsible for the facility, for part of the scientific program at the station (environmental sciences), and for some related scientific programs elsewhere. Other programs at the station reported to the Bedford Institute of Oceanography. The situation was further complicated in 1982 with the creation of the Gulf Region (resulting in program changes and resource transfers) and the beginning of a decade of budgetary reduction. While the reductions resulted in transfers of staff, "fortunately, these cuts did not go so far as to close the Station, a suggestion of an ADM Policy and Planning in that era."[67]

Notable in the history of the St Andrews Biological Station has been the development and growth of science required to manage an increasingly intense and mechanized suite of human activities, and more recently the evolution of a more holistic ecosystem approach to management. The major fisheries, for example, experienced over the past century the dramatic change from the use of small inshore vessels with passive gear (traps, gillnet, hook and line) to widely distributed and highly mobile fisheries using large mechanized vessels with sophisticated fish detection and capture equipment. Responding to this, there was substantial change in the management paradigm. Particularly over the past five decades, an elaborate management scheme has developed. In 1993 Scott Parsons made the point that Canada's marine fisheries have been plagued by recurrent crises, that in just twenty-five years the fisheries went from underdevelopment to overcapacity, and since the 1960s there have been major changes in management.[68] This represents the single most important driver of the evolution of research at the biological station, and indeed the station's history is a rather complete case study of these developments.

A desire to provide the biological basis for management has been apparent from the very beginning of the biological station at St Andrews. Increasing management demands influenced both the programs and the station's structure. The need for a dedicated, year-round staff to provide practical advice was a major element in the change to a permanent staff in the 1920s, and the increase in staff following the Second World War was related to fisheries development needs. A perceived need to have "in-house" expertise to support a more complex management scheme was at least part of the rationale for the change from the Fisheries Research Board to departmental line management in the early 1970s. In recent decades, a considerable fraction of the station's activity has been in support of stock assessment and advice related to ongoing fisheries management. New requirements for more holistic "ecosystem-based" and "integrated management," discussed below, are currently shaping the station's structure and programs.

This emphasis on research for fisheries management of course raises the issue of fundamental versus applied science. While scientists at the St Andrews Biological Station and other federal marine institutions have always undertaken both pure and applied science, the relative importance of the two has been a matter for debate. The need for directed work on applied issues was a prime contributing factor for hiring the permanent staff (as opposed to relying on volunteers) and for founding the very applied "fisheries experimental stations" in the 1920s, but we see ample evidence of ongoing discussion and concern that "applied science would 'grow sterile' without the input of pure science."[69] Although the emphasis has been on applied science, station programs have certainly also undertaken basic biological work that advanced fundamental science. The extensive work on smoltification of salmon by, for example, Saunders and colleagues is a prime example.[70] There has also been a difference over time in the scope for fundamental research, with relatively greater opportunity during the Fisheries Research Board years, and relatively less after the downsizing of the past two decades. The Department of Fisheries and Oceans would maintain that there is an appropriate balance of both applied and fundamental work being undertaken; the fundamental aspects of scientific creativity and influence remain criteria for a research scientist's promotion within the Government of Canada.

The 1977 extension of Canadian jurisdiction over the waters above the continental shelf to 200 nautical miles resulted in an increase in staff to undertake research associated with management of the larger,

200-mile exclusive fishing zone. Environmental research increased rapidly from being incident-driven to research on the structure and function of aquatic ecosystems and to the effects of human activities: some of this is described by Peterson in chapter 9 and Wells in chapter 11.[71] The station remained strong in fisheries, although there was some attrition, especially in the work on invertebrate fisheries, and the transfer of mandate (including for scallops and some lobster areas) to other laboratories. Fisheries programs evolved to support an increasingly demanding and quickly developing management system that included annual quotas established on the basis of peer-reviewed stock assessments. Facilities were improved to allow the consolidation of collections of preserved fish and invertebrates of the north-west Atlantic into a single, well housed, and curated collection in the station's Atlantic Reference Centre. Strong lab-based fish physiology and culture research, strong leadership from the station, and close collaborations led, during this same period, to the development of a finfish aquaculture industry in south-west New Brunswick, the subject of the final chapter in this volume by Robert Cook. While fishery development activities, per se, are a thing of the past, the biological station has continued to undertake science in support of developing sustainable aquaculture (including new species and improved husbandry research), but with a greater emphasis on sustainability.

Wendy Watson-Wright, director from 1992 to 1997, undertook a strategic planning exercise aimed at unifying the "disintegrated groups" at the station. The station was seen as having two major pillars, fisheries and aquaculture, supported by environmental science and oceanography. At the same time, the Department of Fisheries and Oceans entered a period of program review that looked explicitly at the relative relevance of activities to federal priorities. The review identified scenarios of budgetary reductions of 5, 10, and 15 per cent, the last two of which would have seen the station closed. Fortunately, the station survived, and was named the centre for aquaculture research for the Atlantic zone. This started some capital facilities renewal.[72]

During the directorship of T.W. Sephton (1997–2005) the station continued to capitalize on its leadership in aquaculture within a matrix that involved the Bedford Institute of Oceanography and the Gulf Fisheries Centre. It also shored up its position with respect to the fisheries science mandate focused on the Gulf of Maine, which included trans-boundary US and Canadian fisheries and oceanography; its environmental science was by this time heavily involved with the impacts of aquaculture.

Another challenge was planning for the replacement of buildings that were, by now, fraught with an increasing list of maintenance issues associated with an aging seawater system and regulatory shortcomings due to ratcheting health and safety and animal care standards.

More attention has been paid in recent years to "relevance," in a climate of fiscal restraint. There has been ongoing explicit discussion of the department's legitimate role, as well as priorities in allocating research funding, related to earlier issues of "fundamental" versus "applied" science, but going farther. The ongoing climate of restraint has led to the question of what activities or programs one should (or could) eliminate, and to what degree research activities were for public versus private good, when the public expected expenses to be picked up or paid for by the beneficiaries of the science, namely, the fisheries corporations profiting from this work. More than a decade of financial restraint resulted in a complement of staff that was relatively small and nearing what some claimed was minimum critical mass. Station activities were trimmed to focus on only the major activities required of fisheries and oceans management or core deliverables demanded by funding sources. Fiscal restraint forced researchers to seek other sources of funding, or co-funding, and scientific staff became quite entrepreneurial in finding funding sources to allow research. The "entrepreneurial" approach, however, raised concerns that it threatened to lead to research being undertaken towards questions for which funding was available, rather than strategically important ones.

The Oceans Act (1997) and the Species at Risk Act (2002) changed the tone of the science required for management. The Oceans Act, especially, called for a new approach to ocean resource usage, based upon the principles of wise ecosystem-based management. This demanded an enhanced research capacity to provide scientific advice for more integrated management based on the precautionary approach to achieve sustainable development. The station embraced this challenge as an opportunity to reshape research. Arising from the previous strategic planning, scientists sensed that station programs could be unified in an "applied coastal ecosystem science program" to allow research groups of the station to work more closely together, across structural lines. It was recognized that future requests for advice would require the integration of fisheries and aquaculture information with greater oceanographic and environmental considerations, and that the fragmented reporting structure made this difficult.

In 2005, when I became director, Michael Sinclair, director of science for Maritimes Region, unified the station under the director so that it could develop the desired type of cross-disciplinary program and could operate in a more "entrepreneurial" manner. The station faced numerous recognized challenges associated with fiscal restraint, including lack of recent recruitment and potential loss of corporate memory through pending retirements. The rapidly changing landscape of management resulting from the recent Canadian environmental legislation, especially the Species at Risk and Oceans Acts, increased both demands for scientific research, and the diversity of requests for advice, for the most part without additional resources. Further strategic discussions focused on the scientific needs for evolving "ecosystem-based management" and "integrated management" approaches. The station began placing its activities in a framework (figure 4.2) that anticipated several aspects of a new management regime. Individual management plans (for various fisheries, aquaculture, etc.) expanded to include the diverse objectives associated with ecosystem-based management, including conservation objectives that broadened to include productivity, biodiversity, and habitat. Increased consideration also had to be given to social, economic, and institutional (governance) aspects. Consumers were now putting pressure on the fishing and aquaculture industries to comply with market certification for sustainable and environmental practices. There was also a greater need to consider cumulative impacts of the fisheries and other activities. The St Andrews Biological Station recognized that new research was required in support of this new approach, and attempted to reshape itself to undertake the collaborative research needed to provide the tools, information, and case studies required by emerging "ecosystem-based" and "integrated" management approaches.

The St Andrews history symposium of 2008 demonstrated that the uncertainty surrounding the future of the station had been prevalent for more than four decades.[73] At the symposium, previous directors (representing the past forty years) all referred to the threat of closure as a major issue during their time. However, 2008 also marked the beginning of a major project to replace much of the infrastructure of the biological station with a new wet lab, analytical lab, and office buildings, as well as progress on replacement of two research vessels. The station now seemed quite secure relative to past decades.

## Major Themes and Issues over a Century of Aquatic Research in Canada

The preceding sections have demonstrated that there was major development of aquatic science concepts over the past century – much of it

Ecoregion/planning area (umbrella plan)

Ecosystem Assessment

Nested plans for Managed activities

Expanding list of objectives (ecosystem services/values)

| | Fisheries | Aquaculture | Energy | Transport | Other |
|---|---|---|---|---|---|
| Conservation - Productivity | Plans with diverse objectives; Market audit (certification) | | | | |
| - Biodiversity | | | | | |
| - Habitat | | | | | |
| Economic | | | | | |
| Social/cultural | | | | | |
| Institutional/ governance | | | | | |

Audit of cumulative performance

4.2 Proposed practical structure for integrated management

related to the changing needs of science for a rapidly evolving fisheries management regime – and illustrated how the St Andrews Biological Station faced ongoing issues related to the balance of fundamental versus applied research for both developing and restricting human activity. This section addresses several additional major themes, including: the station's role as one of an international constellation of marine and aquatic science stations that collectively advance ocean science, with some discussion of major advances made at St Andrews; the role of the station as a locus of education and its relation with universities; and finally a discussion of the role of public science.

Changing research programs at the St Andrews Biological Station have reflected the enormous development in aquatic science that has occurred both in Canada and internationally in the past century. When the station was established, research undertaken by volunteers from various universities was largely ad hoc. Further, it was inclined to be descriptive, "inventory" science undertaken by generalists (Mills, this volume), but also in keeping with the ethic of "scientific service" by wealthy or gentlemanly naturalists inherited directly from the nineteenth-century British tradition (Hubbard, this volume). The establishment of permanent research stations in St Andrews and Nanaimo resulted in a large increase in the amount of aquatic research, along with the rapid development of new scientific concepts. E.A. Trippel provides a useful list of papers, by decade, which illustrates the diversity of the legacy of station programs and its scientific conceptual heritage.[74] The permanent stations, and especially the move to full-time staff, resulted in increased oceanographic observations and understanding of the natural environment, and the station has produced some of the longest continuous series of observations on the marine environment, as discussed by Chang and Page, Jennifer Martin, and Peterson in different chapters of this volume. Without question, however, the major contribution has been in relation to fisheries ecology.

The Canadian Fisheries Expedition had a substantial impact on the emergence and subsequent evolution of fisheries biology and biological oceanography. Basic concepts and methods in fisheries ecology, including elucidating stock structure, age composition, reproduction, and productivity of populations have all developed over the past century,[75] and the station's scientists contributed prominently in both the conceptual development and the application of these to north-west Atlantic species and populations (see, for example, Caddy, this volume). The development in understanding both fish biology and fisheries is reflected in the taxonomic syntheses of Leim and Scott (1966) and Scott and Scott (1988), the basis of which is housed in the collections of the station's Atlantic Reference Centre.[76] Station programs have also reflected advances in physiological research and culture techniques that have improved our understanding and management of wild populations; aquaculture programs, which once supported hopes of enhancing wild populations, more recently have had practical applications in aquaculture development (Cook, this volume). The station's programs have reflected major developments over the past century in understanding and mitigating the diverse harmful effects of human activities, including pollution, on the environment (Wells, this volume). The period 1908–2008 was a century of rapid advances in scientific methodology, in sampling techniques, and in aquatic environmental experimentation. SABS programs contributed improved technical and experimental methodology (e.g., Wildish and Robinson, this volume; Foulkes, this volume). Using the abundance of high-quality seawater at its doorstep, and the diverse habitats nearby, SABS developed a tradition of strength in lab-based experimentation and coastal field studies.

The aquatic science community is an international one. From the very beginning, even before Hjort's leadership of the Canadian Fisheries

Expedition, the development of Canadian aquatic science has been highly collaborative internationally. This continued with specific studies such as the International Passamaquoddy Fisheries Investigation and the development of an international oceanographic initiative and various international fisheries committees. The station's scientists have, throughout the past century, collaborated widely. For most of the period, Canadian fisheries and aquatic science has enjoyed a high profile internationally, and the journal most closely linked with Canadian aquatic science, the highly regarded *Journal of the Fisheries Research Board of Canada* – now the *Canadian Journal of Fisheries and Aquatic Sciences* – had its origin at the St Andrews Biological Station.

Some of those collaborations have involved Canadian universities. The association of academics with the station has fluctuated. E.E. Prince's express wish for the biological stations from the beginning was "to give an unequalled opportunity to young biologists in the various universities of Canada to carry on original scientific researches,"[77] and indeed the station was originally a facility for academic research. Early work reflected investigators' interests, and while it was diverse and productive, it was neither sustained nor strategic. The department and board needed a sustained effort that could not be delivered by university-based academics, and so hired permanent staff. While this hiring of salaried experts "in house" in the 1930s was good, the university link was greatly reduced. Kenneth Johnstone points out that the reduced access caused little reaction in academic circles for the next two decades,[78] suggesting that excitement about cell biology and related fields put more emphasis on lab-based research.

The station's growth allowed interested researchers to find employment within the Fisheries Research Board, and as has been the case in many countries, the majority of fisheries work was being done in government or dedicated research institutes rather than in universities. However, by the mid-1960s, a renewed academic interest in ecology arose from emerging new conceptual structures and a general increase in environmental concerns. Director John Anderson re-established a formal link between the biological station and the academic community with the Huntsman Marine Science Centre, at the time a consortium of about twenty Canadian universities. Recent directors have attempted to strengthen links with universities yet again, increasing collaboration with the University of New Brunswick in coastal studies. From both the St Andrews Biological Station's and a national science and technology perspective, the collaboration with academia is important for at least two reasons. First, issues of science (for management) are more complex, and require diverse skill sets not always available within government departments. The ecosystem approach and integrated management, for example, require expertise in social sciences, geomatics, and other areas that are not currently strong within the Department of Fisheries and Oceans. Second, Canada has a relatively small research capacity and limited funding available for aquatic research, and it is both logical and strategic to foster increased collaboration. It is interesting to see that recent programs of the Canadian Natural Sciences and Engineering Research Council (NSERC) have encouraged such collaboration.

The St Andrews Biological Station has evolved administratively from an institute of academic investigation to a federal research laboratory of Fisheries and Oceans, through a series of arrangements and federal department affiliations that included the world renowned Fisheries Research Board of Canada. The station's role also changed as it went from being the sole Atlantic laboratory to one of five major Atlantic laboratories of the Department of Fisheries and Oceans. A body of literature (including A.W.H. Needler's article of 1984 and the recent book by Hubbard [2006]) documents the tension that existed, virtually from the beginning, concerning the governance of public marine science. Early conflict between government administrators and the volunteer university scientists of the board played a large part in the 1912 legislation establishing the independent Biological Board of Canada. While this independence was continued when the new legislation (in 1937) changed the name to the Fisheries Research Board of Canada, it was not without proposals that the board's role be reduced to an advisory one; ultimately in 1973, the station's activities became part of the new Department of the Environment, and in 1979 the Department of Fisheries and Oceans.

It is obvious from this review that the context for research over the past century changed enormously, as governmental needs for science associated with the elaborate system of management also evolved. There are some who link the scientific productivity and profile that Canadian marine and fisheries science once enjoyed with the independence of the Fisheries Research Board of Canada. This is certainly worthy of discussion as Canada reviews strategies to improve science and technology.

There is no doubt that the Fisheries Research Board enabled both growth and prominence for Canadian aquatic science internationally.

During this period (1912–73) the board was responsible to the minister concerned with fisheries, but had control of its own affairs. As an organization based on science, the board was made up of eminent university scientists serving in a volunteer capacity, and was, apparently, a successful organization. According to Needler:

> The Board's scientists knew that they worked for a body that appreciated scientific values ... Morale was good and recruitment easy. The Board's scientific publications developed a reputation for scientific excellence. The Board could support long-term and fundamental research as well as immediate and applied. It could discourage waste of time on problems without prospect of scientific solution. The Board, with a majority of scientists but also with Departmental and industrial members, was able to maintain a reasonable balance between the long-term and the immediate and between the fundamental and the practical ... The Board could protect against the tendency for pressure to be greatest for science that has already reached the applied stage as against more fundamental new science on which real progress depends. The board's administration was efficient because of relative freedom from red tape and because of quick access to authority. Scientific know-how could be applied quickly ... Because of few steps between the stations and top authority, administrative decisions could be quick and frustrations kept to a minimum. The independence of the Board was able to avoid the involvement with petty political patronage, which was common in government operations ... Under the Board, recruitment and procurement were based only on merit. One of the most favorable factors was a strong connection with university science at all levels of the Board's operations.[79]

In spite of this success, the Fisheries Research Board was relieved of administrative authority over its facilities and programs in 1973 and became purely an advisory body. A few years later it disappeared altogether.[80]

These changes, leading to the incorporation of fisheries research into the federal department, were well intentioned and part of the larger re-profiling of science following the Royal Commission of Government Organization (Glassco commission) and related initiatives. A forum held in October 1973 to discuss the "The Way Ahead" for the Fisheries Research Board as an advisory body[81] was quite optimistic about the need for science strategy and focus, and better university involvement.

Needler, however, suggested that at this time the stations lost much of their autonomy and identity as administrative units, that they became more regional (rather than national) in scope, and that the increased number of administrative steps between the stations and the top authority in Ottawa resulted in inefficiencies, red tape, frustrations, and loss of morale.[82]

F.R. Hayes, a former chair of the Fisheries Research Board, in his book *The Chaining of Prometheus: Evolution of a Power Structure for Canadian Science*, predicted that the forces at work in Canada at that time would reduce the capacity of government science – and that federal scientists would become mere "expediters" of university and industry science.[83]

The absorption of the Fisheries Research Board into a government department increased the complexity of government management. Former directors R. Brinkhurst and R. Cook experienced the implementation of programs across labs. Cook noted: "I was given … this fisheries environmental sciences division role which had ⅔ of the staff reporting to me here in St Andrews, and the other ⅓ with the Halifax Lab. It meant, from my operational point of view, and to be fair to a bunch of very talented scientists at the Halifax lab, I had to basically sit here and say 'How's everything going in the labs?' and 'Keep me up to date.' So I spent a lot of time burning rubber between St Andrews and the Saint John airport, and between Halifax Airport and the Halifax lab, and back and forth, which was a very inefficient use of my time."[84] The return in 2005 to a reporting relationship in which all station staff report locally (through the director) was an attempt to simplify collaboration and station project development.[85]

Hayes highlighted the dilemma for the effectiveness of science under a governmental structure versus science carried out by an outside agency at arm's length from government. He observed that in various arrangements of agriculture research, a governmental structure has "the tendency of line authority to cleverly slip sawdust into the oats of the research donkey until the animal is becoming moribund," while independent agencies "have the opposite tendency, namely, to drift away from their missions into pure science, which they euphemistically call prerequisite science. The charge against them is not bad research but lack of total relevance." Hayes also noted that applying science to what he calls "adaptive industries" (e.g., power, transportation, and communications) differed from science applied to "industries of necessity" (those associated with natural resources), including fisheries:

> The operation of modern science on a guild, say fishing, requires the dismantling of existing methods as well as the introduction of new ones. Conservatism offers a great resistance to both processes and makes necessary a supporting group of middlemen or interpreters. They have to be dedicated as well as energetic, for they tend to be denounced by academic researchers as inadequately trained, and by the guild as too academic.

He concluded that "the industries of necessity are clearly more difficult to bring under science than adaptive industries which spring directly out of the laboratory."[86]

The management of aquatic science by government has certainly led to increased emphasis on the accountability of research programs with respect to departmental objectives, as noted in the comments of Watson-Wright and other recent directors at the director's panel of 16 October 2008. Although it seems there remains a chronic shortage of science and technology funding in Canada, and a disconnect between the academic and public sectors with respect to aquatic research, the biological stations and other laboratories now within the Department of Fisheries and Oceans remain the national leaders in marine and aquatic research; Canada's international reputation in fisheries science is arguably not as strong as it was in the Fisheries Research Board era, but a recent bibliometric study of aquatic science publications of forty nations for the decade ending 2007 concluded that Fisheries and Oceans research papers are relatively well cited and have high relative impact factors both in traditional areas of research – on fish populations, community productivity, and oceanography – and in more recent areas of research, including aquaculture and ecosystem management.[87] The same study reveals an increasing trend towards international collaboration in papers. It appears that the quality and profile of Canada's governmental aquatic work has been sustained.

The past century of research at SABS has resulted in a great legacy of work of relevance both to Canada's and to the international scientific community's needs. Looking ahead, it is interesting to ask what the future legacy of SABS and of public science might be. There is certainly an exciting and challenging landscape for public marine science. There is an increasing demand for public science in support of new developments in management and increasing public interest in the marine environment. Surely it can be agreed that the underlying responsibility of laboratories such as SABS, which undertake "public science" in support of legislation, is to collect and maintain data, to undertake the scientific investigation that supports the maintenance of ecosystem services, and to provide credible information and advice. Who, if not those involved in public science, will undertake the conceptual development and scientific research necessary to meet the increasing requirements associated with public interest in the marine environment, and with the important evolution of management towards greater sustainability? Important public science elements include monitoring surveys that have been standardized for more than three decades, and represent the most significant source of information on ecosystem state and change; decades of scientific study and understanding of fisheries (and other activities) that form the basis for management; production and safekeeping of data series in fisheries, oceanography, and preserved samples – some extending back almost a century; and finally scientific advance required for implementation of new policies such as those related to species at risk or the ecosystem approach, which will allow "Integrated Management" of human activities.

Needler concluded: "Only a dedicated continuing scientific staff can handle the difficult problem of keeping balance between immediate applied research and the broader, deeper "fundamental" research on which real progress depends. It is, of course, possible to contract out some short-term problems to consultants or universities, but only a strong continuing research staff can properly direct the more important long-term research. That continuing staff, however, profits for breadth of vision from a strong university connection."[88]

As the management of aquatic activities changes to take account of a greater set of ecosystem considerations, the questions being asked of science will continue to become increasingly interdisciplinary and complex. SABS will need to enhance its interaction with academia, with others in the science and technology community and with industry. Restraints within government combined with the increased complexity of management-related science provide incentives for collaborations across traditional university and academic boundaries. SABS is perfectly situated to build on its long tradition, to take advantage of one of the most productive and dynamic marine environments in the world, and to collaborate with industry, academia, and other jurisdictions both nationally and internationally in the continued development of aquatic science. The future will be shaped in large part by the evolution of strategies for science and technology in Canada that include decisions on the balance between public and academic science, on what is considered private versus public good, on long-term data monitoring, and on the overall level of funding of public science.

## NOTES

I am grateful to Eric Mills, Mike Sinclair, Dave Wildish, Blythe Chang, Tim Smith, and Jennifer Hubbard for discussions and advice related to this history. I thank Eva Gavaris, who compiled much of the historical information on which the paper is based. I am extremely grateful to Joanne Cleghorn and Charlotte McAdam of the St Andrews Biological Station library for their ongoing stewardship of the archival material and enthusiasm for the historical research and additional work (including establishment of a photo archive) that accompanied the 100th anniversary of the station. I thank Dave Wildish and Jennifer Hubbard for their efforts in collecting and editing the papers that form this collection.

1 Tim D. Smith, presentation at Workshop on the Evolution of Marine Science in Canada, 16 October 2008.
2 E.D. Gavaris and R.L. Stephenson, "Report of the Commemoration of the 100th Anniversary of the St Andrews Biological Station – 2008," Canada, Manuscript Report of Fisheries and Aquatic Science 2943 (2010): 1–27.
3 Some previous historical accounts include notably the papers arising from previous anniversaries. Among these are J.L. Hart, "Fisheries Research Board of Canada, Biological Station, St Andrews, N.B., 1908–1958: Fifty Years of Research in Aquatic Biology," *Journal of the Fisheries Research Board of Canada* 15 (1958): 1127–61; H.B. Hachey, "History of the Fisheries Research Board of Canada," Fisheries Research Board of Canada Manuscript Report Series (Biological) 843 (1965): 1–499; A.W.H. Needler, "Reflections on the Fisheries Research Board," *Journal of the Fisheries Research Board of Canada* 31 (1974): 1283–4; A.W.H. Needler, "The Seventy-fifth Anniversary of Two Canadian Biological Stations," *Journal of the Fisheries Research Board of Canada* 41 (1984): 216–24; E.A. Trippel, "The First Marine Biological Station in Canada: 100 Years of Scientific Research at St Andrews," *Canadian Journal of Fisheries and Aquatic Science* 56 (1999): 2495–507; and V. Zitko and D.J. Wildish, "The First Marine Biological Station in Canada: Highlights of Environmental Research at St Andrews," *Water Quality Research Journal of Canada* 35 (2000): 809–17. Of great significance

are the books by Kenneth Johnstone, *The Aquatic Explorers* and especially by Jennifer Hubbard, *A Science on the Scales*.

4 A workshop report by E.D. Gavaris and R.L. Stephenson, "Report of the Commemoration of the 100th Anniversary of the St Andrews Biological Station – 2008," and transcripts of the directors' panel and related archival materials are available in the SABS library.

5 Trippel, "The First Marine Biological Station in Canada," and Hubbard, *A Science on the Scales*.

6 D.P. Penhallow, "Report of the Atlantic Biological Station of Canada, St Andrews, N.B. for 1908," *Contributions to Canadian Biology* 1906–10 (1912): 1–21.

7 E.E. Prince, "Presidential Address: The Biological Investigation of Canadian Waters, with Special Reference to the Government Biological Stations," *Transactions of the Royal Society of Canada, Third Series – 1907–1908, vol. 1, section IV (Geological and Biological Sciences)* (1908):71–92; Rigby and Huntsman, "History of the Fisheries Research Board of Canada."

8 Needler, "The Seventy-fifth Anniversary," 217.

9 Johnstone, *Aquatic Explorers*, 30.

10 Needler, "The Seventy-fifth Anniversary," 217.

11 Ibid., 218.

12 Biological Board Act of 1912, cited by Johnstone, *The Aquatic Explorers*, 78.

13 Penhallow, "Report of the Atlantic Biological Station of Canada," 2.

14 E.E. Prince, "Preface," in *Canada, Canadian Fisheries Expedition, 1914–1915, Investigations in the Gulf of St Lawrence and Atlantic Waters of Canada* (Ottawa: King's Printer, 1919), v.

15 Ibid.

16 J. Hjort, "Fluctuations in the Great Fisheries of Northern Europe, Viewed in the Light of Biological Research," *Rapports et Procès-Verbaux des Réunions du Conseil Permanent International pour l'Exploration de la Mer* 20 (1914):1–228.

17 Biological Board of Canada, "Report of the Atlantic Biological Station, St Andrews N.B. for the season 1921," unpublished report (1921), SABS Library, pp. 1– 2.

18 Johnstone, *The Aquatic Explorers*, 101–2.

19 Ibid., 104.

20 Ibid., 7.

21 A.G. Huntsman, "Twenty-five Years of Canadian Fisheries Research," *Canadian Fisherman* 26:9 (1939): 95.

22 A.G. Huntsman, "Fisheries Research in Canada," *Science* 98 (1943): 117–22.

23 A.G. Huntsman, "Fundy Survey: The Herring Fishery," appendix to the "Report of the Atlantic Biological Station for 1931," unpublished report to the Biological Board of Canada (1931), SABS Library.
24 Ibid.
25 A.G. Huntsman, "Report of the Atlantic Biological Station for 1932," unpublished report to the Biological Board of Canada (1932), SABS Library, 11.
26 Johnstone, *The Aquatic Explorers*, 139–40.
27 A.G. Huntsman, "Fundy Survey: The Herring Fishery."
28 Needler, "The Seventy-fifth Anniversary of Two Canadian Biological Stations."
29 Johnstone, *The Aquatic Explorers*, 163.
30 A.W. H. Needler, "Report of the Atlantic Biological Station for 1941," unpublished report to the Fisheries Research Board of Canada, SABS Library (1941), 1.
31 R.H. McGonigle and A.W.H. Needler, "Report of the Atlantic Biological Station for 1940," unpublished report to the Fisheries Research Board of Canada (1940).
32 Needler, "Report of the Atlantic Biological Station for 1941."
33 Hart, "Fisheries Research Board of Canada, Biological Station, St Andrews."
34 A.W.H. Needler, "Report of the Atlantic Biological Station for 1942," unpublished report to the Fisheries Research Board of Canada (1942), SABS Library, 1.
35 Hart, "Fifty Years of Research in Aquatic Biology," 1127–61.
36 A.W.H. Needler, "Report of the Atlantic Biological Station for 1943," unpublished report to the Fisheries Research Board of Canada (1943), SABS Library, 1.
37 Johnstone, *The Aquatic Explorers*, 185.
38 Needler, "The Seventy-fifth Anniversary of Two Canadian Biological Stations," 220.
39 A.W.H. Needler, "Report of the Atlantic Biological Station for 1943."
40 Ibid., 1.
41 A.W.H. Needler, "Report of the Atlantic Biological Station for 1949," unpublished report to the Fisheries Research Board of Canada (1949), SABS Library.
42 Needler, "The Seventy-fifth Anniversary of Two Canadian Biological Stations," 220.
43 Needler, "Report of the Atlantic Biological Station for 1949."

44 Needler, "The Seventy-fifth Anniversary of Two Canadian Biological Stations," 222.
45 Hart, "Fifty Years of Research in Aquatic Biology."
46 A.W.H. Needler, "Report of the Atlantic Biological Station for 1950," unpublished report to the Fisheries Research Board of Canada (1950), SABS Library, 1.
47 A.W.H. Needler, "Report of the Atlantic Biological Station for 1951," unpublished report to the Fisheries Research Board of Canada (1951), SABS Library, 1.
48 Ibid.
49 J.L. Hart, "Report of the Atlantic Biological Station for 1954," unpublished report to the Fisheries Research Board of Canada (1954), SABS Library.
50 A.W.H. Needler, "Report of the Atlantic Biological Station for 1952," unpublished report to the Fisheries Research Board of Canada (1952), SABS Library, 1.
51 Hart, "Fifty Years of Research in Aquatic Biology," 33.
52 J.L. Hart, "Report of the Atlantic Biological Station for 1955," unpublished report to the Fisheries Research Board of Canada (1955), SABS Library, 1.
53 J.L. Hart and D.L. McKernan, "International Passamaquoddy Fisheries Board Fisheries Investigations 1956–59: Introductory Account," *Journal of the Fisheries Research Board of Canada* 17:2 (1960): 1.
54 Fisheries Research Board of Canada, "Biological Station, St Andrews N.B. Annual Report and Investigators' Summaries 1957–58," unpublished report of the Fisheries Research Board of Canada (1958), SABS Library, 33.
55 Johnstone, *The Aquatic Explorers*, 227.
56 Fisheries Research Board of Canada, "Biological Station, St Andrews N.B. Annual Report, 1958."
57 A. Beaulnes, "Government and Science Policy in the 1970s," *Journal of the Fisheries Research Board of Canada* 31 (1974): 1278–80.
58 Hubbard, *A Science on the Scales*, 222–3.
59 Transcript from directors' panel, 16 October 2008, at "Workshop on the Evolution of Marine Science in Canada," in SABS Library.
60 See also V. Zitko and D.J. Wildish, 'The First Marine Biological Station in Canada'.
61 Gavaris and Stephenson, "Commemoration of the 100th Anniversary."
62 R. Halliday, "Career Outline," unpublished manuscript (2004), available in the Bedford Institute of Oceanography Archives, Accession 2006–41.
63 R.L. Stephenson, D.E. Lane, D.G. Aldous, and R. Nowak, "Management of the 4WX Atlantic Herring (*Clupea harengus*) Fishery: An Evaluation of

Recent Events," *Canadian Journal of Fisheries and Aquatic Science* 50 (1993): 2742–57.

64 T.D. Iles and N. Tibbo, "Recent Events in Canadian Atlantic Herring Fisheries," *International Commission of the Northwest Atlantic Fisheries Redbook*, part 3 (1970): 134–47.

65 Stephenson et al., "Management of the 4WX Atlantic Herring."

66 Transcript from directors' panel, 16 October 2008.

67 Ibid.

68 L.S. Parsons, "Management of Marine Fisheries in Canada," *Canadian Bulletin of Fisheries and Aquatic Science* 225 (1993): 1–763.

69 Hubbard, *A Science on the Scales*, 216.

70 R.L. Saunders, E.B. Henderson, B.D. Glebe, and E.J. Loudenslager, "Evidence of a Major Environmental Component in the Determination of the Grilse: Larger Salmon Ratio in Atlantic Salmon (*Salmo salar*)," *Aquaculture* 33 (1983): 107–18; R.L. Saunders, E.B. Henderson, and P.R. Harmon, "Effects of Photoperiod on Juvenile Growth and Smolting of Atlantic Salmon and Subsequent Survival and Growth in Sea Cages," *Aquaculture* 45 (1985): 55–66.

71 See also V. Zitko and D.J. Wildish, "The First Marine Biological Station in Canada."

72 Gavaris and Stephenson, "Commemoration of the 100th Anniversary."

73 Ibid.

74 Trippel, "The First Marine Biological Station in Canada."

75 Smith, *Scaling Fisheries*.

76 A.H. Leim and W.B. Scott, "Fishes of the Atlantic Coast of Canada," *Bulletin of the Fisheries Research Board of Canada* 155 (1966): 1– 485; and W.B. Scott and M.G. Scott, "Atlantic Fishes of Canada," *Canadian Bulletin of Fisheries and Aquatic Science* 219 (1988): 1–731.

77 E.E. Prince, "Marine Biological Station of Canada: Introductory Notes on Its Foundation, Aims and Work," *Contributions to Canadian Biology* (1901): 1–8.

78 Johnston, *Aquatic Explorers*, 138.

79 Needler, "The Seventy-fifth Anniversary of Two Canadian Biological Stations," 218.

80 F. Anderson, "The Demise of the Fisheries Research Board of Canada: A Case Study of Canadian Research Policy," *Journal of the History of Canadian Science, Technology and Medicine* 8 (1984): 151.

81 See the series of perspectives in *Canada, Journal of Fisheries and Aquatic Science* 1974 (31): 1269–322.

82 Needler "The Seventy-fifth Anniversary of Two Canadian Biological Stations."
83 F.R. Hayes, *The Chaining of Prometheus: Evolution of a Power Structure for Canadian Science* (Toronto: University of Toronto Press, 1973).
84 Transcript from directors' panel, 16 October 2008.
85 More recently, Fisheries and Oceans reverted to matrix reporting (reports channelled through project managers overseeing specific projects by teams at different laboratories), with some staff reporting directly to BIO.
86 Hayes, *Chaining of Prometheus*, 33, 138.
87 M. Picard-Aitken, D. Campbell, and G. Côté, *Bibliometric Study in Support of Fisheries and Oceans Canada's International Science Strategy: Report to Fisheries and Oceans Canada by Science-Metrix* (Canada: Fisheries and Oceans Canada, 2009), 1–66.
88 Needler, "The Seventy-fifth Anniversary of Two Canadian Biological Stations," 223.

# 5 Technology in Marine Science at the St Andrews Biological Station, 1908–2008

TIMOTHY JAMES FOULKES

Marine science at the St Andrews Biological Station relied over the years on whatever technologies were then available, but at the same time, the various scientific enterprises also inspired the development of new technology. When scientists were collecting specimens in the early days of marine science, simple hand tools were sufficient for them to study and report on the many marine species found in the Bay of Fundy. As the research programs became more ambitious, the scientists' need to acquire more specimens, and collect them from different environments, depths of water, types of sediment, and so on, led to a steady progression in the development of field equipment for collections, large support vessels, specialized gear aboard them, and laboratory equipment and techniques for scientific experiments and research.

A major technological problem faced in the early days at St Andrews was in dealing with eight-metre tides, twice a day, to obtain a steady and reliable supply of seawater for the laboratory maintenance of live animals. Sea-water supply is usually taken for granted at most marine biological stations. The evolution of the saltwater plumbing system at the station would fill another volume, but it was technology essential to the scientific activities at St Andrews. This chapter presents highlights of developments in technology over the first century at the St Andrews Biological Station from a technologist's perspective, and some of the consequences of this work. However, this presentation does not attempt a comprehensive review of all the technology, especially in the first fifty years. For the earlier technology that was developed the reader is directed to the station's annual reports.

The examples that follow illustrate the diversity and scope of this technology, ranging from the simplest of sampling gear and methods to the more sophisticated systems used for lab work, benthic studies, field surveys, and data management. Some of this technology has had significant consequences, and some of the problems encountered have inspired further developments by others, notably at the Bedford Institute of Oceanography in Bedford, across the body of water called the Bedford Basin from Halifax, Nova Scotia. Engineering projects of the gear research program and the Marine Technology Section from 1964 until 1988 receive most attention in this chapter, because of my direct involvement with them during this period. My focus is mostly on developments by the Marine Technology Section as spin-offs from the work of the Fishing Gear Engineering Research Program from 1964 to 1988. These included various underwater camera vehicles such as BRUTIV. Some developments in the technology needed and used for benthic studies and marine resource stock assessment between 1988 and 2008 in collaboration with the Bedford Institute of Oceanography, are also presented.[1]

Although not the primary focus of this chapter, it is pertinent to touch on specific development projects at the Atlantic Biological Station in the early years. These included the adoption of standardized sampling technology for the Canadian Fisheries Expedition of 1914–15; the Fisheries Technology Program in 1925; the development of cold storage technology and frozen-fish marketing experiments; the development of a pelagic herring trawl specifically for local fisheries; plus the development of fish and bivalve culture (discussed elsewhere in this volume by Robert Cook), enhancement programs, and fish ladders and passes for fish species protection. The advent of steam-powered vessels and the resulting surge in fishing effort raised concerns about declining fish stocks over the years, and led to the development and application, with attendant technology, of marine sciences for fisheries conservation and management throughout Atlantic Canada.

The Canadian Fisheries Expedition was motivated in part by a desire of scientists working for the Biological Board of Canada to locate herring in better condition than the fish habitually taken in the local weir fisheries. These studies provided a large amount of information on plankton, eggs, and larvae of commercial species of fish and herring biology as well as fisheries oceanography. As a result of this expedition, the Atlantic Biological Station's director at the time, Dr A.G. Huntsman, founded in 1925 the Fisheries Technology Program, "to provide scientific research and technical assistance to improve the handling, processing, and quality of fisheries products in Atlantic Canada."

Furthermore, he "was charged by the Biological Board of Canada to … conduct an educational program to ensure that industry personnel were made directly aware of the scientific findings and received training on their application."[2]

An important development in this area was Huntsman's own invention of the method of jacketed cold storage. Dr Huntsman experimented with freezing fish fillets and transporting them to Toronto for sale. The experiment initially went well, but the technology was not effectively passed on to the industry. Consequently, the quality of the fish was compromised, and the product became unacceptable in the marketplace. (One can just imagine the bad smells!) It would be many years before industry would embrace freezing methods again.

In addition to food science, there were also investigations and technological developments connected with the growing commercial fishery, fishing and sampling gear, and concerns for fish stocks that related to their growth, health, and enhancement. During the 1940s, and until the late 1950s, scientists at the biological station collaborated with fishers in developing and testing mid-water trawls for herring. The Atlantic Herring Investigation Committee was formed in 1944 to study herring populations and to determine whether greater use could be made of these stocks by extending the herring fishery to other areas and longer seasons. Previously, the herring fishery had been largely based on inshore fishing methods. Staff from St Andrews associated with this program surveyed the waters for offshore fish populations and experimented with fishing gear not previously used on the Atlantic coast for herring fisheries. The gear they developed was based on existing European drift nets and pelagic trawls. This gear had to be adjusted because Bay of Fundy currents severely damaged European drift nets. Different mesh sizes and various materials were tested until effective and economically feasible solutions were found and eventually adopted and transferred to industry in the late 1950s and early 1960s.

St Andrews scientists also became involved in projects to help anadromous fish get up streams over natural waterfalls. These and the new dams built for various mills at the turn of the century inspired some creative solutions. An example of technology to help fish maintain migration patterns through altered waterways was a fish elevator designed by E.E. Prince and A.P. Knight at the Atlantic Biological Station. This was installed for trial in the Magaguadavic River near St George from 1913–16. It would influence future fish ladders and passes designed to help fish maintain migration patterns through other waterways altered by human activity.

**Development of the St Andrews Biological Station's Workshops**

The scope of marine scientific activities at St Andrews required technological support, so permanent workshops were established early on. By the mid-1950s, the biological station had full-time technical and maintenance staff. The original workshop, built circa 1925, was not replaced until 1958, when a brick building was built in addition to the new main laboratory building. Tradesmen like Willard Ross, Joe Johnson, Frank Langley, Clyde Tucker, Ron Greenlaw, Phillip Green, Hill Brownrigg, Herb Small, and many others on our technical and trades staff helped to create much of the equipment and wet lab facilities that made fisheries research at St Andrews possible during the second half of the century.[3] Workshop facilities expanded from 1966 to 1978 to include marine technology, electronics, and underwater photography.

In the late 1960s, the workshop building just to the south of the main building had a second floor added to it, which was then connected to the main building by an overhead walkway. This new area was dedicated to gear research, electronics, and data management. The latter area was the domain of A. Sreedharran, our gourmet curry cook and data analyst. He became the lord of an early massive, mainframe computer and all its peripherals in his special air-conditioned and dust-free environment. The technology to handle data and punch-cards in that room was, at that time, in the 1960s, very impressive!

Photographic and graphic arts for scientific publications and technical reports were painstakingly provided the old fashioned way with ink pens and photo emulsions for decades by Bill McMullon and Frank Cunningham from their domain in the original 1908 residence building. Bill undertook camera work in the field, much of it underwater, and was noted for his "can do" attitude. Development of these facilities and much of this research would not have been possible without the support of the directors of that period, especially Dr J.L. Hart and Dr J.M. Anderson, to whom the late P.J.G. (John) Carrothers and I remained grateful for their encouragement of our engineering and technology programs beginning in the early sixties. Marine technology and its consequences at St Andrews would probably have been altogether different without the influence of and spin-offs from John

Carrothers's gear research and the mentoring from director Dr John Anderson, who always advocated innovative thinking and a positive approach to problems.

## Fishing Gear Engineering Research, 1962–1987

The fishing gear engineering research project, led by John Carrothers, was created to describe the dynamics and then improve the mechanical performance of commercial fishing gear, and to develop methods for the design of more efficient east coast otter trawls. It was undoubtedly the most ambitious technological endeavour at St Andrews, and the experience gained from it affected many other programs there and elsewhere. The engineering program began in 1962 at the Pacific Biological Station at Nanaimo, BC, where Carrothers undertook the preliminary development of many of the underwater instruments that were later used when the project was moved to St Andrews in 1964. In addition to a camera system that he designed for the headlines of fishing trawls, to be discussed later, a suite of instruments was developed to measure all the operating parameters of full-size commercial groundfish otter trawls. These instruments measured the dimensions of the mouth opening, wing spread, and door spread, the forces on the main bridles and warps, and the dynamic pressure in the mouth of the net (figure 5.1). We used existing instruments to measure angles of the towing warps, ocean currents, vessel speed, and otter trawl velocity.[4] (Twenty-five years later, sophisticated acoustic equipment, e.g., SCANMAR, became available "off the shelf" to measure the dimensions of fishing trawls.) After describing the physical performance of various trawls, the objective was to develop mathematical protocols for their design and improved efficiency. The development of instruments, sea trials, and preliminary data reduction lasted until about 1976, and resulted in a number of technical reports on the physical behaviour of east coast otter trawls.[5] Data analyses and the development of mathematical protocols culminated in a 1986 publication by A.L. Fridman and Carrothers that described the calculations used for fishing gear designs.[6]

There were three main spin-offs from the gear research program, some with significant and enduring effects. These were a large hydraulic flume; the establishment of a new Marine Technology Section; and the further development of underwater photographic equipment, techniques, and vehicles for underwater photography.

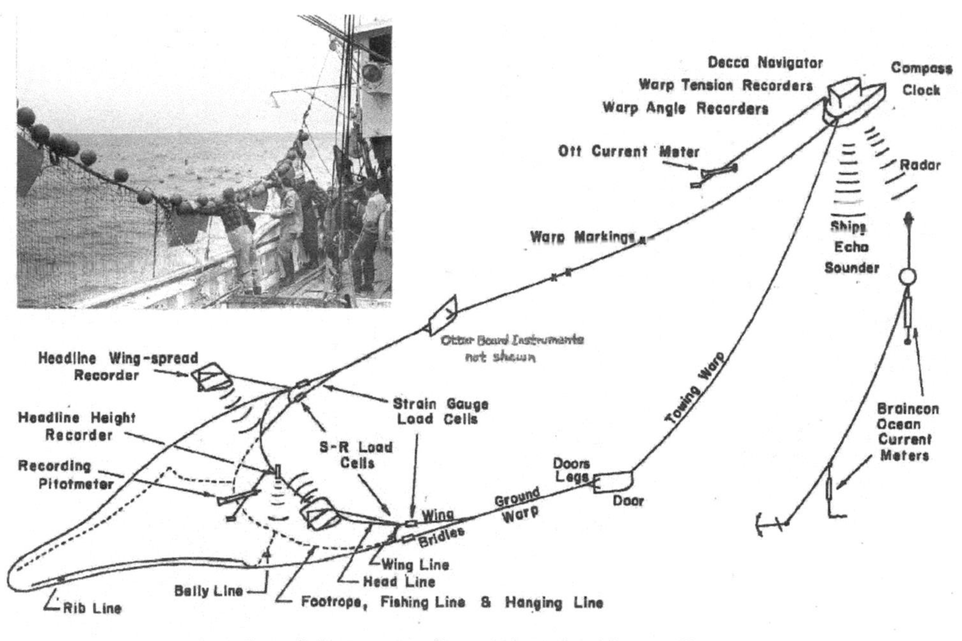


Location of instruments for trawl engineering studies.

5.1 Photo and diagram of instruments on trawl net used in engineering studies

The "Hydraulic Flume" was an instrument test tank and artificial stream which went into operation in 1971. It was originally designed for hydrodynamic tests and calibration of some of the underwater instruments used in the Gear Research Program. It consisted of a sixteen-metre-long, recirculating flow tank, constructed of splined eastern white cedar planks, with acrylic viewing windows. Water was circulated by a variable speed DC electric motor turning a propeller in one corner of the flume tank, and flows could be accelerated to over 200 centimetres per second in the test sections (figure 5.2). The hydraulic flume, housed in the specially built Butler Building, was also used until 1983 by Phil Symons and others as an artificial stream for behavioural studies of juvenile salmon and other fish.[7]

The Marine Technology Section evolved from the gear research program after the latter entered its analytical phase in 1977. In that year extensive premises were established for the new Marine Technology

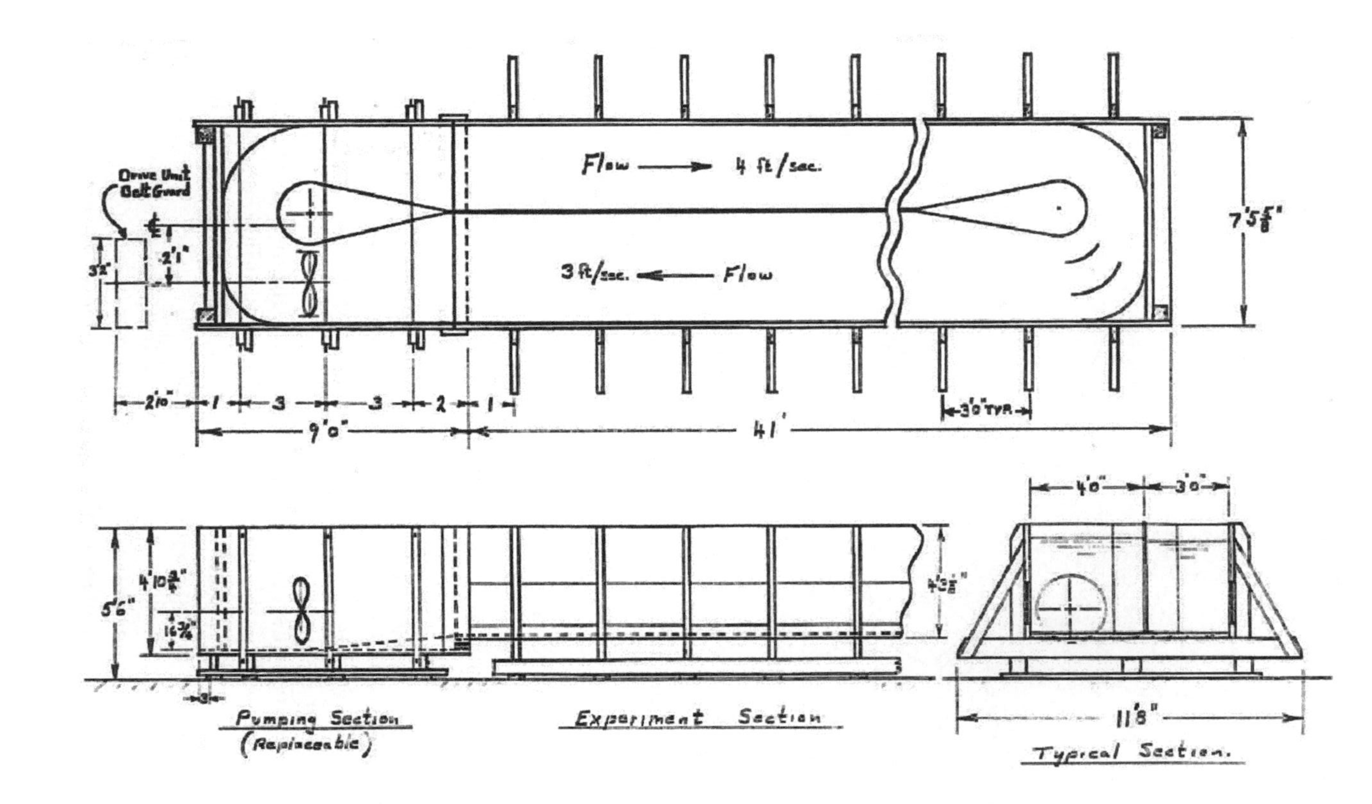


5.2 Diagram, plan, side, and end view of large hydaulic flume

Section, as well as for the biological station's grounds crew and carpentry staff, in a specially designed extension to the warehouse on the hill. This group included myself (usually with a summer student), electronics technologist Sam Polar from Gear Research, mechanic Fred Lord, and electronics technician Marcel Babineau in Sam's electronics department. We felt we could build just about anything, and we just about did. My work with Sam included the gear research instrumentation and other marine technology projects such as the underwater cameras, data telemetry, and control systems for BRUTIV. The work of Sam Polar, and that of his Electronics Section also met the needs of the scientific community for various control, monitoring, and recording systems, and is only mentioned in passing here.

During 1962 and 1964, the station purchased two EG&G 35-mm, underwater cameras and strobe light systems to document fish behaviour in the mouth of the otter trawl net. This led to the development of an extensive assortment of remote vehicles for their use in other applications. For example, the trawl camera case, designed by Carrothers in 1962 for observing fish behaviour in the mouth of the net was meant to be shackled to the headline of a trawl. It was modified for use (unsuccessfully) as a towed body with an experimental elevator in 1967. Later, John Caddy bolted our cameras directly to scallop drags to monitor fishing effects, work he describes elsewhere in this volume. This arrangement worked for short periods until the severe vibration resulted in damage to the camera systems, but useful results were obtained and inspired a gentler camera-vehicle technology. These EG&G camera systems were later extensively modified for work in fisheries research.[8] They were used at SABS for another fourteen years.

Several vehicles and control systems for these and other cameras were designed by the Marine Technology section for projects such as calibrating the acoustic fish counting system, stock assessment surveys, fish and scallop behaviour, and various seabed studies. Later, near the end of the St Andrews station's first century, other scientists like Mike Strong and Peter Lawton would carry on innovating cameras and vehicles using the latest digital technologies for optics, vehicle control, and data. (See "The Last Twenty Years," below.)

### Technology for Stock Assessment Surveys

As the focus of work became directed towards the more commercial aspects of our marine environment, various research groups required stock assessment of commercial species. In addition to traditional methods of trawl surveys and catch statistics for stock assessment, acoustic survey and photographic methods from towed underwater camera vehicles and other technologies were developed, some of which are described here. Scientists also wanted to observe their subjects in their habitat, as well as the effects of commercial fishing at first hand.

Starting in 1968, Director Dr John Anderson arranged for a number of mini-submarines to come to the Maritimes for our use and assessment. With the charter of the Perry *Cubmarine* in 1968 (figure 5.3a), the St Andrews Biological Station became the first institution to use submersibles for fisheries science. This was the submarine that inspired the Beetles' "Yellow Submarine" song, and which was first leased by the Woods Hole Oceanographic Institution for underwater oceanographic research in 1962. Jeff Watson, with Jim Meldrum and Danny Welsh, used it in the Gulf of St Lawrence to study Queen Crabs in situ.[9] It was also used by Udo Buerkle to study fish responses to fishing gear noise. Following the *Cubmarine* in 1968, the Perry *Shelfdiver* (figure 5.3b) was chartered in 1969, and the *Pisces I* in 1970 (figure 5.3c). While submersible technology was not developed at St Andrews, it had a very high profile for a while, and got a lot of researchers interested in exploring the realm of "inner space" at a time when others were exploring the moon with similar enthusiasm. Technology developed at St Andrews for use with *Pisces* included an odometer wheel and camera control system that helped John Caddy quantify scallop behaviour and the various effects of scallop drags.[10] Caddy himself discusses some of this work elsewhere in this volume.

In 1973, the research vessel *E.E. Prince* and the gear research team departed on a mission to Florida for comprehensive studies and to observe otter trawl performance in warm, clear water using our Towed Underwater Research Plane (TURP) and the *Johnson Sea Link*, another manned submersible built by the Harbor Branch Oceanographic Institution in 1971. The mission was aborted while we were off Cape Hatteras when we learned the *Sea Link* had become fatally entangled in the wreckage of the sunken destroyer USS *Fred T. Berry*. The *Sea Link* was recovered by a rescue vessel after twenty-four hours, but two of the four occupants had perished. This, and other close calls, along with their cost, put a damper on future missions with such manned mini-submersibles.[11]

Acoustic survey systems such as ECOLOG, developed by Lloyd Dickie and the acoustics working group at the Marine Ecology Lab at

5.3a Perry *Cubmarine*, 1968

5.3b Perry *Shelfdiver*, 1968

5.3c *Pisces I*, 1970, being inspected by Dr J.R. Weir and Dr J.M. Anderson

the Bedford Institute of Oceanography acoustics working group during the 1970s,[12] yielded lots of data, but it all had to be calibrated for it to be meaningful. At St Andrews, Udo Buerkle directed the development of technology designed to determine the hearing sensitivity of fish, fundamental to the calibration of the acoustic survey system, and their reaction to sounds from trawlers. He also calibrated the target strengths of the echo from fish in relation to their aspect in the sound beam of the transducer,[13] that is, whether the beam was reflected back from the sides of the fish or other angles of their bodies. The Marine Technology Section, under Udo's direction, further developed and refined our towed camera vehicle, BRUTIV, to help verify data from acoustic surveys with photographs of fish in situ (see below) to determine their tilt angle distribution.

In 1974 the Marine Technology Section designed and built a large semi-submersible aluminum tower so that target fish could be manipulated at different angles ten metres below the ECOLOG transducer, which was positioned just below the water surface near the top of the tower. The base of the tower was wide enough to be outside the main lobes of the transducer beam. The tower, anchored just off the biological station's wharf, provided data on the target echo strengths of different species held at different angles to the transducer beam.[14]

A technology called Depth Referencing and Veering Acoustics Body (DRAVAB) was developed for the acoustics group in 1983 to deal with problems from bubbles in the wake of the mother ship which affected the ECOLOG transducer signals. It also obviated the need for heave compensation systems, and allowed a transducer to be streamed off each side of the vessel for greater coverage. Apart from the transducer it carried, DRAVAB was a purely mechanical towed body designed to provide a stable platform for the acoustics transducer, and orient it within a few degrees of vertical regardless of sea state. It also allowed scientists to maintain the transducer at an adjustable set depth and to veer it off to the side of the tow vessel's track. It had positive buoyancy so it would be self-rescuing should the tow cable break. It was not fully de-bugged, and only the one prototype was completed.

## The Birth of Towed Underwater Benthic Camera Sleds at St Andrews Biological Station

In 1965 the Gear Research group fixed the Edgerton camera system to a heavy, suspended steel cage fitted with various devices to trigger the camera at about the right height to photograph the seabed. The cage was later faired to maintain orientation and reduce drag, but researchers found it difficult to maintain its altitude by manual winch control. Although primitive, this was the introduction of remote benthic imaging technology at St Andrews that, nearly forty years later, reached its zenith with the fully automated URCHIN system (see note 25 below).

During the 1970s I was also involved in developing three different sleds for different missions to survey the seabed. The towed Underwater Research Plane (TURP) was designed for Dave Scarratt's SCUBA diver surveys of lobster/Irish moss beds.[15] The Drifting Underwater Collapsible Camera Sled (DUCCS) was designed as a back-up camera vehicle for John Caddy's manned submersible surveys;[16] and the Towed Underwater Automatic Camera Sled (TUACS)[17] was designed for crab surveys conducted by Richard Bailey off Cape Breton. All were designed to be transported easily by pickup truck, and featured inherent stability to provide correct orientation for the camera system, protection of the cameras from vibration, and some sort of rescue system should they encounter a "fatal" obstacle.

These primitive "sleds" had limited usefulness (see "Persistent Problems" below), except for TURP, from which SCUBA divers could observe and manually photograph large areas of the seabed first hand and which was such fun to fly. TURP also led to the development of BRUTIV, an unmanned automated version that will be described and discussed below. BRUTIV saw so much use it became almost legendary at St Andrews, and it was the predecessor of future generations of continental shelf seabed survey vehicles for fisheries science.

All these sleds were towed with a weak link at the vehicle for safe disconnection in case they got entangled in obstructions. DUCCS and TUACS used a safety line and marker buoy, while TURP and BRUTIV had positive buoyancy so they would glide to the surface if the tow-cable weak link broke. BRUTIV also had a self-actuating strobe light and flag, and an EPIRB (emergency position indicating radio-beacon) and tracking receiver for recovery in fog or at night. As a result of this technology, and unlike operations at BIO and elsewhere, St Andrews never lost a camera or a vehicle.

TURP, our diver-controlled Towed Underwater Research Plane) had two-axis (pitch and roll) joystick control for manoeuvrability in water of minimum visibility where boulders and trap lines were regular hazards. Its mission was to enable the study of the effect of raking Irish moss on the lobster fishery and its habitat.[18]

BRUTIV technology (Bottom Referencing Underwater Towed Instrument Vehicle) was developed at St Andrews in 1973 in part to replace the SCUBA divers used by TURP.[19] Our requirement for an automated bottom-following camera vehicle was deemed unfeasible by electronics consultants, so this added challenge was irresistible. Inspired by the Batfish developed at the Bedford Institute of Oceanography (BIO) by Jean-Guy Dessureault,[20] we adapted technology from our Gear Research headline height dimension meter to develop a basic servo-mechanism loop with wing angle feedback, and adopted power for control surfaces from a Batfish hydraulic system. The electronics control system was designed with essential help from Dave McKeown at BIO. We were thus able to develop an automated control system for a stable camera vehicle that could fly at any depth or follow the seabed terrain of the continental shelf at adjustable altitudes down to depths of 100 fathoms.[21]

The first prototype, BRUTIV I (figure 5.4a), used the body of TURP with a pendulum system for roll control, and hydraulics to move the diving planes. This was in use in various configurations until the Mark

5.4a BRUTIV I

2 model was launched in 1977. Common to all prototypes was the analogue altitude control and monitoring system which, from its earliest telemetry system, transmitted height off the seabed as an audible tone to allow operators to "keep an ear" on how well BRUTIV was following the bottom without having to watch some visual indicator for hours on end while the still cameras shot off rolls of 35-mm film. A constant tone meant constant altitude. Electrical interference noises from the strobe lights and film advance motor were also audible and were useful to monitor performance. The Marine Technology Section also developed a low-cost heave compensation system, an arrangement of pulleys and springs on deck which averaged tow-cable tension to allow BRUTIV to follow the seabed more evenly, without the surging which caused nausea in some scientists observing its underwater television when the towing vessel pitched in heavy seas.

Flown in constant depth mode, BRUTIV could fly through and photograph concentrations of herring. Despite its early camouflage colour, fish of all species avoided BRUTIV when it flew closer than two metres during the daytime, but they seemed oblivious to its presence and its lights at night. It was primarily used for studies of groundfish and pelagic fish. Scientists in the Invertebrates and Marine Plants Divisions also used BRUTIV for various offshore lobster, scallop, and crab surveys, and other scientists at BIO used it for geophysical and trawl impact studies.[22]

By 1986, BRUTIV had been tested with acoustic (sonar) sensors. In collaboration with BIO Metrology Division and Seastar Instruments of Halifax, we determined during trials carried out with the *E.E. Prince* on 23–24 March 1986 that BRUTIV could be used with advanced and sensitive fish survey systems when equipped with both side-scan and down-looking sonar, as well as its usual still and video cameras.

BRUTIV II (figure 5.4b) was also used for studies of sea floor dynamics and iceberg scours off Labrador and in Hudson's Bay between 1983 and 1987 by the Atlantic Geoscience Centre at BIO. Atlantic Geoscience Centre researchers were able to combine information from BRUTIV's 35-mm sequential, seabed photos with their side-scan sonar images and estimate the age and frequency of the iceberg scours.[23]

BRUTIV III was launched in 1987 (figure 5.4c). Its first mission with the Atlantic Geoscience Centre was less than successful due to preliminary sea trials at St Andrews being curtailed for want of ship time. After the "bugs" were subsequently worked out at St Andrews in 1988, the entire system was transferred to the Bedford Institute of Oceanography and Sea Star Instruments in 1989, and used until 1999 by Dave McKeown's group at BIO. My last direct involvement with the development of this underwater vehicle technology, after I transferred to the Northwest Atlantic Fisheries Centre (NAFC) in St John's, Newfoundland, in 1987, was the successful sea trials of BRUTIV III in Halifax Harbour in 1992. New side-scan and down-looking sonar systems together with a newly developed surface control unit based on an analogue controller designed and developed at St Andrews were used. In 1994 I arranged, with Seastar Instruments of Halifax, the development of a new surface control unit for use with BRUTIV III.

In spite of almost continuous development of cameras, tow cables, electronics, and data telemetry, our underwater photographs lacked two very important features: altitude and position data. Data management of 35-mm photos was extremely cumbersome, and new data logging and "still video" (digital) cameras were unreliable or not available before 1987. We lacked resources to develop these at St Andrews. Our "leading edge" fibre-optic tow cable and digital control system, tested in 1987, had not worked properly due to software and electronics

5.4b BRUTIV II

5.4c BRUTIV III

problems; and of course Global Positioning systems (GPS) were not widely available for another ten years.

## Technology Development with University Partners

Not all the technology development was done in house at the St Andrews Biological Station. For example, a Blažka respirometer, originally used in swimming performance and oxygen uptake studies in fish by Dick Saunders, was designed and built in the workshops at the University of Toronto. It was subsequently used in the flume lab for other purposes, which are described elsewhere in this volume by Dave Wildish and Shawn Robinson. Flow calibration in the flume lab at St Andrews was originally provided by a velocity calibrator, designed by David Kristmanson of the University of New Brunswick, Chemical Engineering Department, which was built in their workshops in Fredericton.[24] Kristmanson, through a long period of collaboration with Wildish and the St Andrews Biological Station until his retirement (1970–93), provided the design of new flumes that were built in the station's workshops, a collaboration that is also described by Wildish and Robinson in their contribution to this collection.

## The Last Twenty Years

By 1993 staff for the carpentry and plastics workshop facilities had retired. I had moved to Newfoundland to resolve technical problems in that region and responsibilities for technology development at St Andrews Biological Station were officially transferred to the Bedford Institute of Oceanography. With the transfer of responsibility for innovations in electronic and mechanical technology, much less of this type of development was done at St Andrews, but collaborative development of technology within the region continued. At the BIO, BRUTIV morphed into TOWCAM.[25] It used new digital technology solutions, which became commercially available in the 1990s, to solve all the problems perceived earlier in BRUTIV.[26] New video (digital) imaging and digital data logging were the obvious solutions, and were used in TOWCAM, because data management of tens of thousands of seabed photographs was a considerable analytical problem for benthic researchers. TOWCAM video data can now be analysed frame by frame, each with its own associated GPS position and time stamp.[27] TOWCAM has been widely used by benthic ecologists at the BIO[28] and the St Andrews Biological Station.[29] The Underwater Reconnaissance and Coastal Habitat Inventory (URCHIN) was also developed at St Andrews in collaboration with the BIO.[30] This is a geo-referenced remote camera system with associated relational image database for mapping coastal benthic habitat, employing similar methods to those used in TOWCAM.[31]

Collaboration between St Andrews and the BIO also led to a further application of acoustic and digital data logging technology: the development of the Grab Acoustic Positioning System (GAPS), by McKeown and others.[32] The GAPS represents a significant advance in remote benthic sampling from vessels, allowing each sample to be accurately positioned by GPS (and hence revisited at chosen time intervals). GAPS technology was used in one of the first quantitative studies of the influence of pockmarks on benthic communities in soft sediments.[33] In the future, benthic sampling data with GPS coordinates will be of much greater value than sampling lacking precise coordinates. GAPS is a system of hardware, software, and methodology created to provide benthic ecologists, geologists, and engineers with accurate, real-time information about the location of grab and diver-held samplers and towed survey platforms (e.g., TOWCAM and URCHIN). It enables users to predetermine a detailed sample transect across an area of interest and manoeuvre the sampler or diver to within a few metres of each target location before collecting the sample, or to tow a survey platform along a predetermined survey line.

A further example of successful collaboration with BIO technologists is the design and manufacture of two specialized seawater, recirculating flumes for use in the St Andrews flume laboratory, described elsewhere in this volume by Wildish and Robinson. These examples clearly show that the personal contact and synergy that are essential to excellence, and were evident earlier in St Andrews, continued to flourish in these cooperative projects with the BIO team during the last part of the century, and enabled the continued development of technology at the St Andrews Biological Station at the time of the centennial workshop in 2008.

## Conclusion

The applied problems dealt with by marine science in 2008 are more numerous and much more complex than those considered in 1964 when the professional technology program was initiated at the St Andrews

Biological Station. In 1964 one of the most important applied problems was considered to be new and more efficient ways to harvest the sea – hence the Fishing Gear Engineering Research Program. To put this in context, in those days many fisheries biologists considered the fish stocks to be virtually limitless; hence, herring stocks were considered to be the "golden goose" of the sea,[34] although concern, even then, was expressed about their increased exploitation. Since 1970, we have come to witness overfishing in many fish stocks on the east coast of Canada, the beginnings of industrial aquaculture in the Bay of Fundy and the realization of numerous marine environmental problems, some with global dimensions.

To the credit of the Marine Technology Section at the biological station, it changed with the times and was involved with many smaller projects too numerous to include here. Some of these included: a bomb depressor to allow plankton sampling at depth; a newly designed three-metre beam trawl; and a variable buoyancy tagging barge, for tagging herring at a weir. Major achievements were those detailed above, the contributions to acoustic methods for stock assessment, and the development of towed underwater benthic camera sleds, leading to BRUTIV, TOWCAM, and URCHIN. The latter will probably play an important role in future synoptic environmental surveys of the marine, benthic environment.

One could ask, what comes first: technology or the need for it; a good workshop, or the desire to build something? No matter the case, the availability of effective support is a major factor in the success of marine science. During the past century, we and our predecessors were encouraged to seek out new technology, to "boldly go" as it were, and we accomplished great things in an exciting, creative, era. We were fortunate to have many ingenious people in both our technical and scientific staff. We worked effectively as a real team. Even though we lacked some areas of expertise, we never lacked an "esprit de corps," and budgets for our science and technology projects were usually available.

## NOTES

I would like to thank Dr R.L. Stephenson, who was the director of SABS at the time, for inviting me to participate in the centenary celebration, "Workshop on the Evolution of Marine Science in Canada," held in St Andrews from 15–16 October 2008. I am also indebted to the research scientists at SABS who were involved in the technology projects, including many of their technical staff, for their inspiration and cooperation, with particular gratitude for the always willing Sam Polar and his effective electronics work. I would also like to thank Eva Gavaris for her contributions to this paper concerning technology developments in the early days, Frank Langley for his recollections about workshop facilities and activities in the 1950s, and Mike Strong for insights into all the technology that he helped develop during the latter days of SAB's first century. Finally, I am also grateful to Frank Valentine and Dave Wildish for helping to edit, improve, and expand an earlier version of this manuscript and to Jennifer Hubbard of Ryerson University and Ellen Foulkes of MUN (ret.) for helping with this final version.

1 During the preparation of the illustrated talk I gave at the St Andrews biological Station symposium, "The Evolution of Marine Science in Canada," I uncovered many historically interesting images – too many to include here. They may be seen at the SABS web site: www.mar.dfo-mpo.gc.ca/sabs/ (no longer available).
2 J.E. Stewart, "Retrospective of the Halifax Lab," *Bio-Oceans Association Newsletter* 30 (April 2006), at www.bedfordbasin.ca/newsletters/30_APR2006.pdf.
3 Frank Langley, personal communication.
4 P.J.G. Carrothers, "Fishing Gear Engineering Research at Sea" (1966), remastered camcording available at www.mar.dfo-mpo.gc.ca/sabs/ (no longer available).
5 P.G. Carrothers, T.J. Foulkes, M.P. Connors, and A.G. Walker, "Data on Engineering Performance of Canadian East Coast Groundfish Otter Trawls," Fisheries Research Board of Canada Technical Report no. 125 (1969): 1–107; and P.J.G. Carrothers and T.J. Foulkes, "Measured Towing Characteristics of Canadian East Coast Otter Trawls," International Commission for the Northwest Atlantic Fisheries Research Document 71/39 (1971): 1–100.
6 A.L. Fridman and P.J.G. Carrothers, *Calculations for Fishing Gear Design*, FAO fishing manual Fn120, rev. ed. (Farnham, Surry, Eng.: Fishing News Books, 1986), 1–264.
7 See, for example, P.E.K. Symons, "Behaviour and Growth of Juvenile Atlantic Salmon *Salmo salar* and Three Competitors at Two Stream Velocities," *Journal of the Fisheries Research Board of Canada* 33 (1976): 2766–73.
8 T.J. Foulkes, S. Polar, M. Babineau, and P.J.G. Carrothers, "Instruction Manual (for) Modified E G & G model 200A-210 Automatic Camera System," Fisheries Research Board of Canada Manuscript Report 1304 (1974).

9 J.F. Caddy and J. Watson, "Submersibles for Fisheries Research," *Hydrospace* 2 (1969): 12–16.
10 J.F. Caddy, "A Method of Surveying Scallop Populations from a Submersible," *Journal of the Fisheries Research Board of Canada* 27:3 (1970): 535–49.
11 John Caddy, personal communication, 2008.
12 For information about Lloyd Dickie, please see "Dickie, Lloyd Merlin,' in *Canadian Encyclopedia*, http://www.thecanadianencyclopedia.ca/en/article/lloyd-m-dickie/.
13 U. Buerkle, "Detection of Trawling Noise by Atlantic Cod (*Gadus morhua* L.)," *Marine Behavior and Physiology* 4 (1977): 233–42; U. Buerkle and A. Sreedharan, "Acoustic Target Strengths of Cod in Relation to Their Aspect in the Sound Beam," in *Meeting on Hydroacoustical Methods for the Estimation of Marine Fish Populations, Volume II: Contributed Papers, Discussion, and Comments, the Charles Stark Draper Laboratory, June 25–29 1979*, ed. J. Suomala (Cambridge, MA: 1981), 229–47; and U. Buerkle, "Estimation of Fish Length from Acoustic Target Strengths," *Canadian Journal of Fisheries and Aquatic Science* 44 (1987): 1782–5.
14 Buerkle and Sreedharan, "Acoustic Target Strengths of Cod."
15 T.J. Foulkes and D.J. Scarratt, "Design and Performance of 'TURP,' a Diver Controlled Towed Underwater Research Plane," Fisheries Research Board of Canada Technical Report 295 (1971): 1–18.
16 T.J. Foulkes and J.F. Caddy, "'DUCCS,'" a Drifting Underwater Collapsible Camera Sled," Fisheries Research Board of Canada Technical Report 310 (1972): 1–14.
17 T.J. Foulkes, "Design and Development of 'TUACS,' a Towed Underwater Automatic Camera Sled," Fisheries Research Board of Canada Technical Report 292 (1972).
18 D.J. Scarratt, "The Effects of Raking Irish Moss (*Chondrus crispus*) on Lobsters in Prince Edward Island," *Helgoländer Wissenschaftliche Meeresuntersuchungenelgoland* 24 (1973): 415–24.
19 T.J. Foulkes, "The Development of a Bottom-Referencing Underwater Towed Instrument Vehicle, BRUTIV, for Fisheries Research," in *Underwater Photography: Scientific and Engineering Applications*, ed. Paul Ferris Smith (New York: Van Nostrand Reinhold, 1984), 95–107.
20 D.-G. Dessureault, "'Batfish': A Depth Controllable Towed Body for Collecting Oceanographic Data," *Ocean Engineering* 3 (1976): 99–111. The "Batfish" was a streamlined vehicle developed to house fast-responding oceanographic sensors. It is towed behind a ship or small vessel and its depth is controlled from the vessel by a manually or automatically produced command signal. Variable-angle wings controlled its vertical position and a novel control surface assured lateral stability.

21 D.L. McKeown, "Benthic Habitat Studies: An Engineer's Perspective," part I, *Bio-Oceans Association Newsletter* 29 (2006): 2–3, at http://www.bedfordbasin.ca/newsletters/29_JAN2006.pdf.
22 T.W. Rowell, P. Schwinghamer, M. Chin-Yee, K.D. Gilkinson, D.C. Gordon Jr., E. Hartgers, M. Hawryluk, D.L. McKeown, J. Prena, D.P. Reimer, G. Sonnichsen, G. Steeves, W.P. Vass, R. Vine, and P. Woo, "Grand Banks Otter Trawling Impact Experiment: III. Experimental Design and Methodology," Canadian Technical Report of Fisheries and Aquatic Science 2190 (1997): 1–36.
23 H.W. Josenhans and J. Zevenhuizen, "Seafloor Dynamics on the Labrador Shelf," in *Ice Scour Bibliography*, ed. C.R. Goodwin, J.C. Finley, and L.M. Howard (Ottawa: Environmental Studies Revolving Funds Report no. 010, 1985). H.W. Josenhans and J. Zevenhuizen, "Dynamics of the Laurentide Ice Sheet in Hudson Bay, Canada," *Marine Geology* 92 (1990): 1–26.
24 D.J. Wildish and D.D. Kristmanson, *Benthic Suspension Feeders and Flow* (New York: Cambridge University Press, 1997), 409.
25 Centre for Marine Biodiversity, TOWCAM, "Passamaquoddy Bay Pock Marks: Towcam Survey System," 2007. Available from http://www.marinebiodiversity.ca/cmb/research/pockmarks/towcam-survey-system/.
26 Dave McKeown, personal communication, 2008.
27 D.L. McKeown and D.E. Heffler, "Precision Navigation for Benthic Surveying and Sampling," in *Proceedings of the Oceans '97 MTS/IEEE Conference, Halifax, Nova Scotia, Canada, Oct. 6–9, 1997*, vol. 1: 386–90.
28 See, for example, D.C. Gordon Jr., P. Schwinghamer, T.W. Rowell, J. Prena, K. Gilkinson, W.P. Vass, D.L. McKeown, C. Bourbonnais, and K. MacIsaac, "Studies on the Impact of Mobile Fishing Gear on Benthic Habitat and Communities," 2003, at http://www.mar.dfo-mpo.gc.ca/science/review/1996/Gordon/Gordon_e.html.
29 For example, D.J. Wildish, H.M. Akagi, D.L. McKeown and G.W. Pohle, "Pockmarks Influence Benthic Communities in Passamaquoddy Bay, Bay of Fundy, Canada," *Marine Ecology Progress Series* 357 (2008): 51–66.
30 M.B. Strong and P. Lawton, "URCHIN – Manually-Deployed Geo-Referenced Video System for Underwater Reconnaissance and Coastal Habitat Inventory," Canadian Technical Report of Fisheries and Aquatic Science 2553 (2004): 1–28.
31 Somewhat to my chagrin, as I was caught up in recounting my own and earlier work, discussions with Peter Lawton and Mike Strong during the Centennial Workshop in October 2008 revealed these results of collaboration with BIO (e.g., URCHIN AND TOWCAM) after I was no longer involved with technology at St Andrews, which I had overlooked in my presentation. This later work has been described here by Dave Wildish.

32 D.L. McKeown, D.J. Wildish, and H.M. Akagi, "GAPS: Grab Acoustic Positioning System," Canadian Technical Reports of Hydrography and Ocean Sciences 252 (2007): 1–88; at www.dfo-mpo.gc.ca/Library/328342.pdf.
33 Wildish, Akagi, McKeown, and Pohle, "Pockmarks Influence Benthic Communities in Passamaquoddy Bay."
34 S.N. Tibbo, "Herring – the 'Golden Goose' of the Sea," Fisheries Research Board of Canada, Biological Station, St Andrews, NB, General Series Circular no. 55 (1970): 1–5.

# 6 An Overview of Physical Oceanographic Research at the St Andrews Biological Station during Its First Century

BLYTHE D. CHANG AND FRED H. PAGE

Oceanographic research has been a major component of the scientific research program of the Biological Station at St Andrews throughout its history. David Penhallow, its first director, who noted that in the station's first year (1908) attention was to be focused on three major questions – oyster culture, lobster culture, and fish migration – recognized that oceanographic research was required for the study of these issues.[1] Over the years, the oceanographic program has reflected changes in the programs and issues it supported, and in altering political realities, but the emphasis throughout has been on applied research. The Atlantic Biological Station, prior to Newfoundland joining confederation in 1949, was the only major marine research station operated by the federal government in Atlantic Canada. Therefore, oceanographers based at St Andrews conducted research along the entirety of Canada's east coast, including the Bay of Fundy, the Scotian Shelf, the Gulf of St Lawrence, and the eastern Canadian Arctic. Now that there are several marine laboratories operated by the Department of Fisheries and Oceans Canada throughout eastern Canada, the St Andrews Biological Station restricts its oceanographic research mostly to the Bay of Fundy and the Gulf of Maine, although collaborations continue with researchers working in other parts of the Maritime provinces, Newfoundland and Labrador, other parts of Canada, and other countries.

Prior to the station's opening in 1908, only limited oceanographic data had been collected in Atlantic Canada, primarily to improve shipping safety. Eric Mills describes the earlier history of Canadian hydrography in chapter 1. During the first twenty years of research at St Andrews, oceanographic and other research was conducted mainly by university researchers during the summer field season. The goals

6.1 Some of the scientists who have studied physical oceanography at SABS. Top row (L–R): Gordon G. Copeland (1909), James W. Mavor (1914, 1919–20), Alexandre Vachon (1916–20), A.G. Huntsman (1917–33). Middle row: Edmund E. Watson (1926, 1930–2), Harry B. Hachey (1927–40, 1946–59), Alfred W.H. Needler (1929–40), Louis M. Lauzier (1945–70). Bottom row: Ronald W. Trites (1950–60), Neil J. Campbell (1955–60), R. Ian Perry (1984–91), Fred H. Page (from 1991). The numbers in parentheses refer to the years that these researchers were involved in oceanographic research at SABS. Photographs are from SABS archives.

of this early research were to gain a general knowledge of the oceanography in Atlantic Canada, and to relate ocean conditions to fisheries resources and safe navigation, as well as to the potential for oyster aquaculture. Research was conducted primarily in the Bay of Fundy, Gulf of St Lawrence, and the Scotian Shelf. Oceanography became the first year-round program at St Andrews when Harry Hachey was hired in 1928 (Photographs of physical oceanographic researchers affiliated with the station during its first century, including earlier volunteers, Hachey, and the full-time professionals that joined and succeeded him, are shown in figure 6.1). An important issue when Hachey was hired as the station's first full-time employee, was a proposed tidal power project in Passamaquoddy Bay and its possible impacts on fisheries. This required an intensive oceanographic study. Oceanographic research was largely curtailed during the Second World War; however, the need for oceanographic research related to submarine detection led to the establishment of the Atlantic Oceanographic Group (AOG) in 1944, with headquarters at St Andrews, under the leadership of Hachey. After the war ended, the AOG continued at St Andrews, conducting research related to fisheries and naval issues; over time, fisheries issues came to dominate. Considerable research was conducted in 1957–8 related to a revised proposal for tidal power in Passamaquoddy Bay. Oceanographic research was also conducted during this period in the Gulf of St Lawrence, the Atlantic waters off Nova Scotia, and the eastern Arctic. When the AOG was moved to Halifax in 1960, a fisheries oceanography program remained at SABS, led by Louis Lauzier, until his transfer to Ottawa in 1970. A period of relative inactivity ended when Ian Perry was hired in 1984 to head a new oceanography program, with a primary goal of strengthening links between fisheries biology and oceanography. Perry left in 1991 and was replaced by Fred Page, who initially continued the fisheries oceanography work, and later expanded the oceanography program, with an emphasis on the near-shore marine environment. St Andrews researchers are currently studying oceanography in relation to many issues, including fisheries, shipping safety, aquaculture, tidal power generation, Marine Protected Areas, harmful marine phytoplankton, and invasive species.

## Physical Oceanographers and Their Work at the Atlantic Biological Station

During the station's first twenty years (1908–27) scientific research was mainly conducted by "volunteer workers"; there were no paid scientific staff at St Andrews until 1916, when A.G. Huntsman was hired as curator (later director); a second staff scientist was not hired until 1924. For the volunteer workers – mostly professors and students affiliated with universities in eastern Canada and the north-eastern United States, as well as some government scientists – St Andrews provided laboratory facilities, boats, equipment, accommodations, and occasionally a small honorarium. While research was conducted only during the summer season up until 1928, some year-round activities, including collecting water temperature and salinity data, commenced in the early 1920s.

W. Bell Dawson, the founder and head of the Tidal and Current Survey – whose contributions are described in chapter 1 by Eric Mills – was the first physical oceanographer associated with the Atlantic Biological Station. He visited in 1908, but did not conduct any research at that time. He did, however, agree to prepare a preliminary report on water temperatures and densities in the Bay of Fundy, based on existing data. The preliminary report was to be the basis for collecting additional oceanographic data.[2]

Before 1908, Dawson had been active in measuring sea levels and currents throughout Atlantic Canada. His appointment in 1893 as the first head of the Tidal Survey marked the beginning of systematic tidal and current surveys in Canadian waters. His first current surveys were conducted in the Strait of Belle Isle area, Cabot Strait, and the St Lawrence estuary from 1894 to 1896.[3] In the summers of 1904 and 1907, Dawson conducted current surveys at a series of locations in the outer Bay of Fundy.[4] These current surveys included the deployment of current meters and data collected on water temperature and density. The stations were mainly offshore, chosen to lie along the major shipping lanes, since this work was intended for the production of tide and current tables to aid marine navigation. However, Dawson noted that knowledge of water currents was also of importance to understanding the distribution of fish and other marine life, including spawning and survival.[5] Dawson returned to St Andrews as a visiting scientist, with his assistant A.R. Lee, in the summer of 1917; they installed a tide gauge at the end of the station wharf.

The first oceanographic work done by researchers formally based at the Atlantic Biological Station appears to have been in July–August 1909, when water temperature and density data were collected by Gordon Copeland of the University of Toronto (assisted by D.L. McDonald of McGill University), to examine the potential for oyster culture in the

Passamaquoddy Bay area.[6] Based on his findings, he concluded that Passamaquoddy Bay was too cold for oyster reproduction.

Most physical oceanographic research during the era of volunteer workers was not conducted by oceanographic specialists, but rather by biologists (table 6.1). The methods they used to study physical oceanography at St Andrews were considerably different from later methods (figure 6.2). The earliest researchers could measure temperatures, densities, and salinities only at discrete times and depths. Surface temperatures were measured with thermometers and temperatures at depths were measured using reversing thermometers. They collected water samples at depth using Nansen reversing bottles or other similar bottles, and measured the density (specific gravity) of water samples using hydrometers. They calculated the salinity from the chlorinity, measured by silver nitrate titration.[7] The titration method was used until 1978, when a laboratory salinometer was purchased.

In August 1910, A. Brooker Klugh, of Queen's University, collected some water temperature and density (specific gravity) data in the lower Saint John River and Kennebecasis Bay.[8] Additional temperature and density data were collected in August 1914 by E. Horne Craigie of the University of Toronto, as part of a project led by James W. Mavor of the University of Wisconsin that examined the potential for oyster culture in bays between the St Croix River and Saint John.[9] In the summers of 1914 and 1915, Craigie, directed by Mavor, and assisted by W.H. Chase of Acadia University in 1915, collected more temperature and density data in Passamaquoddy Bay, the St Croix estuary, St Marys Bay, and the Bay of Fundy.[10]

Early researchers at the Atlantic Biological Station lamented their inability to measure currents directly. Copeland noted of his 1909 work, "I should have liked to have made an exact study of the currents in several localities, but lack of instruments, and of opportunity prevented it."[11] Craigie expressed similar thoughts: "It is also regrettable that no current-meter of any kind was to be had, as some observations with such an instrument would undoubtedly throw much light upon the subject by indicating the direction and strength, as well as the fluctuations of the currents at various points."[12] This is not because such devices did not exist, but probably reflected a lack of funds: the earliest reported current meter measurements in Atlantic Canada were by W. Bell Dawson of the Tidal and Current Survey, starting in 1894; he used meters with either anemometer- or propeller-type rotors.[13]

Table 6.1 "Volunteer workers" and other non-SABS employees who studied physical oceanography at SABS, 1908–34*

| Researcher and affiliation | Year(s) and project |
|---|---|
| Gordon G. Copeland (U. Toronto) | 1909: hydrography and oyster culture potential (Passamaquoddy Bay) |
| D.L. McDonald (McGill U.) | 1909: assisted Copeland |
| A. Brooker Klugh (Queen's U.) | 1910: hydrography (lower Saint John River) |
| E. Horne Craigie (U. Toronto) | 1914–15: hydrography (Bay of Fundy, Passamaquoddy Bay) |
| James W. Mavor (U. of Wisconsin; Union College, Schenectady, NY) | 1914, 1919–20: hydrography and drifter releases (Bay of Fundy) |
| Albert D. Robertson (U. Toronto) | 1914–15: hydrography and oyster culture (Malpeque Bay) |
| Johan Hjort (Norwegian Directorate of Fisheries) | 1914–15: Canadian Fisheries Expedition (Gulf of St Lawrence, Scotian Shelf) |
| Paul Bjerkan (Norwegian Directorate of Fisheries) | 1915: Canadian Fisheries Expedition (Gulf of St Lawrence, Scotian Shelf) |
| W.H. Chase (Acadia U.) | 1915: assisted Craigie |
| Julius Nelson (New Jersey Agricultural Experiment Station) | 1915: hydrography and oyster culture (Malpeque Bay) |
| Alexandre Vachon (U. Laval) | 1916–20: salinity determinations (Bay of Fundy, Gulf of St Lawrence) |
| W. Bell Dawson (Tidal and Current Survey) | 1917: tide gauge installation (SABS) |
| A.R. Lee (Tidal and Current Survey) | 1917: assistant to Dawson |
| George F. Sleggs (Dalhousie U.) | 1922: drifter releases (Cabot Strait) |
| W.M. Anderson (U. Toronto) | 1924: hydrographic assistant (SABS) |
| H.M. Allan (U. Toronto) | 1924: warm water lobster survey (Shelburne) |
| Harry H. Bell (Dalhousie U.) | 1924–5: warm water lobster survey (Lunenberg) |
| Arthur F. Chaisson (St Francis Xavier U.; Harvard U.) | 1924–8: warm water lobster survey (Halifax and Tusket-Pubnico areas) |
| W.G. Jones (U. New Brunswick) | 1924: warm water lobster survey (Chaleur) |
| Edmund E. Watson (Queen's U.; International Passamaquoddy Fisheries Board) | 1926, 1930–2: hydrography and tidal power proposal (Passamaquoddy Bay, Bay of Fundy) |
| Harry B. Hachey (U. New Brunswick) | 1927: drifter releases (Passamaquoddy Bay) |
| Frits Johansen (Ottawa) | 1927, 1929: hydrography and drifter releases (Labrador, Hudson Bay, Hudson Strait) |
| B.W. Taylor (McGill U.) | 1928: hydrography and drifter releases (Hudson Strait) |
| Charles J.A. Hughes (International Passamaquoddy Fisheries Board) | 1931–2: hydrographical assistant to Watson |
| J. Wallwin Fisher (U. Toronto) | 1932–3: hydrography and oyster culture (Shediac) |

* Volunteer workers were mostly university professors and students who came to SABS to conduct research during the summer season.

6.2 Some vessels and equipment used to study physical oceanography at SABS. Top row (L–R), vessels: the *E.E. Prince* (1913–32); the *CCGS Pandalus III* (background; since 1986) and the *Vector* (since 2007). Second row, water sampling equipment: Nansen reversing water bottle (Canadian Fisheries Expedition, 1915); Sea-Bird Electronics conductivity-temperature-depth profiler (since 1990s). Third row, drifters: glass drifter bottle (1960s); seabed drifter (1961–73); Seimac Convertible Accurate Surface Trackers (since 1996). Bottom row, current meters and circulation model: Gurley meter (1950s); Teledyne RD Instruments Acoustic Doppler Current Profiler in subsurface flotation device (since 1990s); example circulation model output (since 2000). All photos are from SABS archives, except the Gurley meter (courtesy of Canadian Hydrographic Service, Ottawa).

While at this time most of the St Andrews oceanographic research was directed towards biological questions, researchers were also interested in understanding physical features of the water bodies they were investigating. Craigie stated:

> The object of this work was to obtain as much information as possible not only about the actual temperatures and densities of the water, but also about the nature of the currents of warm and cold water, how these are affected by the tides, etc. Such observations, besides being of importance and interest in themselves, are valuable on account of their bearing upon the haunts and habits of fish frequenting the waters studied, or passing through these waters in their migrations.[14]

The results enabled the scientists to provide descriptions of the general circulation patterns. They determined that for Passamaquoddy Bay, "warm water passes out through the Western Passage, cold water from the outside enters through Letite passage."[15] For the Bay of Fundy they found

> a clear indication that water on the Nova Scotia side of the lower part of the bay is, on the whole, warmer than on the New Brunswick side, and the plan of the surface temperatures suggests a current of warm surface water from the Atlantic flowing in along the south shore and then turning north about half way up the bay, so that its influence is not visible in the higher profiles. All other evidence, however, indicates a simple tongue of cold water up the middle of the bay.[16]

In 1914 and 1915 Albert Robertson of the University of Toronto – assisted in 1915 by Julius Nelson of the New Jersey Agricultural Experiment Station – collected temperature and salinity data in Malpeque Bay, Prince Edward Island, in an investigation into the causes of major mortalities in oyster populations and the potential for oyster culture.[17]

This was a precursor to later oyster research led by Alfred Needler. In 1929, Needler collected temperature, salinity, and pH data in Malpeque Bay. This led to the establishment in the following year of the Ellerslie substation of the Atlantic Biological Station, under Needler's leadership. The Ellerslie substation was the centre for Atlantic Canadian oyster investigations, which included the collection of additional temperature and salinity data in Malpeque Bay and other areas in the Gulf of St Lawrence.[18] Needler served as the head of the Ellerslie substation until he became director of the Biological Station at St Andrews from 1941 to 1954.

The Canadian Fisheries Expedition of 1914–15, led by the Norwegian scientist Johan Hjort, was designed to study the herring resources of Atlantic Canada, primarily in the Gulf of St Lawrence and the Scotian Shelf, but a major component was the collection of hydrographic data. The expedition's final report included two chapters on the oceanography of Canadian Atlantic waters, written by Norwegian scientists J.W. Sandström and Paul Bjerkan.[19] Many years later, Hachey noted: "The very complete report of the Canadian Fisheries Expedition provided a very fine reference book on various phases of oceanography. The treatise on hydrodynamics, which occupied a prominent part of the report, has for 50 years continued to be valued as a text. Further, the methods of investigation introduced into Canadian oceanography were to be adopted and developed by Canadians in the years that followed."[20]

Alexandre Vachon, of Laval University, worked at St Andrews from 1916 to 1920. His main activity was conducting salinity analyses (by titration) of water samples collected in various surveys in the Bay of Fundy and the Gulf of St Lawrence. He noted that

> as the plankton, which regulates, to a great extent, the migrations of the fish, is itself at the mercy of the chemical, physical and mechanical conditions of the sea, it is easily understood of what economical importance a correct knowledge of these conditions will prove. We speak of the migrations of the herrings and sardines; they are the same as those of the plankton which serve as food for them, and the presence of the plankton is ruled by depth, light, temperature, salinity, pressure and density.[21]

Vachon's 1918 report includes temperature and salinity data collected in July–October 1916 at several "Prince" monitoring stations in the Bay of Fundy (figure 6.3), including Prince 5, located between Campobello Island and The Wolves.[22] This is the first published data from these stations, which were named after the Atlantic Biological Station vessel the *E.E. Prince*, the first ship used to collect data from these stations; the vessel was named after the first commissioner of fisheries for Canada. Most researchers considered data from this station to be indicative of conditions generally in the Bay of Fundy.

James Mavor (by this time affiliated with Union College, Schenectady, New York) made the first direct observations of water currents at St Andrews using drifters. In 1919 and 1920 he conducted research on the

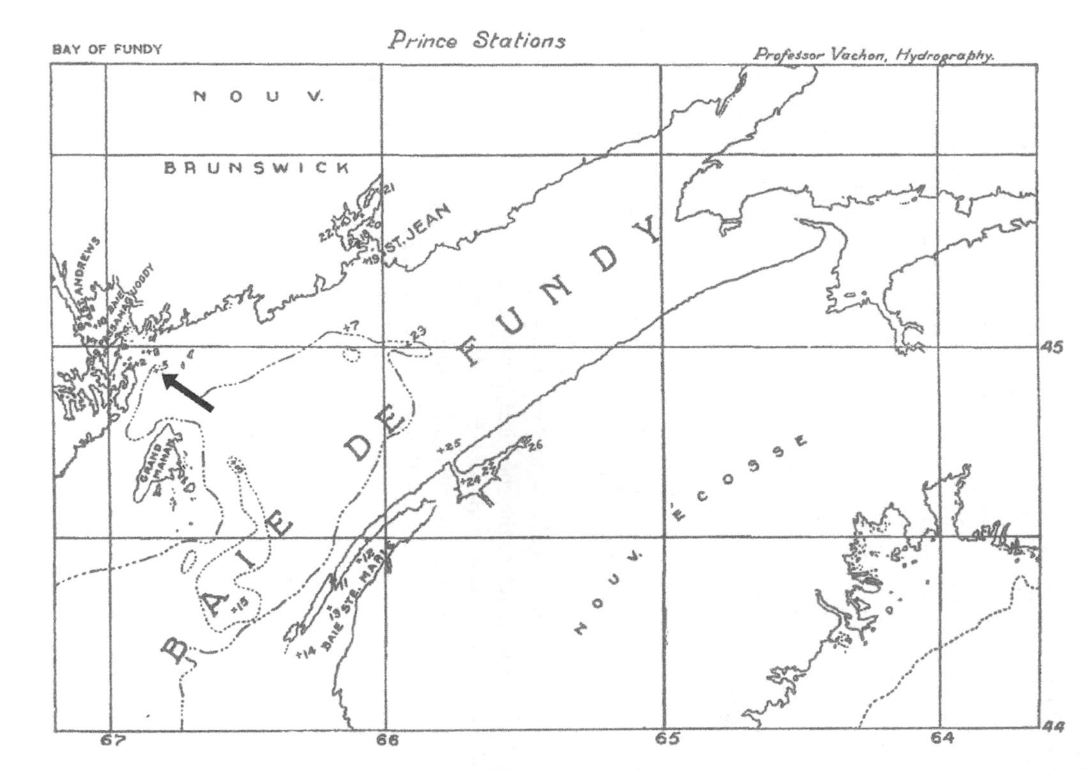


6.3 The first published map showing the locations of the original "Prince" hydrographic stations, including Prince 5, located just east of Campobello Island (indicated with an arrow). This map shows stations sampled in 1916. From A. Vachon, *Contributions to Canadian Biology 1917–1918* (1918): 295–328.

currents in the lower Bay of Fundy "'to determine what, if any, general movement of the water other than the tides occurs in the Bay of Fundy.'"[23] The original drifters were floating glass bottles; until the 1960s similar drifters were used to track surface currents. They could only provide data on the date and location of release and the date and location of recovery; the path taken between release and recovery had to be guessed, and the time taken for a drifter to reach the recovery point was poorly known. Mavor published several articles on the circulation in the Bay of Fundy.[24] Almost 400 drift bottles (2–10 oz glass bottles) were released in the summer of 1919 from locations in the lower Bay of Fundy; Mavor combined an analysis of their movements with temperature, salinity, and density data; he also analysed hydrographic data collected between 1916 and 1918, and re-analysed Dawson's current meter data from 1904 and 1907. His description of the general circulation pattern agreed with that made earlier by Craigie and Chase:[25]

> Water enters the bay on its eastern side and flows northeast along the coast of Nova Scotia; it crosses the bay to the New Brunswick side and flows southwestward out of the bay, the bulk of the water probably passing to the east of Grand Manan. The rate at which the water flows is probably somewhere between 5 and 10 nautical miles per day, so that the complete circuit probably takes from 20 to 40 days.[26]

He also published a note on the path of the Gulf Stream, based on four drifters released in the Bay of Fundy in 1919 and recovered on the European side of the Atlantic Ocean in 1920–1.[27] Mavor noted that knowledge of water movements for "any consideration of the life history of marine organisms is obvious; eggs, larvae or post-larval stages may be transported far from the place of breeding to waters either suited or unsuited to their survival."[28]

A.G. Huntsman, who was primarily a biologist, was also as the station's director chiefly responsible for developing the oceanographic program at St Andrews, including the drift bottle program. Between 1922 and 1930, St Andrews scientists released more than 8000 drift bottles off Canada's east coast, including in Passamaquoddy Bay, the Bay of Fundy, the Scotian Shelf, the Gulf of St Lawrence, Cabot Strait, the Strait of Belle Isle, the Labrador coast, Hudson Strait, and Hudson Bay. Much of this work was part of an international cooperative program to study North America's eastern coastal circulation, organized by the International Committee on Deep-Sea Fisheries Investigations (later renamed the

North American Council on Fishery Investigations), whose members included Canada, the United States, Newfoundland, and France.[29]

Huntsman also led expeditions to collect oceanographic data in the Gulf of St Lawrence in 1917 and 1918, and off south-western Nova Scotia in 1921. In 1923 he led the Strait of Belle Isle expedition in 1923, which collected biological and oceanographic data (temperature, salinity, and density data; current meter records; and drift bottle releases); the data were later used to describe the area's physical oceanography.[30] Huntsman also described the circulation in and outside Halifax harbour, based on temperature and salinity data collected in 1922 and 1924, and drift bottle releases in 1923.[31]

Other early Atlantic Biological Station oceanographic work included the collection of water temperature and salinity data during a faunistic survey in St Marys Bay in 1919 and a survey to determine the extent of warm water suitable for lobsters in the Bay of Chaleur area of New Brunswick and the outer coast of Nova Scotia in 1924–5. Fisheries expeditions to Hudson Strait and Hudson Bay in 1927, 1928, and 1929 collected surface temperature data and water samples, and released drift bottles.

Atlantic Biological Station annual reports indicate that monthly or weekly hydrographic sampling started in 1913, although the exact locations are unclear; for the period before 1924 we have found Prince 5 data only for some months in 1916 and 1918. Since 1924, there are monthly temperature and salinity records from Prince 5 near the surface, at mid-depth, and near the sea floor, with the exception of some missed months during and immediately after the Second World War and some other missed months likely due to bad weather or vessel problems (figure 6.4). Since 1989, temperature and salinity data have been recorded at Prince 5 throughout the water column using a conductivity-temperature-depth profiler.[32] The Prince 5 station appears to have been chosen for long-term sampling because it is located in an area with strong currents and good depth (about 100 m), in close proximity to the biological station. Watson in 1936 declared that Prince 5 is "certainly not representative of the bay of Fundy."[33] However, Hachey noted that the monthly data collected at Prince 5 are fairly representative of conditions at other stations in the Bay of Fundy.[34]

The total number of volunteer researchers at the Atlantic Biological Station peaked in the late 1920s. In 1928, the number of paid scientific staff at the Station reached eight, including a new full-time hydrographer, Harry Hachey. After 1934, there were almost no volunteer workers at St Andrews, due to a policy change regarding volunteer workers in 1931, a shortage of funds during the economic depression of the 1930s, and an increase in the size of the scientific staff at St Andrews starting in the late 1920s.[35]

Harry Hachey's was the first year-round research program at St Andrews. He first joined the station in the summer of 1927, while affiliated with the University of New Brunswick. In that summer, he released drift bottles to study the water circulation in Passamaquoddy Bay. The results, not published at the time, suggested a cyclonic circulation in a counter-clockwise direction.[36] The following year, he was appointed station hydrographer, the first year-round scientist employed at the Atlantic Biological Station. This marked the beginning of a dedicated oceanographic research program at St Andrews.

Hachey's early work focused on the Bay of Fundy, including research related to a tidal power project in Passamaquoddy Bay (see below) and on tidal mixing in the Saint John River estuary.[37] In 1929 he organized a series of coastal stations in Atlantic Canada, where surface water and air temperatures were taken regularly. In 1930, he was put in charge of the Hudson Bay Fisheries Expedition, which released drift bottles and collected temperature, salinity, and density data; the data were used to estimate the circulation pattern of surface waters in Hudson Bay.[38] Hachey's 1931 manuscript on the oceanography of the Bay of Fundy, based on data collected in 1929 and 1930, was not published; a revised version was produced by Hachey and Bailey in 1952.[39]

### The 1930s: The Cooper Power Project Proposal, a Fire, and the Great Depression

In 1926, the Dexter P. Cooper Co. proposed the construction of several dams across Passamaquoddy and Cobscook Bays to extract power from the tides (figure 6.5). The International Passamaquoddy Fisheries Commission (IPFC) was formed in 1931 to investigate this project's probable effects on the area's fisheries. During 1931 and 1932, the IPFC scientific investigations were based at St Andrews and the *E.E. Prince* was loaned to the IPFC. Edmund Watson of Queen's University, who had conducted some preliminary research at St Andrews in the summers of 1926 and 1930, collected temperature and salinity data for the IPFC along several transects across the Bay of Fundy in 1931 and 1932. He was assisted in this work by Charles Hughes. The 1931 data were lost in a fire (see below) – the

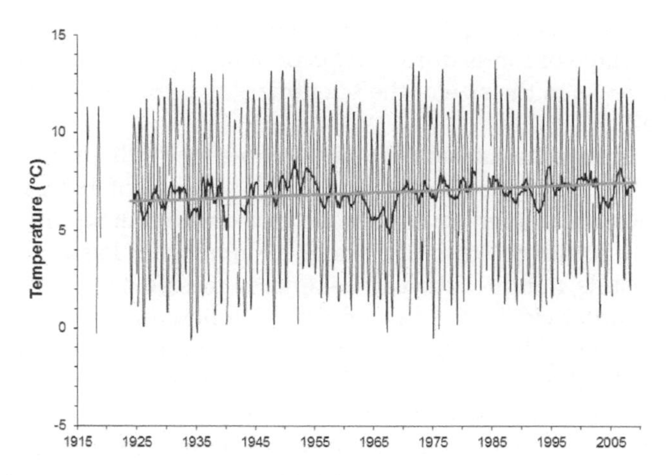


6.4 Surface seawater temperatures at the Prince 5 station, east of Campobello Island, 1916–2008. The earliest data records are for 1916 and 1918. Regular monthly records are available from 1924 on, with the exception of some months during and immediately after the Second World War (1940–1 and 1945–6), some months in the winters of 1982–3 and 1983–4, and the occasional month missed due to weather and/or equipment problems. The thin black line shows all temperature records, usually taken near the middle of each month; the thick black line shows the 12-month running average; the straight grey line indicates the long-term trend.

implications of which are discussed by Rob Stephenson in chapter 4. Watson's report concluded that the mixing produced by the Quoddy passages was important locally, but only amounted to about 6 per cent of the total mixing in the Bay of Fundy, and was therefore not likely to have a major influence on general circulation of the Bay of Fundy and Gulf of Maine.[40] Hachey also conducted research related to the Cooper project, using tidal gauges and experimental tanks. He concluded that closing off Passamaquoddy Bay would not affect the natural period of oscillation of the waters of the Bay of Fundy, and would increase the amplitude of the Bay of Fundy tides by 1.7 per cent, which would increase spring tides by about 14 cm at Saint John and 21–26 cm at the head of the bay.[41]

The IPFC report concluded that the project's physical oceanographic impacts would not extend far beyond the outer Quoddy region.[42] The report, however, also predicted considerable changes to the waters inside

Passamaquoddy and Cobscook Bays (e.g., the tides in Passamaquoddy Bay would be reduced to a fifth of natural levels), and that the herring fishery inside Passamaquoddy Bay would be virtually eliminated. The Cooper project did not proceed, largely because of the predicted impacts on the herring fishery. A smaller project, involving the damming of Cobscook Bay alone, was subsequently begun, but was not completed due to lack of funding.[43]

The fire that destroyed the Atlantic Biological Station's main laboratory building in March 1932 resulted in a loss of facilities, equipment, the library, and some oceanographic data. A new laboratory was operating the following year, but the disruption in activities, combined with staff and budget cuts related to the Great Depression, resulted in a period of reduced oceanographic research. Temperature and salinity data continued to be collected at fixed stations in the Bay of Fundy, but the chief oceanographic investigations were on the Atlantic coast of Nova Scotia.[44] In 1930, Edmund Watson had developed a method for measuring currents for the IPFC, using surface and subsurface drags; he measured the time taken for the drags to travel a certain distance, and used this information to estimate current velocities in Bocabec Bay. In 1933 Hachey studied the mixing of stratified waters using experimental tanks, and made current velocity measurements in Passamaquoddy Bay using surface and depth drags.[45] He studied the hydrography of Saint John harbour in 1934–5.[46] Also during the years from 1932 to 1938 Hachey analysed thermograph data from Canadian National Steamships sailing on the Boston-Bermuda route to examine temporary migrations of Gulf Stream water.[47]

Additional temperature and salinity data were collected by the Ellerslie substation to examine the potential for oyster culture in various locations in the Gulf of St Lawrence and Bras d'Or Lakes. Water samples were also collected from various stations in the Gulf of St Lawrence, in connection with cod, haddock, and lobster investigations; drift bottles were released in the Gulf of St Lawrence in 1936 and 1938.[48]

## The Atlantic Oceanographic Group, 1944–60

In 1940–4, during the Second World War, Hachey was on leave with the Canadian military, and the Atlantic Biological Station's oceanographic research program became inactive. The station's ship, *Zoarces*, used for deep-sea observations, was taken over by the Royal Canadian Navy. Some temperature and salinity monitoring did continue at St Andrews, however, and some oceanographic work was conducted by other programs. While studying larval smelt migration, Russell McKenzie released drift bottles in the Miramichi Bay area in the summers of 1943 and 1944.

Hachey returned in 1944 when the Atlantic Oceanographic Group (AOG) was formed and located at the Atlantic Biological Station. The AOG was jointly supported by the Royal Canadian Navy, the Fisheries Research Board of Canada, and the National Research Council. The initial purpose of the AOG was to work on problems of underwater sound transmission related to the detection and destruction of submarines during the war. The AOG was initially headed by Hachey, on assignment with the Royal Canadian Navy. In 1946 Hachey returned to the biological station, and the AOG continued operating at St Andrews, responsible for furthering oceanographic activities in waters contiguous to the Atlantic and Arctic coasts of Canada. In 1946, the Joint Committee on Oceanography (JCO) was formed to coordinate oceanographic research in Canada. Hachey served as its secretary and senior (later chief) oceanographer, and continued in these positions when the JCO was reorganized into the Canadian Committee on Oceanography in 1959. The AOG research program was designed to meet Fisheries Research Board and Navy requirements, although over time fisheries issues came to dominate. The naval research was highly classified;[49] this probably explains the absence of published reports on this work.

The AOG was based at St Andrews until 1960. Hachey led the AOG until mid-1954, and remained at St Andrews as chief oceanographer for Canada until he was transferred to Ottawa in 1959. The number of oceanographers at St Andrews and the AOG grew rapidly, peaking in 1957–9, when considerable research was conducted in relation to another proposed tidal power project in Passamaquoddy Bay. Most of the St Andrews Biological Station's employees who have conducted physical oceanographic research are included in table 6.2; however, due to incomplete records, the table does not include seasonal and term employees. The list also does not include researchers from other programs who collected physical oceanographic data during their fieldwork, nor does it include ships' crews who assisted in oceanographic data collection. Oceanographers working at the AOG during 1944–60 included Louis Lauzier (from 1945); Hugh McLellan (1947–57); William Bailey (from 1948); Ronald Trites (from 1950); Donald MacGregor (from 1950, summers only); Neil Campbell (from 1955); Arthur Collin (1955–9); Fabian Forgeron (from 1957); and Roland Chevrier (from 1957). The scientists

Table 6.2 Physical oceanographers and support staff at SABS

| Name | Years at SABS |
|---|---|
| **Research scientists** | |
| A.G. Huntsman | 1917–33 |
| Harry B. Hachey | 1928–40, 1946–59 |
| Alfred W.H. Needler | 1929–40 (Ellerslie) |
| Louis M. Lauzier | 1945–70 |
| Hugh J. McLellan | 1947–57 |
| William B. Bailey | 1948–60 |
| Ronald W. Trites | 1950–60 |
| Donald G. MacGregor | 1950–61 (summers only) |
| Arthur E.H. Collin | 1955–8 |
| Neil J. Campbell | 1955–60 |
| J. Roland Chevrier | 1957–60 |
| Fabian D. Forgeron | 1957–60 |
| R. Ian Perry | 1984–91 |
| Fred H. Page | 1991–present |
| Michael Dowd | 1999–2003 |
| **Other scientific staff** | |
| E. Lloyd Graham | 1948–51 |
| Leslie H. Brownrigg | 1949–52 |
| Carl C. Cunningham | 1950–60 |
| John A. Sullivan | 1951–6 |
| George B. Taylor | 1952–60 |
| John H. Hull | 1952–83 |
| J. Graham Clark | 1954–70 |
| Reginald K. Robicheau | 1956–9 |
| Thomas A. Grant | 1957–60 |
| Arthur W. Brown | 1963–74 |
| Randy J. Losier | 1984–present |
| Blythe D. Chang | 2003–present |
| E. Paul McCurdy | 2004–present |
| John C.E. Reid | 2008–present |
| **Administrative staff** | |
| Helen E. Rigby | 1944–9 |
| Madelyn M. Irwin | 1950–60 |
| Shaun Smith-Gray | 1999–2001 |
| Brenda J. Best | 2001–7 |
| Sheila B. Gidney | 2007–present |

*Notes*: Before 1928, most oceanographic research was conducted by "volunteer" workers. The list does not include term and seasonal employees, students, and ships' crews. The years refer to the period when the individuals were working in oceanography at SABS; A.G. Huntsman and A.W.H. Needler were not primarily physical oceanographers, but conducted some physical oceanographic research during the years indicated.

were assisted by several technicians, including Lloyd Graham (1948–51); Leslie Brownrigg (1949–52); Carl Cunningham (from 1950); John Sullivan (1951–6); George Taylor (from 1952); John Hull (from 1952); Graham Clark (from 1954); Reginald Robicheau (1956–9); and Thomas Grant (from 1957).

The primary effort in Atlantic oceanography was directed towards collecting and analysing basic oceanographic data over a large portion of the western North Atlantic Ocean. Starting in the late 1940s, bathythermographs were used by St Andrews researchers to collect temperature data throughout the water column. An instrument to record vertical profiles of temperature and salinity was first used at St Andrews in 1948, when an STD (salinity-temperature-depth) recorder was obtained on loan through the Royal Canadian Navy and the US Navy. However, reversing thermometers and water samples were still being used to collect temperature and salinity data at depths into the 1980s. Conductivity-temperature-depth (CTD) profilers are now used to collect such data. Time series of temperature and salinity data can now be obtained using continuously recording instruments deployed at fixed stations.

The collection of temperature and salinity data by the AOG was a major component of the Atlantic Herring Investigation in the Gulf of St Lawrence from 1945–9;[50] the oceanographic work was led by Lauzier. Seasonal oceanographic surveys in the Bay of Fundy, Scotian Shelf, and Gulf of St Lawrence were conducted starting in 1950. Manuscript reports described the oceanography of the Bay of Fundy and Passamaquoddy Bay.[51] Additional surveys were conducted around Newfoundland (including the Grand Banks), Labrador, and the eastern Arctic (including Hudson Strait and Hudson Bay). McLellan used temperature-salinity relations to study mixing and circulation on the Scotian Shelf.[52] Another research focus was the "slope water" between the continental shelf and the Gulf Stream.[53] Lauzier and Trites investigated the deep waters in the Laurentian Channel, using temperature and salinity data collected since 1915.[54] Studies related to oyster culture in the Gulf of St Lawrence required additional temperature and salinity data. In 1952–3 data was collected to study the impacts of constructing the Canso Causeway between Cape Breton Island and mainland Nova Scotia.[55] In the 1950s, the AOG also researched the formation and distribution of ice in the Gulf of St Lawrence and the Canadian Arctic.[56]

Oceanographic research was also conducted in Atlantic Canada by other research groups, often in collaboration with the AOG. Fisheries

Research Board scientists collected temperature and salinity data during fisheries surveys; the Naval Research Establishment of the Navy continued conducting oceanographic research related to defence issues; and the Canadian Hydrographic Service collected current meter data to aid safe navigation. The Naval Research Establishment assisted the AOG in a study of currents in the Grand Manan Channel using a Geomagnetic Electrokinetograph (GEK).[57] The GEK measured the electromotive force between two electrodes of a two-wire cable towed behind a ship. A collaborative AOG–Naval Research Establishment project released over 800 drifter bottles in 1954 to examine Scotian Shelf circulation.[58]

In 1951, Bostwick Ketchum of the Woods Hole Oceanographic Institution was invited to SABS to analyse salinity data from the Bay of Fundy and Passamaquoddy Bay. In the resulting report on the exchanges of fresh and salt waters in these bays, the flushing time for the Bay of Fundy was estimated to be seventy-six days (using average freshwater discharge rates for the Saint John River in April–July) and for Passamaquoddy Bay sixteen days.[59] In later research, Bailey estimated the flushing rates for the Bay of Fundy to average about 86 days during the spring-summer freshet and 173 days at other times.[60]

Other collaborative work involving the AOG included collecting data on temperatures (using bathythermographs) and surface currents (using a GEK), as part of "Operation Cabot," a Canada-USA multi-ship study of the northern edge of the Gulf Stream in 1950. MacGregor and McLellan used a GEK to study the currents in Grand Manan Channel.[61] The AOG also collected surface and subsurface temperature, salinity, and other chemical data from Bermuda to Baffin Bay, as part of the International Geophysical Year's "Project Deep Water Circulation" in 1958.

## Research on a Revised Tidal Power Project in Passamaquoddy Bay, 1957–8

A revised tidal power project in Passamaquoddy and Cobscook Bays by the Canadian and US governments in 1956 proposed to construct a dam similar to that of the earlier Cooper project (figure 6.5), but modified to enclose a slightly larger area outside of Cobscook Bay. The International Passamaquoddy Fisheries Board (IPFB) was established to examine the potential impacts of this project on fisheries. The AOG conducted research for the IPFB in 1957–8 on the circulation of the Quoddy

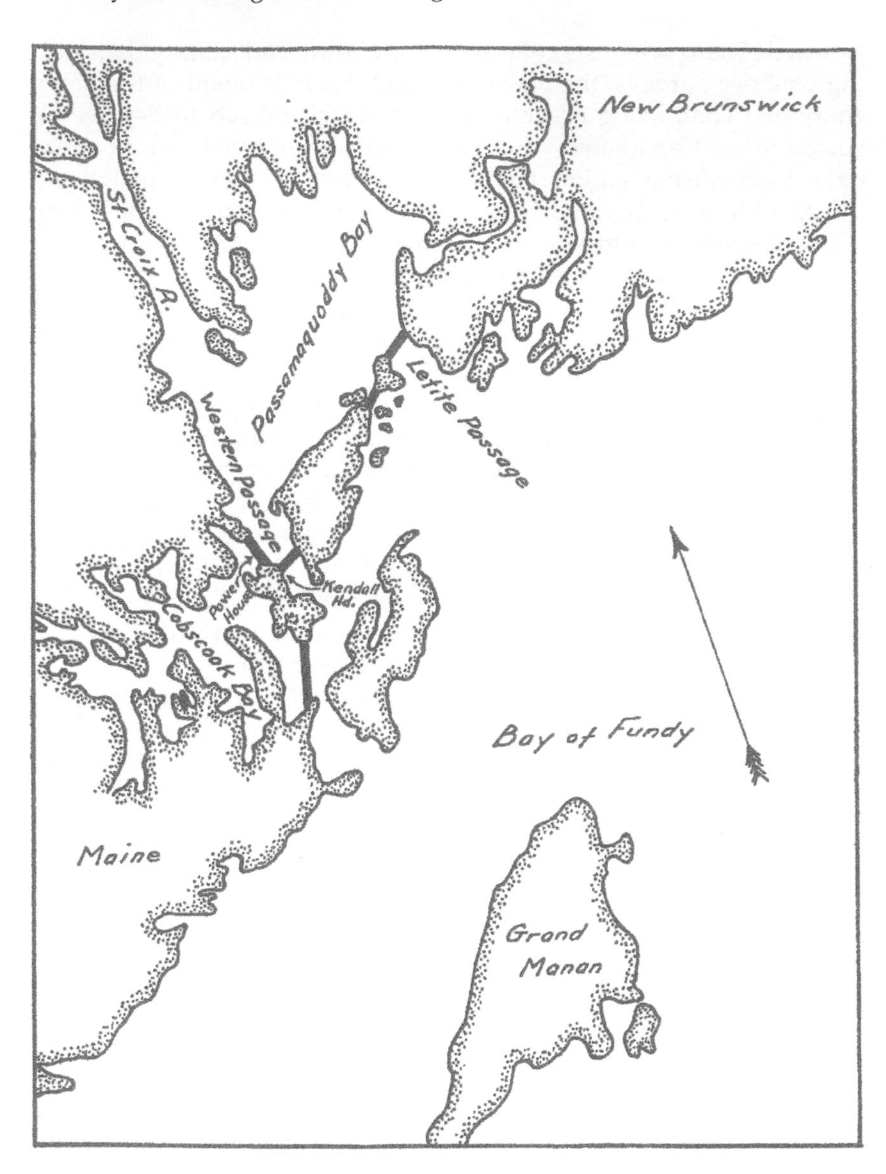


6.5 The Cooper power project, originally proposed in 1926. Thick lines indicate locations of proposed dams. From H.B. Hachey, "The Probable Effect of Tidal Power Development on Bay of Fundy Tides," *Journal of the Franklin Institute* 217 (1934): 747–56.

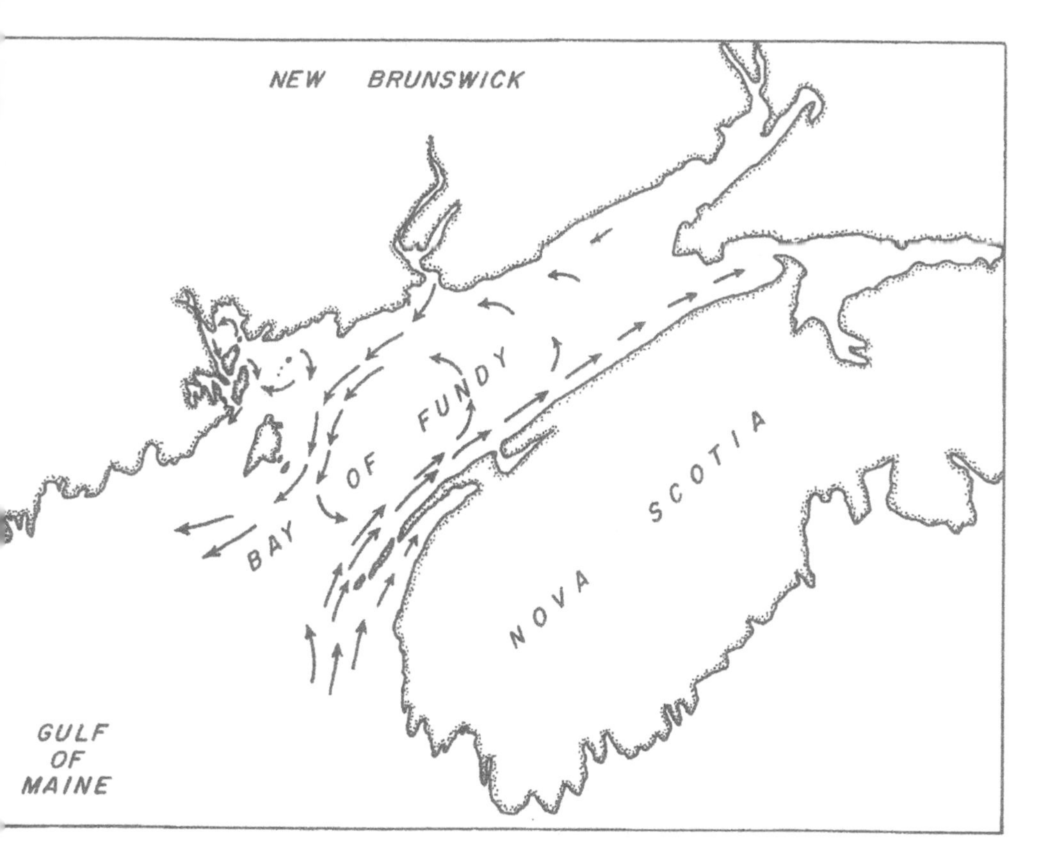


6.6 Non-tidal surface circulation in the Bay of Fundy. From R.W. Trites, "Probable Effects of Proposed Passamaquoddy Power Project on Oceanographic Conditions," *Journal of the Fisheries Research Board of Canada* 18 (1961): 163–201.

Region. The results were published in a report to the International Joint Commission,[62] as well as in scientific journals. AOG scientists involved in the oceanographic work included Joseph R. Chevrier, F.D. Forgeron, Donald MacGregor, and Ron Trites. AOG scientists released more than 8,000 drift bottles, collected temperature and salinity data, and measured water currents.[63] Current meter data were also collected by Warren Forrester of the Canadian Hydrographic Service, in collaboration with the AOG.[64] He measured currents in 1957–8 primarily using an Ekman (propellor-type) current meter, supplemented by using a Chesapeake Bay Institute drag and a Gurley (bucket wheel) current meter;[65] Trites and MacGregor made other current measurements using an electromagnetic induction method.[66]

The data from this project, together with previous research, were used to describe the general surface circulation in the Bay of Fundy and the Quoddy Region. The surface circulation in Passamaquoddy Bay was found to be, on average, counterclockwise, with variable flow in Letete Passage, outflow through Western Passage, and, in Head Harbour Passage, outflow on the Campobello Island side and inflow on the Deer Island side.[67] Using temperature and salinity data, the flushing time in Passamaquoddy Bay was estimated to vary from eight to seventeen days,[68] similar to the earlier estimate of sixteen days by Ketchum and Keen.[69] Trites described the Bay of Fundy's general circulation pattern as a counterclockwise gyre: water flowed in from the open ocean along the Nova Scotia coast, then flowed out of the bay to the south-east of Grand Manan Island and from there either along the Maine coast or across the bay to the Nova Scotia coast (figure 6.6).[70] Considerable seasonal variation in the circulation patterns was observed in Passamaquoddy Bay and the Bay of Fundy.

The International Passamaquoddy Fisheries Board's conclusions were similar to those for the earlier Cooper project: the power project would cause major changes to the oceanography inside Passamaquoddy and Cobscook Bays, and immediately outside the dams; the effects outside of Head Harbour and Bliss Island were predicted to be insignificant, with an increase in the tidal range in the Bay of Fundy of about 1 per cent.[71] The maximum tidal change was predicted to occur at the head of the bay, where the mean tidal range would increase by 18–21 cm;[72] these predictions were similar to those made by Hachey in 1934[73] for the earlier Cooper tidal power proposal. The revised power project was determined to be not feasible economically, and did not proceed.[74]

## Fisheries Oceanography, 1960–1970, and a Reduced Oceanographic Program, 1970–1984

The Atlantic Oceanographic Group was transferred to Halifax in 1960, and then to the new Bedford Institute of Oceanography (BIO) in Dartmouth in 1962 – although until 1964 the group still reported to the St Andrews Biological Station. This left only a small fisheries oceanography program at St Andrews. The AOG continued to collaborate with St Andrews researchers, especially on Bay of Fundy issues. The fisheries oceanography program at St Andrews was led by Louis Lauzier, assisted by technicians Graham Clark, John Hull, and Arthur Brown. This program's purpose was to study the ocean environment to explain the annual and seasonal fluctuations in the production of commercial fish, and to forecast variations in the environment. The program expanded drift-bottle releases to study water circulation in the Cabot Strait, Northumberland Strait, the Scotian Shelf, the Gulf of Maine, and the Bay of Fundy.[75] Drift bottle recovery data were used to describe the surface circulation pattern in Northumberland Strait.[76] In the 1960s, seabed drifters were developed to study currents near the sea floor, but these too only provided information on the release and recovery points. Seabed drifters were used to determine the bottom circulation patterns on the continental shelf of Atlantic Canada.[77] Lauzier also coauthored an atlas of surface water circulation on the continental shelf of eastern North America[78] and collaborated with fisheries scientists to collect temperature and salinity data and drifter data to investigate the distribution of larval swordfish[79] and herring. [80]

After Lauzier transferred to Ottawa in 1970, there was a period of very limited oceanographic research at the St Andrews Biological Station until 1984. A reduced program was carried out by technicians Clark, Hull, and Brown, who continued the drift-bottle release program initiated by Lauzier until 1973: a total of 114,000 surface drifters and 51,000 seabed drifters were released during 1960–73, of which 22,000 surface drifters and 13,000 seabed drifters were recovered by the end of 1973.[81] Also, hydrographic data continued to be collected from fixed stations (including Prince 5 and the Biological Station's wharf) and during fisheries survey cruises.

During this period, physical oceanographic studies related to some major project proposals in the Bay of Fundy were led by researchers based at BIO. In 1973, a large oil refinery was proposed at Eastport, Maine, in Cobscook Bay, although it did not proceed. To help assess this project's potential impacts, existing physical oceanographic data was summarized, additional current meter data were collected, and the dispersion of spilled oil was predicted.[82] In 1984 David Greenberg predicted the effects on physical oceanography of tidal power barriers in the upper Bay of Fundy, which were proposed in the late 1970s and early 1980s;[83] no such tidal power barriers were constructed, although there is now renewed interest in extracting tidal power in the Bay of Fundy (see below). One major project that did proceed in the Bay of Fundy was the Point Lepreau Generating Station – the only nuclear power plant in Atlantic Canada – constructed in 1975–81. Bedford Institute of Oceanography scientists collected hydrographic data and conducted a drifter release program before the plant began operating, and monitoring continued after operations began, to assess the impact of radioactive, thermal, and chemical releases.[84]

## A Renewed Oceanographic Research Program, 1984 to Present

Oceanographic research was re-initiated at St Andrews in 1984, when Ian Perry was hired to create stronger connections between fisheries biology and oceanography. Perry and his technician Randy Losier were situated within the SABS groundfish assessment and research section. This new SABS oceanography program was focused on determining the best ways of incorporating environmental and oceanographic data (which was collected during groundfish surveys) into stock assessments. This research led to the recognition that some of the inter-annual variations in groundfish bottom-trawl survey results could be due to environment-driven changes in the spatial density of fish stocks.[85]

The SABS oceanography program, with colleagues at BIO, participated in the Fisheries Ecology Program. This program focused on the ecology of the haddock stock off the south-western Scotian Shelf, and its goal was to provide sound advice for the stock's management. The program included research on factors affecting the initiation of spring phytoplankton blooms[86] and the drift of cod and haddock larvae from the Browns Bank spawning grounds. Perry later participated in multidisciplinary research on Georges Bank, studying the diel (daily) vertical migrations of juvenile cod and haddock[87] and the advection of zooplankton and fish larvae onto and off of the bank.[88] This led to SABS involvement in the GLOBEC (Global Ocean Ecosystem Dynamics) program begun by US and Canadian researchers on Georges Bank, in particular with modelling studies of circulation and transport and the recruitment of haddock larvae.[89] The recognition that climate change could affect ocean conditions led to collaborative research on the impacts of carbon dioxide-induced climate change on invertebrate and fish stocks in the north-west Atlantic.[90] Climate change research programs have recognized the importance of maintaining long-term monitoring programs, such as the Prince 5 station.

Perry transferred to the Pacific Biological Station in 1991, and was replaced by Fred Page. Page had been involved in the Fisheries Ecology Program as a graduate student, studying the influence of transport and dispersal processes on the recruitment dynamics of haddock.[91] He had also been recruited as a postdoctoral fellow to the GLOBEC program to provide a linkage between the physical and biological oceanographers. Page and Losier initially continued to focus on fisheries oceanography within the GLOBEC program; they led the temperature and salinity sampling component of the DFO Maritimes Region Fisheries Groundfish Survey Program.[92] They continued the work of Perry to associate fish distribution and oceanographic conditions, including the transport and dispersal of the early life stages of commercially important fish species.[93]

Since the mid-1990s, St Andrews oceanographers have used newer drifters equipped with GPS (Global Positioning System); these drifters can record locations at regular time intervals, thus providing continuous data on the track and speed of a drifter from release until recovery. They also transmit their data to satellite systems, so researchers can track drifter positions on their computers and target drifter recoveries by searching in the area of the last recorded position and using direction-finding equipment to home in on the transmitter signal.

In 1996, under Page's leadership, the oceanography program at SABS once again became a separate research section at St Andrews. The oceanography group grew to include Page, Losier, Michael Dowd (from 1999 to 2003), Blythe Chang, Paul McCurdy, and John Reid. The SABS physical oceanographic research program as of 2008 measured and modelled water circulation and the transport and dispersal of organisms and substances in the water. The program expanded from a fisheries focus into other coastal issues with oceanographic connections; the primary focus became aquaculture-related oceanography.

New tidal power projects in the Bay of Fundy have been proposed recently, using newer technologies, such as in-stream turbines, instead of dams. While the greatest potential appears to be in the upper Bay of Fundy's Minas Basin, interest in smaller tidal power projects in the lower Bay of Fundy, such as in the Head Harbour Passage between Deer and Campobello Islands, led to the SABS oceanography group collecting data related to this issue.

The SABS oceanography group also continues to monitor physical oceanographic conditions in the lower Bay of Fundy; the eighty-five-year record from Prince 5 since 1924 shows a distinct seasonal cycle, with a range in monthly mean surface temperature of about 10° C (figure 6.4); this was also noted in 1954 by Bailey, MacGregor, and Hachey based on Prince 5 data from 1924 to 1952.[94] Comparison of data from different depths indicates that waters at this location are relatively well mixed. In general, it appears that temperature trends at Prince 5 are indicative of low frequency trends throughout the Bay of Fundy and Gulf of Maine.[95] There appears to be a long-term warming trend of about 0.01°C per year (about 1°C over the 85-year period), but with considerable variation over time. Further analysis indicates that the warming trend has been mainly in winter months.

Surface water temperatures have also been collected at the biological station's wharf since 1921. Data were originally recorded twice daily (early morning and late afternoon); since the 1970s continuously recording thermographs have been used. The data have been analysed by Hachey in 1939, Hachey and McLellan in 1948, Lauzier in 1966, and Lauzier and Hull in 1969.[96] For the period from 1921 to 1962, Lauzier observed an overall increase in average temperature, but with considerable inter-annual variation;[97] also, the trend was more pronounced in the annual minimum temperatures, as was observed at Prince 5. Oceanographic monitoring data continue to be collected at these stations by station staff as part of the Atlantic Zone Monitoring Program (AZMP). These long-term data sets are an invaluable resource for research on issues such as climate change.

The current meters now in use can record current speed and direction at frequent time intervals (e.g., five to ten minutes) for continuous periods of days to months. These instruments record the data internally. One commonly used type of current meter, the InterOcean S4, measures the movement of water through a magnetic field around the meter housing. Another type, the acoustic Doppler current meter, emits acoustic signals which are used to estimate the current speed of particles suspended in the water; Acoustic Doppler Current Profilers (ADCPs) can provide vertical or horizontal profiles of current velocities. Such instruments have been extremely useful in recent work.

Models have also been used over the years to study water circulation, often in relation to proposed tidal dams or other construction proposals. Early models consisted of water-filled tanks. Hachey constructed a tank to study the mixing of fresh and salt water in an estuary.[98] He also used an oscillating system of tanks to investigate possible effects of the proposed tidal power project in Passamaquoddy Bay in the early 1930s.[99] Latterly, computer models for the Bay of Fundy and Gulf of Maine area have been developed in collaboration with scientists at the Bedford Institute of Oceanography.[100] Such computer models can be used to predict current velocities between points where actual field data are available, as well as to predict the effects on water circulation of proposed developments. Circulation models, together with current meter deployments and drifter releases, have been used to estimate water circulation in the vicinity of salmon farms, and then to predict the potential for disease spread among farms;[101] the potential for interactions between farms and other activities;[102] and the movements of phytoplankton blooms in the vicinity of salmon farms.[103] For example, Page led colleagues in developing a model to examine dissolved oxygen levels at salmon farms[104] and Michael Dowd used models to study the relationship between oceanographic conditions and shellfish aquaculture production in Prince Edward Island.[105]

In addition to the work conducted by the oceanography group, the staff from other programs at the St Andrews Biological Station has continuously collected oceanographic data, including hydrographic data collected during fisheries survey cruises in the Bay of Fundy and Scotian Shelf. Temperature and salinity data have been collected in the lower Bay of Fundy by David Wildish for research on the environmental impacts of salmon aquaculture, by Jennifer Martin for a phytoplankton monitoring program, and by Shawn Robinson in research on scallop spat distribution. Their research is discussed separately in this volume in the chapter by Martin and the co-authored chapter by Wildish and Robinson. During the 1990s, issues related to salmon and other aquaculture – the history of which is described by Robert Cook in chapter 12 – became the focus of the oceanography group. Oceanographic research has contributed to aquaculture issues such as farm siting, environmental monitoring, fish health management, the implementation of bay management areas, integrated multitrophic aquaculture, and offshore aquaculture.

## Conclusion

The oceanography group's ability to address issues currently affecting the Bay of Fundy and other Atlantic Canada waters builds on the considerable amount of research conducted by station researchers and oceanographers since its inception. For over a century, volunteer and professional oceanographers at St Andrews have made essential contributions to understanding the hydrography and marine conditions of the Bay of Fundy and more distant waters. This research has been primarily applied, driven by the need for information related to specific issues. Fisheries-related issues have been important in defining oceanographic research programs throughout most of the biological station's history. Oceanographic data has been collected as part of fisheries surveys, beginning with the Canadian Fisheries Expedition of 1914–15. Investigations into the relationships between oceanographic parameters and the distribution and abundance of commercial fish and invertebrate species have provided insights into the dispersal of early life stages, stock structure, and recruitment variability.

In addition to their important contributions to fisheries and aquaculture research, oceanographers at the St Andrews Biological Station have augmented local and international science through significant work relating to other biological and physical oceanographic issues. They have assisted climate change studies, by maintaining the longest hydrographic sampling effort on Canada's east coast. Their contributions to tidal power investigations included assessing the potential for ecosystem effects of tidal barrages in the twentieth century and in-stream tidal turbines in the early twenty-first century, as well as the potential amount of power that could be generated. They have augmented our understanding of harmful algal blooms by monitoring physical conditions associated with plankton monitoring programs and helping to predict the formation of blooms and their subsequent transport and dispersal. Oceanographers at SABS have also contributed to understanding issues related to coastal zone management, search and rescue, oil spill response, aquatic invasive species, and Marine Protected Areas.

NOTES

We thank Joanne Cleghorn and Charlotte McAdam of the St Andrews Biological Station (SABS) Library, Ilona Monahan of the Canadian Hydrographic Service, Ian Perry of the Pacific Biological Station, Randy Losier and Paul McCurdy of the SABS Coastal Ocean and Ecosystem Section, and SABS summer students Robyn Tidd and Caleigh Dunfield for their assistance. Reports on the history of the Fisheries Research Board of Canada and on oceanographic research in Canada have been extremely helpful, especially those of Rigby and Huntsman (1958), Hachey (1961, 1965), Campbell (1976), and Johnstone (1977). Information was also obtained from the annual reports of SABS (formerly known as the Atlantic Biological Station), the Atlantic Oceanographic

Group, and the Fisheries Research Board of Canada. We also thank John Smith (Bedford Institute of Oceanography) for reviewing the draft manuscript and providing valuable comments.

1 D.P. Penhallow, "Report on the Atlantic Biological Station of Canada, St Andrews, N.B., for 1908," *Contributions to Canadian Biology, 1906–10* (1912): 1–21.
2 Ibid.
3 B.J. Tait and L.F. Ku, "A Historical Review of Tidal, Current and Water Level Surveying in the Canadian Hydrographic Service," in *From Leadline to Laser: Centennial Conference of the Canadian Hydrographic Service, Canadian Special Publication of Fisheries and Aquatic Science* 67 (1983): 51–61.
4 W.B. Dawson, *The Currents at the Entrance to the Bay of Fundy and on the Steamship Routes in Its Approaches off Southern Nova Scotia from Investigations of the Tidal and Current Survey in the Season of 1904* (Ottawa: Canada, Dept. of Marine and Fisheries,1905); W.B. Dawson, *Tables of Hourly Direction and Velocity of the Currents and Time of Slack Water in the Bay of Fundy and Its Approaches as far as Cape Sable: From Investigations of the Tidal and Current Survey in the Seasons of 1904 and 1907* (Ottawa: Canada, Dept. of Marine and Fisheries, 1908); W.B. Dawson, *Temperatures and Densities of the Waters of Eastern Canada, Including the Atlantic from the Bay of Fundy to Newfoundland, the Gulf of St Lawrence, and the Straits Connecting It with the Ocean: From Investigations by the Tidal and Current Survey in the Seasons of 1894 to 1896, and 1903 to 1911* (Ottawa: Canada, Department of the Naval Service, 1922), 1–102.
5 Dawson, *Temperatures and Densities of the Waters of Eastern Canada.*
6 G.G. Copeland, "The Temperatures and Densities and Allied Subjects of Passamaquoddy Bay and Its Environs: Their Bearing on the Oyster Industry," *Contributions to Canadian Biology 1906–1910* (1912): 281–94.
7 P. Bjerkan, "Results of the Hydrographical Observations Made by Dr. John Hjort in the Canadian Atlantic Waters during the Year 1915," in *Canadian Fisheries Expedition, 1914–1915, in the Gulf of St Lawrence and Atlantic Waters of Canada* (Ottawa: Canada, Dept. of the Naval Service, 1919), 348–403.
8 A.B. Klugh, "A Hydrographic Reconnaissance of the Lower St. John and the Kennebecasis, New Brunswick," *Biological Board of Canada Manuscript Report* 48 (1910), 1–4.
9 J.W. Mavor, E.H. Craigie, and J.D. Detweiler, "Investigation of the Bays of the Southern Coast of New Brunswick with a View to Their Use for Oyster Culture," *Contributions to Canadian Biology 1914–1915* (1916): 145–9.
10 E.H. Craigie "Hydrographic Investigations in the St Croix River and Passamaquoddy Bay in 1914," *Contributions to Canadian Biology 1914–1915*

(1916): 151–61; E.H. Craigie, "A Hydrographic Section of the Bay of Fundy in 1914," *Contributions to Canadian Biology 1914–1915* (1916): 163–7; E.H. Craigie and W.H. Chase, "Further Hydrographic Investigations in the Bay of Fundy," *Contributions to Canadian Biology 1917–1918* (1918): 127–48.

11 Copeland, "The Temperatures and Densities and Allied Subjects of Passamaquoddy Bay and Its Environs," 282.

12 Craigie, "Hydrographic Investigations," 151.

13 Tait and Ku, "A Historical Review."

14 Craigie, "Hydrographic Investigations," 133.

15 Ibid., 157.

16 Craigie and Chase, "Further Hydrographic Investigations in the Bay of Fundy," 133.

17 A.D. Robertson, "First Report on the 'Barren Oyster Bottoms' Investigation, Richmond Bay, P.E.I.," *Contributions to Canadian Biology 1914–1915* (1916): 55–71; and J. Nelson, "An Investigation of Oyster Propagation in Richmond Bay, P.E.I., during 1915," *Contributions to Canadian Biology 1915–1916* (1917): 53–78.

18 A.W.H. Needler, "The Oysters of Malpeque Bay," *Bulletin of the Fisheries Research Board of Canada* 22 (1931): 1–30.

19 Bjerkan, "Results of the Hydrographical Observations Made by Dr. John Hjort"; and W.J. Sandström, "The Hydrodynamics of Canadian Atlantic Waters," in *Canadian Fisheries Expedition, 1914–1915, in the Gulf of St Lawrence and Atlantic Waters of Canada* (Ottawa: Dept. of the Naval Service, 1919), 221–347.

20 H.B. Hachey, "History of the Fisheries Research Board of Canada," Fisheries Research Board of Canada Manuscript Report 843 (1965): 292.

21 A. Vachon, "Hydrography in Passamaquoddy Bay and Vicinity, New Brunswick," *Contributions to Canadian Biology 1917–1918* (1918): 295–328.

22 Ibid.

23 J.W. Mavor, "The Circulation of the Water in the Bay of Fundy. Part I: Introduction and Drift Bottle Experiments," *Contributions to Canadian Biology*, new series, 1 (1922): 103.

24 J.W. Mavor, "Drift Bottles as Indicating a Superficial Circulation in the Gulf of Maine," *Science* 52 (1920): 442–3; J.W. Mavor, "Circulation of the Water in Bay of Fundy and Gulf of Maine," *Transactions of the American Fisheries Society* 50 (1921): 334–44; Mavor, "The Circulation of the Water in the Bay of Fundy. Part I."; J.W. Mavor, "The Circulation of the Water in the Bay of Fundy. Part II: The Distribution of Temperature, Salinity and Density in 1919 and Movements of Water Which They Indicate in the Bay of Fundy," *Contributions to Canadian Biology*, new series, 1 (1923): 353–75.

25 Craigie and Chase, "Further Hydrographic Investigations."
26 Mavor, "Circulation of the Water in Bay of Fundy," 103.
27 J.W. Mavor, "The Course of the Gulf Stream in 1919–21 as Shown by Drift Bottles," *Science* 57 (1923): 14–15.
28 Mavor, "Circulation of the Water in the Bay of Fundy," 103.
29 North American Council on Fishery Investigations, *Proceedings 1921–1930 (No. 1)* (Ottawa: Printer to the King's Most Excellent Majesty, 1932).
30 A.G. Huntsman, W.B. Bailey, and H.B. Hachey, "The General Oceanography of the Strait of Belle Isle," *Journal of Fisheries Research Board of Canada* 11 (1954): 198–260.
31 A.G. Huntsman, "Circulation and Pollution of Water in and near Halifax Harbour," *Contributions to Canadian Biology*, new series, 2 (1924): 69–80.
32 F.H. Page, B.D. Chang, and J.L. Martin, "A Century of Oceanographic Monitoring at the St Andrews Biological Station," *Gulf of Maine News (Regional Association for Research on the Gulf of Maine)* 7:2 (2000): 1–5.
33 E.E. Watson, "Mixing and Residual Currents in the Tidal Waters as Illustrated in the Bay of Fundy," *Journal of the Biological Board of Canada* 2 (1936): 189.
34 H.B. Hachey, "The Replacement of Bay of Fundy Waters," *Journal of the Biological Board of Canada* 1 (1934): 121–31.
35 Hachey, "History of the Fisheries Research Board of Canada."
36 W.B. Bailey, "Some Features of the Oceanography of the Passamaquoddy Region," Fisheries Research Board of Canada Manuscript Report (Oceanography and Limnology) 2 (1957): 1–30.
37 H.B. Hachey, "Tidal Mixing in an Estuary," *Journal of the Biological Board of Canada* 1 (1935): 171–8.
38 H.B. Hachey, "The General Hydrography and Hydrodynamics of the Waters of the Hudson Bay Region," *Contributions to Canadian Biology*, new series, 7 (1932): 91–118; H.B. Hachey, "The Circulation of Hudson Bay Water as Indicated by Drift Bottles," *Science* 82 (1935): 275–6.
39 H.B. Hachey and W.B. Bailey, "The General Hydrography of the Waters of the Bay of Fundy," Fisheries Research Board of Canada Manuscript Report 455 (1952): 1–100.
40 Watson, "Mixing and Residual Currents in the Bay of Fundy."
41 H.B. Hachey, "The Probable Effect of Tidal Power Development on Bay of Fundy Tides," *Journal of the Franklin Institute* 217 (1934): 747–56.
42 International Passamaquoddy Fisheries Commission, "Report of the International Passamaquoddy Fisheries Commission," Washington, DC, U.S House of Representatives, 73rd Congress, 2nd Session, Document no. 200 (1934): 1-24.

43 A.G. Huntsman, "International Passamaquoddy Fishery Investigations," *Journal du Conseil International pour l'Exploration de la Mer* 13 (1938): 357–69.
44 H.B. Hachey, "The Waters of the Scotian Shelf," *Journal of the Fisheries Research Board of Canada* 5 (1942): 377–97; H.B. Hachey, "Water Transports and Current Patterns for the Scotian Shelf," *Journal of the Fisheries Research Board of Canada* 7 (1947): 1–16.
45 H.B. Hachey, "Movements Resulting from Mixing of Stratified Waters," *Journal of the Biological Board of Canada* 1 (1934): 133–43.
46 H.B. Hachey, "Hydrographic Features of the Waters of Saint John Harbour," *Journal of the Fisheries Research Board of Canada* 4 (1939): 424–40.
47 Hachey, "Movements Resulting from Mixing of Stratified Waters."
48 Atlantic Oceanographic Group. 'Canadian Drift Bottle Data, Atlantic Coast," Fisheries Research Board of Canada Manuscript Report (Oceanography and Limnology) 15 (1958): 1–80.
49 N.J. Campbell, "An Historical Sketch of Physical Oceanography in Canada," *Journal of the Fisheries Research Board of Canada* 32 (1976): 2155–67.
50 A.H. Leim, S.N. Tibbo, L.R. Day, L. Lauzier, R.W. Trites, H.B. Hachey, and W.B. Bailey, "Report of the Atlantic Herring Investigation Committee," *Bulletin of the Fisheries Research Board of Canada* 111 (1957): 317 pp.
51 Hachey and Bailey, "The General Hydrography of the Waters of the Bay of Fundy"; W.B. Bailey, "Some Features of the Oceanography of the Passamaquoddy Region"; and W.B. Bailey, "Oceanographic Observations in the Bay of Fundy since 1952," Fisheries Research Board of Canada Manuscript Report of the Biological Stations 633 (1957): 1–72.
52 H.J. McLellan, "Temperature–salinity Relations and Mixing on the Scotian Shelf," *Journal of the Fisheries Research Board of Canada* 11 (1954): 419–30.
53 H.J. McLellan, L. Lauzier, and W.B. Bailey, "The Slope Water off the Scotian Shelf," *Journal of the Fisheries Research Board of Canada* 10 (1953): 155–76; and H.J. McLellan, "On the Distinctness and Origin of the Slope Water off the Scotian Shelf and Its Easterly Flow South of the Grand Banks," *Journal of the Fisheries Research Board of Canada* 14 (1957): 213–39.
54 L.M. Lauzier and R.W. Trites, "The Deep Waters in the Laurentian Channel," *Journal of the Fisheries Research Board of Canada* 15 (1958): 1247–57.
55 D.G. MacGregor, "Water Conditions in the Strait of Canso," Fisheries Research Board of Canada, Manuscript Report of the Biological Stations 552 (1954): 1–30.
56 N.J. Campbell and L.M. Lauzier, "Ice Studies of the Atlantic Oceanographic Group," Fisheries Research Board of Canada, Manuscript Report (Oceanography and Limnology) 60 (1960): 1–12.

57 D.G. MacGregor and H.J. McLellan, "Current Measurements in the Grand Manan Channel," *Journal of the Fisheries Research Board of Canada* 9 (1952): 213–22.
58 R.W. Trites and R.E. Banks, "Circulation on the Scotian Shelf as Indicated by Drift Bottles," *Journal of the Fisheries Research Board of Canada* 15 (1958): 79–89.
59 B.H. Ketchum and D.J. Keen, "The Exchanges of Fresh and Salt Waters in the Bay of Fundy and in Passamaquoddy Bay," *Journal of the Fisheries Research Board of Canada* 10 (1953): 97–124.
60 W.B. Bailey, "Oceanographic Observations in the Bay of Fundy since 1952," Fisheries Research Board of Canada, Manuscript Report of the Biological Stations 633 (1957): 1–375.
61 MacGregor and McLellan, "Current measurements in the Grand Manan Channel."
62 D.F. Bumpus, J.R. Chevrier, F.D. Forgeron, W.D. Forrester, D.G. MacGregor, and R.W. Trites, "International Passamaquoddy Fisheries Board Report to the International Joint Commission. Appendix 1, Oceanography" (1959).
63 J.R. Chevrier and R.W. Trites, "Drift-bottle Experiments in the Quoddy Region, Bay of Fundy," *Journal of the Fisheries Research Board of Canada* 17 (1960): 743–62; R.W. Trites, "Temperature and Salinity in the Quoddy Region of the Bay of Fundy," *Journal of the Fisheries Research Board of Canada* 19 (1962): 975–8; and R.W. Trites and D.G. MacGregor, "Flow of Water in the Passages of Passamaquoddy Bay Measured by the Electromagnetic Method," *Journal of the Fisheries Research Board of Canada* 19 (1962): 895–919.
64 W.D. Forrester, "Current Measurements in Passamaquoddy Bay and the Bay of Fundy 1957 and 1958," *Journal of the Fisheries Research Board of Canada* 17 (1960): 727–9.
65 Ibid.
66 Trites and MacGregor, "Flow of Water in the Passages of Passamaquoddy Bay."
67 J.R. Chevrier and R.W. Trites, "Drift-bottle Experiments in the Quoddy Region, Bay of Fundy," *Journal of the Fisheries Research Board of Canada* 17 (1960): 743–62.
68 R.W. Trites, "Probable Effects of Proposed Passamaquoddy Power Project on Oceanographic Conditions," *Journal of the Fisheries Research Board of Canada* 18 (1962): 163–201.
69 B.H. Ketchum and D.J. Keen, "The Exchanges of Fresh and Salt Waters in the Bay of Fundy and in Passamaquoddy Bay," *Journal of the Fisheries Research Board of Canada* 10 (1953): 97–124.

70 R.W. Trites, "Probable Effects of Proposed Passamaquoddy Power Project on Oceanographic Conditions," *Journal of the Fisheries Research Board of Canada* 18 (1961): 163–201.

71 Ibid.

72 H.J. McLellan, "Energy Considerations in the Bay of Fundy System," *Journal of the Fisheries Research Board of Canada* 15 (1958): 115–34.

73 H.B. Hachey, "The Probable Effect of Tidal Power Development on Bay of Fundy Tides," *Journal of the Franklin Institute* 217 (1934): 747–56.

74 International Joint Commission, United States and Canada, "Report of the International Joint Commission on the International Passamaquoddy Tidal Power Project" (Washington, DC, and Ottawa, ON: International Joint Commission, 1961).

75 L.M. Lauzier, J.G. Clark, and A.W. Brown, "Canadian Drift Bottle Program, 1960–1963, Atlantic Coast," Fisheries Research Board of Canada Manuscript Report (Oceanography and Limnology) 178 (1964): 1–24.

76 L.M. Lauzier, "Drift Bottle Observations in Northumberland Strait, Gulf of St Lawrence," *Journal of the Fisheries Research Board of Canada* 22 (1965): 353–68.

77 D.F. Bumpus and L.M. Lauzier, *Serial Atlas of the Marine Environment. Folio 7: Surface Circulation on the Continental Shelf off Eastern North America between Newfoundland and Florida* (New York: American Geographical Society of New York, 1965).

78 L.M. Lauzier, "Bottom Residual Drift on the Continental Shelf Area of the Canadian Atlantic Coast," *Journal of the Fisheries Research Board of Canada* 24 (1967): 1845–59.

79 S.N. Tibbo and L.M. Lauzier, "Larval Swordfish (*Xiphias gladius*) from Three Localities in the Western Atlantic," *Journal of the Fisheries Research Board of Canada* 26 (1969): 3248–51.

80 S.N. Messieh, S.N. Tibbo, and L.M. Lauzier, "Distribution, Abundance and Growth of Larval Herring (*Clupea harengus L.*) in Bay of Fundy–Gulf of Maine Area," Fisheries Research Board of Canada Technical Report 277 (1971).

81 A.W. White and H.M. Akagi, "A Compilation of Total Releases and Recoveries of Drift Bottles and Sea-bed Drifters in Continental Shelf Waters of the Canadian Atlantic Coast from 1960 through 1973," Fisheries Research Board of Canada Manuscript Report 1281 (1974): 1–49.

82 R.H. Loucks, R.W. Trites, K.F. Drinkwater, and D.J. Lawrence, "Summary of Physical, Biological, Socio-economic and Other Factors Relevant to Potential Oil Spills in the Passamaquoddy Region of the Bay of Fundy. Section 1: Physical Oceanographic Characteristics," Fisheries Research

Board of Canada Technical Report 428 (1974): 1–59; R.H. Loucks, D.J. Lawrence, and D.V. Ingraham, "Summary of Physical, Biological, Socio-economic and Other Factors Relevant to Potential Oil Spills in the Passamaquoddy Region of the Bay of Fundy. Section 2: Dispersion of Spilled Oil," Fisheries Research Board of Canada Technical Report 428 (1974): 61–101; and R.W. Trites, "Comments on Flushing Time for Passamaquoddy Bay and Wind-generated Waves in the Quoddy Region," Fisheries Research Board of Canada Technical Report 901 (1979): 3–7.

83 D.A. Greenberg, "The Effects of Tidal Power Development on the Physical Oceanography of the Bay of Fundy and Gulf of Maine," Canadian Technical Report of Fisheries and Aquatic Sciences 1256 (1984): 349–69.

84 G.L. Bugden, "Oceanographic Observations from the Bay of Fundy for the Pre-operational Environmental Monitoring Program for the Point Lepreau, N.B. Nuclear Generating Station," *Canadian Data Report of Hydrography and Ocean Sciences* 27 (1985): 1–41; and R.W.P. Nelson, K.M. Ellis, and J.N. Smith, "Environmental Monitoring Report for the Point Lepreau, N.B. Nuclear Generating Station – 1991 to 1994," *Canadian Technical Report of Hydrography and Ocean Sciences* 211 (2001): 1–125.

85 R.I. Perry and S.J. Smith, "Identifying Habitat Associations of Marine Fishes Using Survey Data: An Application to the Northwest Atlantic," *Canadian Journal of Fisheries and Aquatic Science* 51 (1994): 589–602.

86 R.I. Perry, P.C.F. Hurley, P.C. Smith, J.A. Koslow, and R.O. Fournier, "Modelling the Initiation of Spring Phytoplankton Blooms: A Synthesis of Physical and Biological Interannual Variability off Southwest Nova Scotia, 1983–85," *Canadian Journal of Fisheries and Aquatic Science* 46 (suppl. 1) (1989): 183–99.

87 R.I. Perry and J.D. Neilson, "Vertical Distributions and Trophic Interactions of Age-0 Atlantic Cod and Haddock in Mixed and Stratified Waters of Georges Bank," *Marine Ecology Progress*, series 49 (1988): 199–214.

88 R.I. Perry, G.C. Harding, J.W. Loder, M.J. Tremblay, M.M. Sinclair, and K.F. Drinkwater, "Zooplankton Distributions at the Georges Bank Frontal System: Retention or Dispersion?" *Continental Shelf Research* 13 (1993): 357–83.

89 F.E. Werner, F.H. Page, D.R. Lynch, J.W. Loder, R.G. Lough, R.I. Perry, D.A. Greenberg, and M.M. Sinclair, "Influences of Mean Advection and Simple Behavior on the Distribution of Cod and Haddock Early Life Stages on Georges Bank," *Fisheries Oceanography* 2 (1993): 43–64.

90 K.T. Frank, R.I. Perry, and K.F. Drinkwater, "Predicted Response of Northwest Atlantic Invertebrate and Fish Stocks to CO2-induced Climate Change," *Transactions of the American Fisheries Society* 119 (1990): 353–65.

91 F.H. Page and P.C. Smith, "Particle Dispersion in the Surface Layer off Southwest Nova Scotia: Description and Evaluation of a Model," *Canadian Journal of Fisheries and Aquatic Science* 46 (suppl. 1) (1989): 21–43.
92 F.H. Page and R.J. Losier, "Temperature Variability during Canadian Bottom-trawl Summer Surveys Conducted in NAFO Division 4VWX, 1970–1992," *International Council for the Exploration of the Sea Marine Science Symposium* 198 (1994): 323–31.
93 S.J. Smith and F.H. Page, "Associations between Atlantic Cod (*Gadus morhua*) and Hydrographic Variables: Implications for the Management of the 4VsW Stock," *International Council for the Exploration of the Sea, Journal of Marine Science* 53 (1996): 597–614; and F.H. Page, M. Sinclair, C.E. Naimie, J.W. Loder, R.J. Losier, P.L. Berrien, and R.G. Lough, "Cod and Haddock Spawning on Georges Bank in Relation to Water Residence Times," *Fisheries Oceanography* 8 (1999): 212–26.
94 W.B. Bailey, D.G. MacGregor, and H.B. Hachey, "Annual Variations of Temperature and Salinity in the Bay of Fundy," *Journal of the Fisheries Research Board of Canada* 11 (1954): 32–47.
95 F.H. Page, B.D. Chang, and J.L. Martin, "A Century of Oceanographic Monitoring at the St Andrews Biological Station," *Gulf of Maine News (Regional Association for Research on the Gulf of Maine)* 7:2 (2000): 1–5.
96 H.B. Hachey, "Surface Water Temperatures of the Canadian Atlantic Coast," *Journal of the Fisheries Research Board of Canada* 4 (1939): 378–91; H.B. Hachey and H.J. McLellan, "Trends and Cycles in Surface Temperatures of the Canadian Atlantic," *Journal of the Fisheries Research Board of Canada* 7 (1948): 355–62; L.M. Lauzier, "Foreshadowing of Surface Water Temperatures at St Andrews, N.B.," *International Commission for the Northwest Atlantic Fisheries Special Publication* 6 (1966): 859–67; and L.M. Lauzier and J.H. Hull, "Coastal Station Data Temperatures along the Canadian Atlantic Coast 1921–1969," Fisheries Research Board of Canada Technical Report 150 (1969): 1–25.
97 Lauzier, "Foreshadowing of Surface Water Temperatures at St Andrews."
98 H.B. Hachey, "The Replacement of Bay of Fundy Waters," *Journal of the Biological Board of Canada* 1 (1934): 121–31.
99 Hachey, "The Probable Effect of Tidal Power Development."
100 See, for example, D.A. Greenberg, J.A. Shore, F.H. Page, and M. Dowd, "A Finite Element Circulation Model for Embayments with Drying Intertidal Areas and Its Application to the Quoddy Region of the Bay of Fundy," *Ocean Modelling* 10 (2005): 211–31.
101 B.D. Chang, F.H. Page, R.J. Losier, D.A. Greenberg, D.J. Chaffey, and E.P. McCurdy, "Application of a Tidal Circulation Model for Fish Health

Management of Infectious Salmon Anemia in the Grand Manan Island Area, Bay of Fundy," *Bulletin of the Aquaculture Association of Canada* 105:1 (2005): 22–33.

102 B.D. Chang, F.H. Page, and B.W.H. Hill, "Preliminary Analysis of Coastal Marine Resource Use and the Development of Open Ocean Aquaculture in the Bay of Fundy," Canadian Technical Report of Fisheries and Aquatic Sciences 2585 (2005): 1–36.

103 B.D. Chang, R.J. Losier, F.H. Page, D.A. Greenberg, and D.J. Chaffey, "Use of a Water Circulation Model to Predict the Movements of Phytoplankton Blooms Affecting Salmon Farms in the Grand Manan Island Area, Southwestern New Brunswick," Canadian Technical Report of Fisheries and Aquatic Science 2703 (2007): 1–64.

104 F.H. Page, R. Losier, P. McCurdy, D. Greenberg, J. Chaffey, and B. Chang, "Dissolved Oxygen and Salmon Cage Culture in the Southwestern New Brunswick Portion of the Bay of Fundy," in *Environmental Effects of Marine Finfish Aquaculture, Handbook of Environmental Chemistry, Vol. 5*, ed. B.T. Hargrave (Berlin: Springer-Verlag, 2005), part M: 1–28.

105 M. Dowd, "Oceanography and Shellfish Production: A Bio-physical Synthesis Using a Simple Model," *Bulletin of the Aquaculture Association of Canada* 100:2 (2000): 3–9; and M. Dowd, "Seston Dynamics in a Tidal Inlet with Shellfish Aquaculture: A Model Study Using Tracer Equations," *Estuarine, Coastal and Shelf Science* 57 (2003): 523–37.

# 7 Experimental Flow Studies at St Andrews Biological Station

DAVID J. WILDISH AND SHAWN M.C. ROBINSON

For over half a century scientists at the St Andrews Biological Station undertook experimental ecological, physiological, and behavioural studies to investigate how the flow of seawater and fresh water drives biological factors. These studies were initiated in response to the needs of important Maritime industries including the fisheries, fish culture, and bivalve culture, and also encompassed inquiries into the environmental side-effects of industries such as forestry, which aerially sprayed forests and streams with pesticides, the pulp and paper industry, and mining resulting in heavy metal pollution in streams. In aquatic environments, water flow is particularly important to those mobile organisms, such as fish, which swim in moving water and which delivers food to them. Water flow is also vital for generally sessile organisms, such as bivalves, which rely on flow to carry food particles to them and carry away their wastes and bio-products, such as carbon dioxide, ammonia, pheromones, gametes, and faeces. In this chapter a historical account of the flow simulators used at St Andrews beginning in the 1960s is presented. Until 1993, flow simulators were constructed at St Andrews, but in that year the manned workshop facilities were shut down, and the station no longer had the ability to create experimental devices. Thereafter, flume construction was undertaken at the Bedford Institute of Oceanography. These flow simulators were used to conduct freshwater experimental physiological and behavioural research, predominantly with salmonid fishes. Marine flow studies included ecological, physiological, and behavioural studies of mussels, scallops, and oysters, particularly in relation to feeding and growth. The legacy of this research is basic information on the biological responses of fish and bivalves to water flow, which allows for the construction of predictive models of great value in fish and bivalve culture, and assists in managing and protecting the marine environment.

This chapter reviews the history of field and laboratory experimental research at St Andrews Biological Station that is related to two applied research areas: freshwater fish physiology and behaviour and marine bivalve ecology, physiology, and behaviour, where flow was an important variable. The focus here is on research publications that elucidated the effects of flow as a key environmental variable, as determined by testing contrasted null and alternate hypotheses. Some experimental tests were conducted in the field; those that were laboratory based often required the design and construction of a flow simulator.

In a larger historical context isolated flow studies go back at least to Charles Darwin, who wrote a monograph on the suspension-feeding cirripedes, or barnacles, concerning their taxonomic structure and function, and on the formation, structure, and functioning of suspension-feeding coral reefs. Such references and many which followed did not incorporate a modern understanding of the physics of flow – in other words, of hydrodynamics. Hydrodynamic practice and theory was developed by practical engineers early in the twentieth century and summarized in a book published in German by Herman Schlichting in 1951. The first English edition of *Boundary Layer Theory* was published in 1955. C.B. Jorgensen's 1966 book *Biology of Suspension Feeding* was the first to incorporate hydrodynamics and benthic biology as a coherent interdisciplinary field of study and was updated in 1990. Other summarizing books dealing with various aspects of the interdisciplinary field of biology and hydrodynamics came along much later and include: S. Vogel's *Life in Moving Fluids: The Physical Biology of Flow*, first published in 1981; Mark W. Denny's *Air and Water: The Biology and Physics of Life's Media*, published in 1993; K.H. Mann and J.R.N. Lazier's 1991 monograph *Dynamics of Marine Ecosystems: Biological–Physical Interactions in the Oceans*; and David J. Wildish and David D. Kristmanson's *Benthic Suspension Feeders and Flow*, published in 1997.[1]

From the beginning, St Andrews Biological Station scientists concentrated on scientific questions derived from the marine, estuarine, and freshwater-based industries which developed in the local area. Historically, the areas of relevance to the station included the Bay of Fundy, Gulf of Maine, Scotian Shelf, and Northumberland Straits and their watersheds. The industries established there in the last century include many types of fishing, including sport fishing, aquaculture, and several industrial activities that use the aquatic environment directly, such as for the disposal of wastes. For many of the industrially centred scientific questions researched at St Andrews, water movement, whether as marine tidal or wind-generated currents or freshwater flow rate, was recognized to be a key environmental variable. It is fitting that the biological station, situated as it is at the mouth of the Bay of Fundy – a macrotidal estuary, where the semi-diurnal spring tides may exceed eight metres – became involved with flow-related research.

Many of the initial flow studies were field based, observational, oceanographic ones designed to understand the background environment for finfish and bivalve fisheries, as well as for environmental conservation. Elsewhere in this volume, Chang and Page describe the physical oceanographic field studies of flow connected with this research and their historical development. The earliest experiment featuring flow was a field study conducted by C.J. Kerswill, which compared bivalve growth at low and high flow rates. These experiments were part of Kerswill's Ph.D. research, and originally appeared in his 1941 thesis; reports of the results were finally published in 1949.[2]

## Flow Simulators and Laboratory Experiments: The Machines and Methods Needed to Understand Fish and Bivalve Physiology and Behaviour

In a river or the sea, the fluid flowing over the bottom is slowed by its contact with the solid surface and the relative degree of friction it exerts on the flow. The discrete layer, so formed, is called the benthic boundary layer. Within it, velocity decreases logarithmically towards the bottom. Above the benthic boundary layer is the free stream flow, unhindered by contact with the bottom. In some of the flow simulators to be described here, a benthic boundary layer and free stream flow is created. In most studies the mean free-stream velocity was measured, although in some, attempts were made to measure the velocity near the experimental subject, even if this was within the benthic boundary layer.[3]

Flow simulator designs that have been used at St Andrews are shown in table 7.1. The earliest were for studies of the physiology of juvenile freshwater fish. In 1964 R.L. Saunders devised and built a neutral buoyancy flume at St Andrews that was capable of flows up to twenty-six centimetres per second. Its working section was fifty centimetres in length and it was powered by a small electric submersible pump. In this flume Saunders could measure the density of small experimental fish by determining their neutral buoyancy point, that is, the water pressure

Table 7.1 Flow simulators used at St Andrews Biological Station in the period up to 2008

| Simulator | | | Volume capacity | |
|---|---|---|---|---|
| no. | Name | Type | L (depth cm) | Reference |
| 1 | Neutral buoyancy flume | Fresh-water recirculating | 2 (7.2) | Saunders (1964)<br>Neave et al. (1966) |
| 2 | Avoidance trough | Fresh-water flow-through | 6.5 (8.5) | Sprague (1964)<br>Sprague et al. (1965) |
| 3 | Artificial stream | Freshwater recirculating | 186, 200 (70) | Symons (1976) |
| 4 | Single-channel flume | Seawater flow-through | 500 (20)<br>1125 (full) | Wildish and Kristmanson (1984) |
| 5 | Kirby-Smith growth tubes | Seawater flow-through tubes | 165 (1 tube)<br>1319 (8) | Kirby-Smith (1972)<br>Wildish and Kristmanson (1985) |
| 6 | Multiple-channel flume | Seawater flow-through | 740 (20 cm)<br>1670 (full) | Wildish et al. (1987)<br>Wildish and Kristmanson (1988)<br>Wildish et al. (2008) |
| 7 | Blažka respirometer | Fresh water or Seawater recirculating | 50 | Byrne et al. (1972)<br>Saunders and Kutty (1973)<br>Peterson (1974)<br>Beamish (1978)<br>Holmberg and Saunders (1979)<br>Wildish et al. (1987) |
| 8 | Modified Vogel flume | Seawater recirculating | 90 | Vogel (1981)<br>Wildish and Miyares (1990)<br>Wildish et al. (1992) |
| 9 | Saunders and Hubbard mini flow tank | Seawater recirculating | 205 | Saunders and Hubbard (1944)<br>Wildish and Saulnier (1993)<br>Wildish et al. (1998)<br>Newell et al. (2001)<br>Wildish et al (2000) |
| 10 | Annular flume | Seawater recirculating | 450 | |

at which fish just began to float.[4] Other physiological experiments in which flow was a variable were conducted in a Blažka respirometer, to be described below.

In 1964 J.B. Sprague designed an avoidance trough for measuring the avoidance behaviour of small freshwater fish exposed to toxic substances within the apparatus. The avoidance trough was 114 centimetres long and consisted of a 14.6 cm internal diameter Plexiglas™ tube filled to a depth of 8.5 centimetres. Clean fresh water was introduced at either end of the trough and exited at its midpoint in a series of plastic drainage pipes. Toxicant solution could be introduced at either end.[5] In the early 1970s, P.E.K. Symons developed a smaller artificial stream (not included in table 7.1) for studying Atlantic salmon behaviour.[6] The definitive device to study stream velocity effects on freshwater fish behaviour, which appears as item number 3 in table 7.1, was originally designed and constructed by John Carrothers and Tim Foulkes of the St Andrews Biological Station for instrumentation calibration purposes, and its development is described by Foulkes elsewhere in this volume. It was a 16-metre-long, recirculating raceway with a fixed-pitch propeller driven by a 15 horsepower motor. It had a water volume capacity of 186,200 litres, and was capable of free-stream flows greater than one hundred centimetres per second. It should be acknowledged here that flow studies would be impossible without the technical expertise of those who designed and built flow simulators at the St Andrews Biological Station over the years: Joe Johnson, Frank Langley, Hill Brownrigg, Ron Greenlaw, Clyde Tucker, Herb Small, John Carrothers, Tim Foulkes, Sam Polar, Art Carson, Brian Kohler, and Richard Carson.

The earliest laboratory experiments on bivalves and flow were carried out at the ecological level, that is, using populations of animals. Since seston depletion effects were the main interest, the devices were of flow-through design. It would be wise at this point to explain the meaning and significance of "seston." Seston is the name given to the solid matter suspended in sea- or fresh water, inclusive of all living, dead, or inorganic particles. Seston concentration and quality are important to the many benthic animals which are suspension feeders and which depend on seston being carried by water currents to them as food. A measure of total seston content – for example, as dried weight on a filter – will usually be a poor measure of its quality, because only a variable fraction can be used as food by the suspension feeder. Therefore, many alternative measures have been tried. These include measuring organic carbon after ignition at high temperature, chlorophyl *a* content as a measure of living phytoplankton, cell counts of a pure strain of phytoplankton, or adenosine triphosphate (ATP) content as a measure of all living particles, inclusive of bacteria. None are entirely satisfactory and a definitive test of a particular seston depends on the growth response it elicits in a named suspension feeder, which can be measured in laboratory experiments.

These experiments on bivalves and flow were conducted by Wildish and Dave Kristmanson, but this collaboration was an outgrowth of an earlier research focus. The following is Wildish's account of how this collaboration grew: Soon after his arrival at the station a new pulp mill became operational in 1971. It drew fresh water from Lake Utopia and discharged effluent into the L'Etang Inlet. This body of seawater extended fourteen kilometres inland and connected to the Bay of Fundy. The station director, Dr John Anderson, put Wildish in charge of field research to determine what effect the pulp mill effluent might have on the L'Etang ecosystem. He was able to recruit an old friend to help: Nigel Poole, who was a lecturer in microbiology at the University of Aberdeen, and who was a microbial ecologist with previous experience of pulp-mill pollution in Scotland. His supervisor, Dr Vlado Zitko, introduced him to a professor in chemical engineering at the Fredericton campus of the University of New Brunswick, who had skills in chemical mixing problems that could readily be applied to pulp-mill effluent mixing in the L'Etang Inlet. One result of their teamwork was the realization that all benthic animals (Wildish's study area) were influenced by the hydrodynamics of the environment (Kristmanson's study area) they lived in (as well as by pulp-mill effluent at sufficiently high levels!). During the course of the joint studies in L'Etang Inlet Kristmanson and Wildish became close friends. The latter had formulated an overarching trophic group mutual exclusion theory based on the L'Etang work, which obviously required many more experimental tests to reject or validate it. Kristmanson and Wildish determined to work together to achieve this, and thus began a twenty-five-year collaboration. Their modus operandi began with a theoretical discussion on how the hydrodynamics might influence suspension-feeding animals and progressed to designing experimental tests with simulated flows to test a formulated hypothesis. Such discussions often took place in the Faculty Club at the University of New Brunswick or at Wildish's home in St Andrews or latterly in Bayside.

Kristmanson designed a single-channel flume that was constructed at St Andrews and placed near extreme high-water spring tidal height at the biological station. A submerged pump near the station's wharf supplied natural seawater. Free-stream flows of up to ten centimetres per second were the maximum available with the pump and head used and hence limited its further use. The next simulator came from a design by W.W. Kirby-Smith, a researcher at the Duke University Marine Laboratory at Beaufort, North Carolina.[7] It was constructed at the St Andrews Biological Station of eight Plexiglas™ tubes. Seawater supply to the tubes, from a constant head box, provided average flow rates of greater than 5 centimetres per second, according to calculations based on measured volumetric flows divided by the tube cross-section area. The Kirby-Smith growth tubes allowed growth measurements to be made on blue mussels (*Mytilus edulis*) and demonstrated seston depletion effects as tissue growth. However, the results were clearly apparatus specific and could not be used to extrapolate to the field, because of the complex flow profiles within each tube. To overcome this problem, Professor Kristmanson designed a new simulator, the multiple-channel flume, which was built at the biological station, with four individual 15–20 centimetre-wide channels in which a benthic boundary layer could be induced. Free-stream flows of up to 15 centimetres per second were obtained with a submersible pump of 8 horsepower.

To examine physiological trophic (i.e., feeding) responses of bivalves, we required a flow simulator which enabled faster flows and measurement of seston depletion with one or a few bivalves as subjects. These specifications dictated a small-volume, recirculating device in which we could place the subjects so that their filtration activity did not interfere with other subjects, and in which the slow decline in seston concentration indicated the feeding rate over short time periods of a few hours. The first device used for this purpose was the Blažka respirometer, constructed at the University of Toronto for fish physiologists. Subjects in this device could be subjected to flow rates up to around 100 centimetres per second. Complex tube flows in the Blažka respirometer were replaced with more realistic flows, with a benthic boundary layer, in the modified Vogel flume (also constructed in the biological station workshops from a design in Vogel's 1981 book). The problem with this design and its sealed-shaft impellor was that it could produce flows of only up to approximately 40 centimetres per second. The definitive device for measuring physiological effects of flow on bivalve feeding was the Mini Flow tank (figure 7.1) designed and built of Plexiglas™ at the Bedford Institute of Oceanography in Dartmouth, Nova Scotia, by M. Chin-Yee of the mechanical engineering department. Improvements

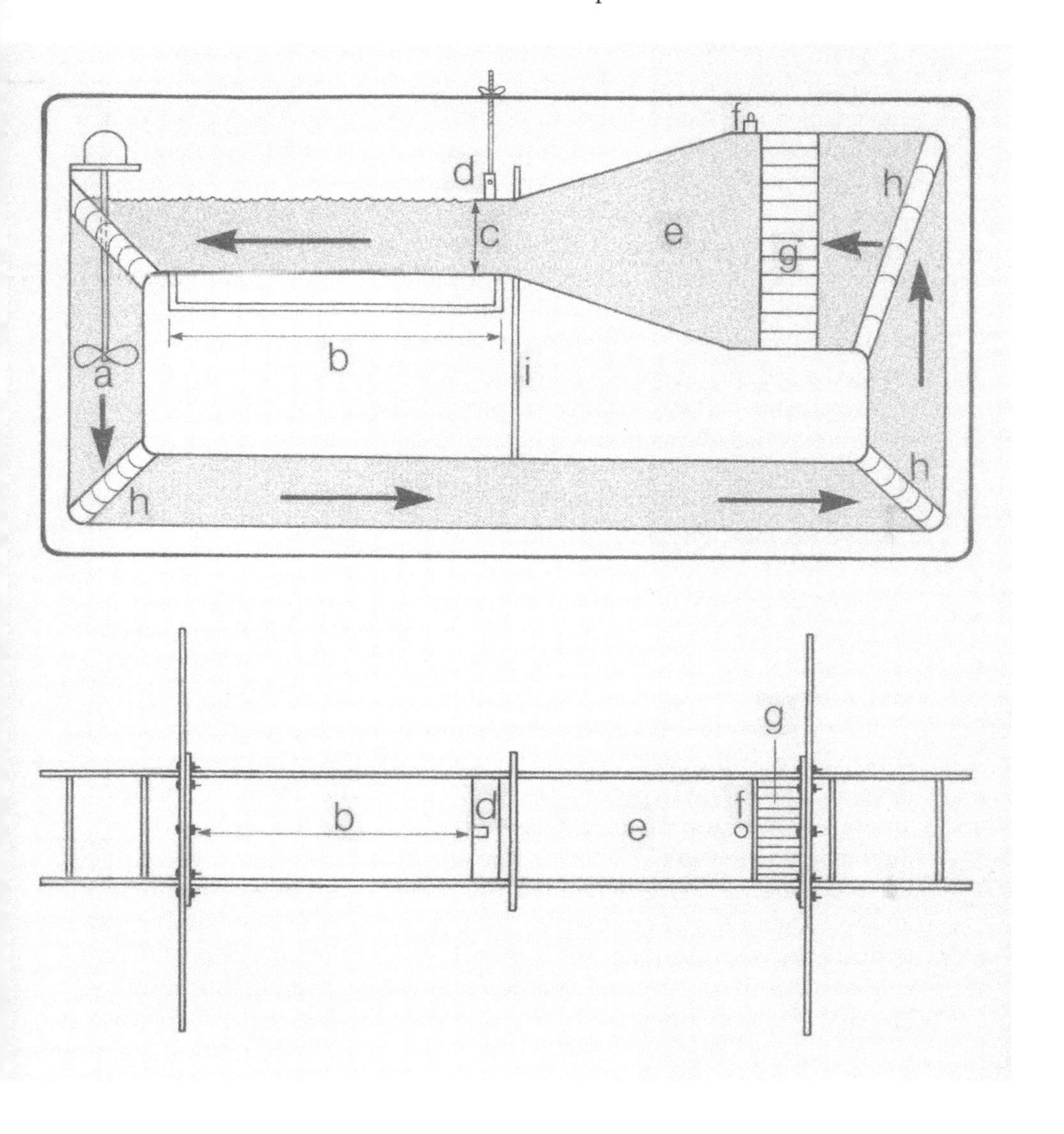


7.1 The Saunders and Hubbard Mini Flow Tank (Wildish and Saulnier, 1993)

for natural flow forming in the Mini Flow tank were incorporated from an earlier (1944) naval architecture design of Saunders and Hubbard that enabled free stream flows in excess of 100 centimetres per second.[8] The impellor was driven by compressed air generated remotely, thus allowing the Mini Flow tank to be vibration and noise free.

A Plexiglass™ annular flume was also designed and constructed by George Steeves and his staff at the mechanical engineering department of the Bedford Institute of Oceanography. Flow in the annular flume was induced by surface paddles with a maximum mainstream flow speed of approximately 45 centimetres per second. The annular flume incorporated a sediment core which could be recessed in the flume floor to investigate flow effects of those suspension-feeding bivalves that dwell in the sea-floor sediments and that need to be in burrows before feeding can occur.

## Bivalve Ecology, Physiology, and Behavioural Studies

The scientific study of marine benthic biology may be dated from the first quantitative study by the Danish investigator C.G.J. Petersen published in 1911.[9] Benthic biology is the study of organisms that live on or under the surface of the sea floor, including organisms that are attached to bottom substrates. Because so few subsequent researchers have been able to devote all their time to this subject, its theory has been slow to develop. The ruling hypothesis in the 1970s was D.C. Rhoads and D.K. Young's trophic group amensalism hypothesis.[10] Amensalism indicates interactions between two species, one of which inhibits the other, while the inhibiting species itself remains unaffected by the other. Rhoads and Young's theory sought to explain why suspension feeders are limited to firm substrates, and deposit feeders to soft sediments. Deposit feeders, they theorized, "process" the sea-floor while feeding, thus limiting larval settlement and growth of suspension feeders. In 1977 Wildish proposed an alternative hypothesis, called trophic group mutual exclusion, in which the proportion of suspension feeders to deposit feeders was determined primarily by the current speed spectra in the local habitat as it affected feeding and maintenance of position. Hence, slower flows favoured deposit feeding as it ensured high sedimentation and low erosions rates, while faster flows favoured transport of seston to suspension-feeding macrofauna. This hypothesis was part of an overarching multiple limiting factor theory. The theory stated that benthic community composition, biomass, and productivity were controlled by interactive physical factors – for example, temperature, salinity, hydrodynamics, dissolved oxygen, sediments, and sediment exchanges with water – in addition to biological factors such as food supply, supply of colonizing larvae, behavioural responses, and competition. The most important of the many ecological limiting factors was arguably hydrodynamics, and so a first step was to experimentally test the effect of flow on the trophic responses of a range of suspension feeders.

In the 1940s C.J. Kerswill conducted experiments with oysters and quahaugs in submerged trays near the Ellerslie Field Station, PEI. His studies showed that those trays with open ends and access to more sestonic food grew faster than those with substantial flow restriction. The demonstration of shell growth enhanced by higher flows was mirrored by growth of quahaugs at two sites in the Bideford River, PEI, where the fastest growth was at the higher flow site.[11] Despite Kerswill's demonstration of the efficacy of velocity in stimulating bivalve growth, his experiments did not determine why faster flows stimulated growth and were site specific. Nor did they provide a quantitative prediction that could be used to choose the best locations for growing bivalves based on flow characteristics. One purpose of subsequent work at the St Andrews Biological Station was to answer these questions with universally applicable results which allowed for the creation of predictive models. Such models provide a rationale for choosing bivalve culture locations, or for making environmental predictions. For example, they can be used to determine how populations of suspension feeders in eutrophic estuaries – estuaries in which there is enhanced production or even algal blooms due to excess nutrients – will consume or respond to enhanced phytoplankton production.

For populations of any suspension feeder, seston depletion effects are critically important and have previously been observed in many groups of species in field studies by measuring seston concentration. Wildish and Kristmanson in 1984 were the first to simulate seston depletion effects in a realistic flume environment using simulator number 4 in table 7.1 – a single-channel flow-through flume. Seston depletion involves a downstream loss of seston along a blue mussel reef community due to the feeding behaviour of all suspension feeders present, and the inability of turbulent processes within the benthic boundary layer to replenish seston at a rate equal to that being removed. A dimensionless index, the seston depletion index, or SDI, was introduced by Wildish and Kristmanson in 1997 as a simple way to indicate whether depletion was likely at a specific location.[12] The SDI compares the total suspension-feeding potential in a unit area to the rate at which turbulence within the benthic boundary layer can deliver seston to the same area. Where the seston depletion index is greater than one, seston depletion is likely. Wildish and Kristmanson demonstrated the reality of the seston depletion effect using the flow-through Kirby-Smith growth tubes – simulator number 5 in table 7.1 – in which downstream blue mussels grew more slowly than upstream ones.[13] In 1988, growth experiments with giant scallops in simulator number 6 showed that periodicity of flow rates, as occurs during the regular flood-ebb tidal cycles, suggested that optimum flows for seston uptake and feeding were needed for more than one-third and up to two-thirds of the time, in order to compensate for periods spent in restrictive high flows.[14] Skimming flow develops over a mussel reef, or any other element of roughness such as a bed of cobbles where a sheared layer develops, which might hinder exchange of seston across it, thereby affecting the flux of seston within the boundary layer. In 2008 Wildish, Kristmanson, and S.M.C. Robinson examined this possibility with horse mussel growth experiments in a multiple-channel flume – flow simulator number 6 in table 7.1 – and suggested that skimming flow would not reduce seston levels enough to affect population growth.[15]

Feeding rates in bivalves can be observed by measuring the loss of seston with time either by cell counting, chlorophyll *a*, or particle number determinations. Physiological level feeding experiments conducted by Wildish and Kristmanson in 1987, with giant scallops, using both a multiple-channel flume and a Blažka respirometer – flow simulators numbers 6 and 7 in table 7.1 –showed for the first time that seston uptake – or feeding – and growth was operative over a range of flow speeds with a central mode (or unimodal). W.W. Kirby-Smith's earlier work in 1972 with the bay scallop also fitted this pattern, rather than the inverse relationship with flow he had proposed.[16] The unimodal response in giant scallops was in four stages in their response to velocity,[17] and this pattern is probably a general one for many suspension feeders. As figure 7.2 shows, seston uptake rates are a function of both velocity and seston concentration in the giant scallop, which interact in determining feeding rates.[18]

In 1993, Wildish and Alinne M. Saulnier proposed an integrated environmental-physiological model of feeding and growth in the giant scallop. Their model was based on experiments done in a Saunders and Hubbard Mini Flow Tank – flow simulator number 9 in table 7.1 – investigating important environmental controls of bivalve feeding: the

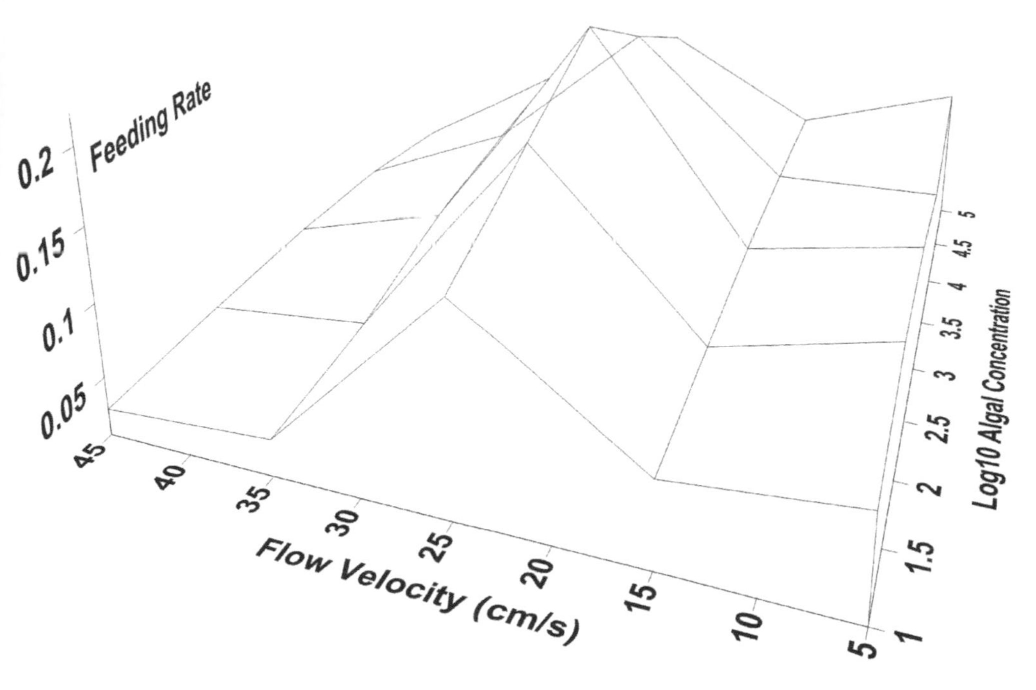


7.2 Feeding rate of giant scallops

concentration and quality of seston and the velocity and direction of flow relative to the exhalant opening. Stage "a" feeding inhibition occurs when flow velocity is low enough to allow dilution of the scallops' inhalant stream with exhaled seawater, which is already cleared of seston. Such conditions occurred during some earlier filtration-feeding experiments in which bivalves were placed in chambers that were too small, and flow mixing was inadequate. Stage "b" filtration occurs at an optimum velocity where small flow increases do not affect feeding. Wildish and M.P. Miyares of the Universidad de Oviedo, Spain, in their 1990 experiment demonstrated stage "c" filtratation in blue mussels, in which velocity inhibited feeding and growth. They measured feeding rates at optimum seston concentrations, using a modified Vogel flume with recirculating seawater (flow simulator number 8 in table 7.1).[19] Carter R. Newell, Wildish, and B.A. MacDonald in 2001 also showed stage "c" filtration using the Saunders and Hubbard Mini Flow Tank

– flow simulator number 9 – to measure feeding rates in blue mussels. Both studies gave a negative slope between 10 and 35 centimetres per second when flow was measured near the inhalant siphon. At flows of greater than 35 centimetres per second, feeding was inhibited (stage "d"). Newell, Wildish, and MacDonald experimentally established that the feeding rate, as influenced both by seston concentration and velocity, was directly indicated by the exhalant opening area.[20] This finding had practical value in confirming that the feeding rate of mussels in natural conditions could be monitored by measuring the exhalant area, for example, by using remote video.

In preliminary experiments done in 2000 with horse mussels in the Saunders and Hubbard Mini-Flow Tank – flow simulator number 9 – Wildish, Hugh M. Akagi, and Natalie Hamilton used the exhalant siphon open area to indicate feeding at an optimum seston concentration. They found that horse mussels could feed optimally at up to 35 centimetres per second – stage "b" – and at reduced rates up to 55 centimetres per second – stage "c" – a result consistent with horse mussels being present where ambient flows are high.[21] This provides an explanation for the Wildish and Kristmanson "paradox" based on comparing velocity-controlled blue mussel feeding discussed above and horse mussel production related to velocity at field locations in the Bay of Fundy.[22] The paradox was based on the false assumption that feeding responses to velocity were the same in blue and horse mussels. If the horse mussel feeding and growth responses to velocity are confirmed, it will suggest that each suspension feeder has a characteristically different unimodal feeding and growth response to velocity.

Another area of interest in the study of bivalve physiological feeding responses is how feeding is affected by the presence of toxic microorganisms. Some strains of the dinoflagellate microalga *Alexandrium sp.* produce potent neurotoxins when ingested in sufficient quantities by vertebrate animals, including fish, which result in neurotoxic symptoms, and in humans cause paralytic shellfish poisoning with possible fatal results. This is discussed by Jennifer Martin in chapter 10, and of course, it is a problem that was first noticed in the Bay of Fundy. Martin, Wildish, and Saulnier in St Andrews collaborated with Patrick Lassus and Michèle Bardouil, their colleagues from IFREMER, Nantes, France, to conduct experiments using a universally cultured bivalve, the Pacific oyster, *Crassostrea gigas*, to determine its feeding responses to *Alexandrium sp.* Physiological feeding rates were determined for each oyster by coupling the seawater in a Saunders and Hubbard Mini Flow

Tank – number 9 in table 7.1 – to a flow fluorometer to measure in real time the disappearance of chlorophyll *a* caused by oyster feeding. The research group included velocity as a variable in preliminary experiments because the unimodal response to velocity of *Crassostrea gigas* had not previously been determined. They found that with a control microalga, *Isochrysis sp.*, as food, the "b" part of the unimodal response was greater than 24 centimetres per second; consequently, they ran further experiments at optimum velocities of approximately 15 centimetres per second. Further results showed that when either toxic or non-toxic strains of *Alexandrium sp.* were present, feeding was inhibited by a stop-and-start mechanism that resulted in no, or reduced, feeding by *Crassostrea gigas.*[23]

## Fish Physiology and Behavioural Studies

Freshwater fish, such as salmon parr, behaviourally select moderately fast flows in streams in order to optimize feeding opportunities in the form of drift insects, and to receive well-oxygenated water. As a consequence, they face flows which may be sufficient to sweep them away. One of the fundamental features of fish anatomy is their gas-filled swim bladders, which enable fish to maintain or change their positions vertically, to reach or maintain a chosen depth, and to maintain stability. In 1964 Richard L. Saunders conducted experiments in a neutral-buoyancy flume constructed at the St Andrews Biological Station – flow simulator number 1 in table 7.1.

He placed anaesthetised salmon parr and brook trout individually in an airtight flume, as shown in figure 7.3. A partial vacuum was applied to the system and the reduction in pressure was recorded manometrically at the point at which the immobilized fish just floated up from the bottom, using the Cartesian diver principle. He set up the experiments by pre-treating the fish either in still water or at flows of 46 centimetres per second, the latter treatment causing a reduction in swim-bladder volume, with a consequent reduction in the fish's buoyancy. The higher density of less buoyant fish allowed them to maintain their position within the reduced velocities of the benthic boundary layer near the stream bed.[24] Saunders followed up these experiments in 1966, this time in collaboration with N.M. Neave, C.L. Dilworth, and J.G. Eales.[25] They focused their experiments on finding out how adjustments in buoyancy were triggered by changes in water velocity and discovering the time required to make them. In these experiments velocity was a continuous variable.

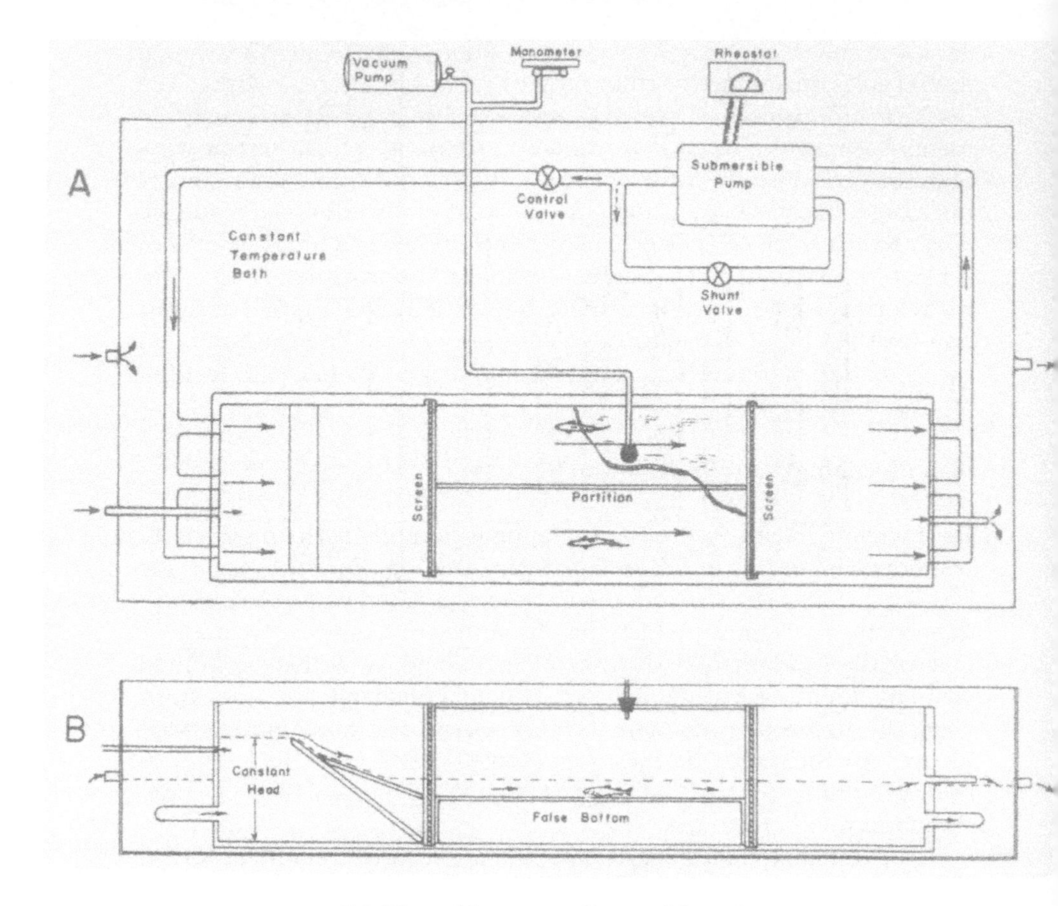


7.3 Neutral buoyancy flume of Saunders

In 1972, John M. Byrne, H. Beamish, and Saunders researched the effects of salinity, temperature, and exercise in post-smolt Atlantic salmon.[26] Treatment smolts were forced to swim in a Blažka respirometer (figure 7.4) initially at low flows, with step-wise increases until they were swimming at flows of 70–80 centimetres per second for two hours. By contrast, controls were unexercised. The researchers examined the effect of pre-treatment at different combinations of salinity and temperature by determining blood plasma osmolality and ionic concentration, that is, concentrations of salts and salt ions. Unexercised, control fish were able to osmoregulate, or control their blood plasma salt and ion

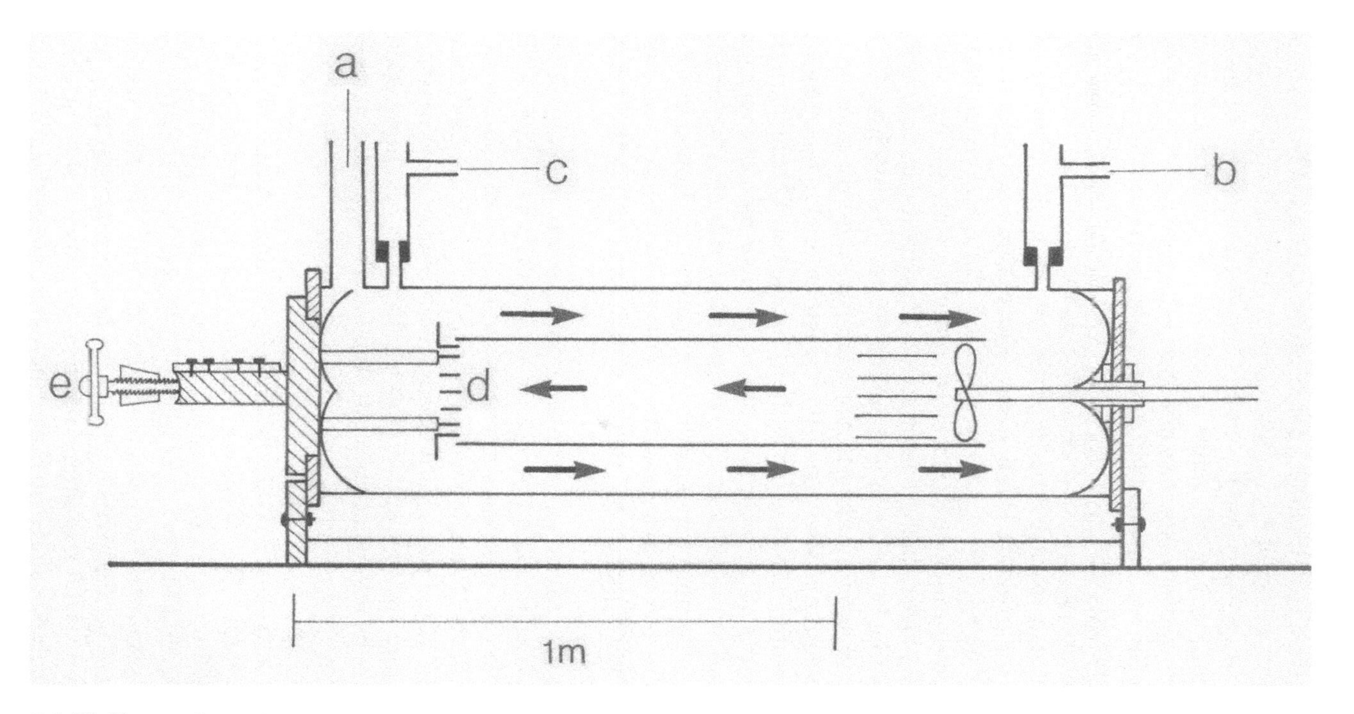


7.4 Blažka respirometer

concentrations, at all treatment combinations, but exercised fish showed marked increases in blood osmolality over controls in full-strength seawater. This suggested that the smolts absorbed seawater, but were unable to osmoregulate efficiently in seawater. A detailed knowledge of the physiology of smolting in salmonids is important if they are subjects of culture. The fundamental knowledge obtained in earlier experiments by Saunders and collaborators proved to be of considerable utility when salmon culture got under way – see Cook, this volume.

In larval fish rearing, it is important to know the flow speeds required by the growing fish to allow for optimal feeding and to avoid the larval fish being swept away. In 2001 Richard H. Peterson and Paul Harmon determined the critical swimming velocity (the swimming velocity just before being swept away) of striped bass larvae in a tubular, Plexiglas™ swimming chamber (30 centimetres long by 6 centimetres in diameter). Flow to the chamber could be adjusted with a submersible pump. They used dye measurements to measure velocity profiles in initial experiments without fish larvae. During experiments with live larvae they recorded volumetric flow measurements, which were related to velocity at the centre of the tube by the empirically derived relationship between the two methods of flow measurement. Striped bass larvae increased their critical swimming velocity from 0.54 centimetres per second on day one to 2.64 centimetres per second on day five.[27]

A number of physiological or behavioural experimental studies have included pollutants and flow. Those presented here include studies of organic pollution that leads to reduced dissolved oxygen levels; studies of the effects of aerial forest spraying with pesticides and the effects of use of the wood preservative pentachlorophenol, both of which find their way into streams; and studies of the effects of mining operations which cause heavy-metal runoff in streams. These studies have helped to build our understanding of how anthropogenic contaminants affect fish physiology.

Velocity and dissolved oxygen levels are important ecological limiting factors for Atlantic salmon smolts. Saunders and M.N. Kutty in 1973 studied the effect of step-wise lowering of dissolved oxygen during exercise in a Blažka respirometer in which the smolts were forced to swim at a constant 55 centimetres per second. The critical dissolved oxygen level beyond which smolts could not maintain their position in the flow was 4.5 milligrams per litre. If velocity was increased the critical dissolved oxygen level also increased, suggesting an interaction between the two. Atlantic salmon were found to have more stringent dissolved oxygen requirements than other freshwater fishes.[28]

Behavioural experiments with Atlantic salmon parr by Philip E.K. Symons in 1973 were conducted in an earlier and smaller version of an artificial stream (see flow simulator number 3, table 7.1). He created velocities of 20 to 80 centimetres per second in the artificial stream and placed in it fish that were pretreated for 15 to 16 hours by being kept in water containing either 1000 or 100 micrograms per litre of fenitrothion, an organophosphate insecticide. Treated fish showed a decrease in the number that held territories of 50 per cent at 1000 micrograms per litre and 20 per cent at 100 micrigrams per litre.[29] Treated parr recovered after two to three weeks and reclaimed their territories. The following year, in a physiological experiment with brook trout, *Salvelinus fontinalis*, Richard Peterson tested the effects of fenitrothion on their swimming performance in a Blažka respirometer. The trout were pretreated with fenitrothion in a concentration range from 150 to 1500 micrograms per litre for a fixed period of 24 hours. He used step-wise increases of velocity to determine the critical swimming velocity for each treated trout. The no-effect level of fenitrothion in this test lay between 150 and 500 micrograms per litre. The critical swimming velocity for control brook trout was 47.5 centimetres per second, which was reduced to 35 centimetres per second for fenitrothion-treated trout at 1500 micrograms per litre.[30] In another swimming performance study in a Blažka respirometer, Bo Holmberg and Saunders in 1979 tested the American eel, *Anguilla rostrata*, exposed to the wood preservative pentachlorophenol. Eels were pretreated at 100 micrograms per litre of pentachlorophenol for four days, followed by four days of recovery in running clean water. The eels' swimming performance was tested in the respirometer at a low and a higher (35 centimetre per second) flow. Eels pretreated with pentachlorophenol showed an increase in active and standard rates of oxygen consumption over untreated controls.[31] This indicated that the chemical created physiological stress.

John B. Sprague in 1964 designed avoidance experiments with Atlantic salmon parr. He wanted to address the issue of copper and zinc pollution originating from base-metal mining in the watershed of the Northwest Miramichi River, New Brunswick (see Sprague et al. 1965).[32] Heavy-metal solutions were pumped to a randomly selected end of the avoidance trough – flow simulator number 2 in table 7.1 – with the other end acting as the control. The movements of a single fish were continuously recorded by pencil and kymograph drum over a ten-minute period by an observer. Analytical results were based on the time spent in the treated versus the control side of the trough. The avoidance bioassay was not designed to predict the behaviour of salmon parr in a stream, but the results did show that the parr could recognize copper and zinc and their mixtures at sublethal concentrations.

Experiments to compare the behaviours of different fish species at slow flows – with a mean of 20 centimetres per second from a range of flows from 10 to 50 centimetres per second – and fast flows – with a mean of 40 centimetres per second and a range from 10 to 100 centimetres per second in the artificial stream – simulator number 3 in table 7.1 – were undertaken by Symons in 1976. The null hypothesis he was testing was that juvenile Atlantic salmon did not have a competitive advantage over other coarse fish present. He stocked the artificial stream with five fish taxa – juvenile Atlantic salmon, white suckers, common shiner, blacknose dace, and creek chub – as found commonly in local streams. They were fed a fixed ration level consisting of brine shrimp and mealworms. The fast-flow treatment resulted in decreased growth when compared to the slow treatment for all fish. Blacknose dace and white suckers, but not common shiner, grew slower at fast flows if territorial salmon were present. This supports the alternate hypothesis that Atlantic salmon have a competitive advantage in fast flows, due to the aggressive territorial behaviour they show to other fish taxa.

## How Flow Experiments Revealed the Importance of Flow in the Environment

Scientists at the St Andrews Biological Station, since the time of C.J. Kerswill's pioneering investigations in the 1940s, conducted a wide range of research studies that confirmed that flow is an important variable in many fundamental and applied problems in aquatic biology. The apparent disparate nature of the research discussed here is due to the nature of the applied problems that stimulated them. However, if we look closely we can find many commonalities in the biological research results obtained. As an example, consider how fish and bivalves deal adaptively with flows that tend to sweep them away. A fish or bivalve resting on or near the bottom within the benthic boundary layer will experience lift due to Bernouilli's principle. This is because of its position within the benthic boundary layer, where flow will be locally higher on the upper surface of the animal, and the resultant pressure difference will generate a lift force. The lift force is resisted by the negative buoyancy of the animal body, and in freshwater fish, Saunders showed in 1964 that they were able to increase negative buoyancy in response to velocity increases by reducing swim-bladder volume.[33] No such option is available to bivalves and they must use other adaptive methods to avoid being swept away. Mussels and smaller scallops use fine sticky threads, called byssus, to attach and anchor themselves to the substrate. Byssus production per mussel is proportionately higher at locations where the velocity is also higher. Older giant scallops employ a different behavioural method to avoid being swept away by directing jets of seawater to excavate a recessed pit in the substrate. Settlement in the pits isolates them from faster flows, thereby reducing lift. Consistent with this idea, Wildish and Saulnier in 1993 showed that it was larger scallops – those greater than 8 centimetres in valve height – during valve gaping, that is, when they were feeding, that were the most vulnerable to lift forces.[34] This may partially explain why we find only the larger scallops recessed in the substrate. We suggested that there may be a trade-off for scallops: increased feeding at fast flows, but increased likelihood of being swept away at such flows. Following from this, we would expect larger scallops to be present in faster flows than smaller ones.

Criticism is an integral part of the scientific enterprise and valid criticism stimulates better field and laboratory methods of experimentation. Flow experiments may be criticized by using variants of one of the following two general arguments: first, flow simulator experiments in the laboratory provide unrealistic environmental conditions; and second, field flow experiments may provide realistic environmental conditions, but may be selective and unable to provide the full range of environmental conditions. Some of the studies discussed here are open to both types of criticism. An example of the first is the tube flow laboratory experiments in simulators 2, 5, and 7 in table 7.1, and also the larval swimming chamber. Flow within a tube is complex with the cross-sectional flow profile being cone shaped, with the point of the cone at the tube centre. This results because a benthic boundary layer develops on every solid surface that the flow contacts. Such conditions rarely occur in the field (except where mussels clog intake pipes); in addition, it is problematic to specify the velocity that the subject actually experiences. It is also not possible to relate tube flow characteristics to free-stream flows. Sprague's 1964 avoidance experiments incorporated an additional unrealistic environmental feature: the tangential flow separating the toxicant from the non-toxicant sides of the avoidance trough. The second type of criticism can be levelled at the field experiments done by Kerswill in the 1940s with oysters and quahaugs in Prince Edward Island. In the light of current knowledge (some of it gained at the St Andrews Biological Station) we can appreciate that Kerswill's experiments were correct for the two sites chosen, but clearly site specific. Thus, if different pairs of sites had been chosen, different results might have occurred. Faster velocities do not necessarily enhance, and some may inhibit, feeding and growth. This results because feeding is unimodally linked, in a taxon-specific way, to velocity, and interactively with seston concentration. At the ecological scale, faster velocities do increase the flux rate of seston reaching the bivalve reef and, providing velocity-induced inhibition is absent, increase both the population growth rate and the path length of the supported reef.

Despite these apparent criticisms of the validity of the data gathered from laboratory systems, it quickly becomes obvious that the early flow studies formed the basis of the cyclical relationship between field observations, testing hypotheses in the lab under controlled conditions, and then further experimentation in field conditions. The basic understanding acquired by this process, and the results of early experiments on flow, allow a conceptual and scientific basis to broader concepts that are currently being pursued, such as carrying capacity, pathways of effects of various anthropogenic compounds, and ecosystem-based management policies.

A long-standing goal of bivalve culture research at St Andrews was to enable the creation of predictive models. From the literature it can be seen that further development of the first "turbulent mass transfer of seston" model published by Wildish and Kristmanson in 1979 has occurred elsewhere.[35] Models are now presented to predict the carrying capacity of commercially important bivalves, as well as to assess material fluxes between the pelagos and benthos for environmental management purposes. Commercial models, such as MUSMOD, are now available which can predict blue mussel carrying capacity at a given location.[36] Subsequent developments have been towards more sophisticated physical oceanographic input to the models, but each requires considerable site-specific input data to produce useful predictions. As far as we are aware, there have been no attempts to incorporate the taxon-specific unimodal responses of bivalves to velocity and seston concentration.

Central to the progress of science is communication, and one of the channels of communication for benthic science has been a series of "Benthic Workshops" that have allowed scientists interested in benthic biology to meet and share ideas. The benthic workshops at the St

Andrews Biological Station began as a one-off event to celebrate the 75th anniversary of the station in 1983. The director in 1983, Dr Bob Cook, gave Wildish a small sum of money to organize a scientific meeting. With it he was able to pay the travel costs of invited speakers whom he had selected based on a shared interest in benthic biology and hydrodynamics. The speakers at the first Benthic Workshop held in 1983 were Don Rhoads, of the Department of Geology and Geophysics at Yale University; John Roff of the Department of Zoology at the University of Guelph; Richard Newell from the Horn Point Environmental Laboratories, University of Maryland; Dave Kristmanson of the Chemical Engineering Department of the University of New Brunswick; Dave Wildish; Jon Grant of the Department of Oceanography, Dalhousie University; and Ulrich Lobsiger of Lobsiger Associates Ltd., Halifax. Richard Newell was the son of Wildish's late thesis supervisor, Prof. Gordon E. Newell of Queen Mary College, London University. The successful meeting was very stimulating to the participants, so there was pressure to repeat it. Wildish consulted with local colleagues and friends, particularly Jon Grant and Barry Hargrave of the Bedford Institute of Oceanography, with the result that the benthic workshop meetings became a biennial event, with one meeting skipped in 1987, so today they are even-year events. Although each meeting has a theme, presentations were not limited to this subject.[37] These benthic workshops have served to keep local scientists from the Maritimes, Quebec, and Maine in touch with each other's latest work, to provide graduate students with a chance to practise their communication skills and enable the one-to-one contacts so necessary to the development of any human intellectual project. One reason for the continued success of the meetings is the provision by the Huntsman Marine Science Centre of reasonably priced accommodations and meals to all participants during the workshop. Following the retirement of Dave Wildish in 2005 the benthic workshop has been locally organized by Dr Shawn Robinson of the St Andrews Biological Station.

In addition to communication, St Andrews Biological Station scientists have continued a tradition of assisting in educating scientists, some of whom go on in the field of benthic biology and flow studies. Educational contacts were and continue to be both formal and informal. For example, as Wildish comments, "Although not formally appointed as a supervisor for the following one successful Ph.D. and one M.Sc thesis I believe that my work was inspirational to … two graduate students, not just through my published work but as a result of contact with, including invited lectures given, at both universities."[38] In addition Wildish was more formally associated with a number of graduates as supervisor,[39] and served as an external examiner of many M.Sc., and Ph.D. examinations at the University of New Brunswick, Dalhousie University, Guelph University, Université du Québec, Carleton University, and Oslo University.

The future would appear bright for flow-related work in St Andrews. The biological station has just completed (2012) a new state-of-the-art science complex, including a large wet laboratory that will be supplied with very high-quality seawater from the mouth of the St Croix estuary. There are provisions for a new flume laboratory. In addition, testing of hypotheses on flow in the field derived from lab experiments has been aided by the development of new hydrodynamic, three-dimensional finite-element flow models within the local area.[40] Such models will allow researchers to choose a geographic area for field experiments, where specific flow patterns exist in the Bay of Fundy. Tools are now available to measure flow at much finer temporal and spatial scales, reflecting the dramatic increase in their sophistication over the last few decades. This will allow investigation of processes at scales previously unattainable, but that are relevant to present-day management decisions, such as regarding the environmental impact from organic loading of various anthropogenic activities.[41] A new flume laboratory and sophisticated measurement systems, coupled with a departmental philosophy centred around ecosystem-based management objectives, should ensure that research on flow-related problems will be relevant, interesting, and challenging for research scientists working at St Andrews over the next century.

We may have given the impression that the work reviewed here was accomplished in splendid isolation. (St Andrews is a small town with idyllic land- and seascapes.) This is clearly not the case: many researchers whose work is included here were or are part of a network or "invisible college," linked to international colleagues who shared their scientific interests and sub-disciplines. Face-to-face contact is invaluable: the importance of encouraging the cross-fertilization of ideas, where the exchange is in both directions, cannot be overemphasized if the goal is to support world-class science at St Andrews and within Fisheries and Oceans Canada. International meetings attended by St Andrews scientists have been a means of disseminating, as well as gathering, information from within these collegial groups.[42]

Since the time of C.J. Kerswill, St Andrews scientists have been contributing to world-wide research efforts that have increased the scientific understanding of the physiological, environmental, and ecological effects of flow in aquatic ecosystems. This is a field which, despite its apparent theoretical focus, has been shown here to have grown from and contributed to practical issues encountered in the Bay of Fundy region, such as fisheries management and aquaculture, and identifying and mitigating pollution problems. Understanding the ecosystem with respect to flow is an exercise in the understanding of scales. Impacts on individual organisms in which small-scale hydrodynamic forces affect the ability of the animal to maintain itself in the flow and to capture food particles contained within a particular parcel of water marks the first level of understanding. The organism-scale forces combine into micro-scale forces that affect the organism and other members of the same species in the local area, whether that be a mussel reef or a rock surface. This building of scales through the organism, from micro to meso and macro scales, allows us to understand where the effects are experienced at successive community, population, and ecosystem levels. As a consequence of this fractal pattern, it is essential to understand the basic functions of the organism and how it interacts with its flow environment. Lacking such information, it is impossible to advance our understanding to successively higher ecosystem levels. The legacy of earlier work on flow at the St Andrews Biological Stations and elsewhere has placed the current generation of scientists in a much better position to assist in the construction of rational, conservation-minded management plans, based on this continuum of scale-related processes.

## NOTES

We thank the SABS library staff, Charlotte Gibson and Joanne Cleghorn, and student Robyn Tidd for bibliographic assistance. On behalf of all the authors cited in this chapter we thank the following for technical assistance: Eugene Henderson, Paul Harmon, Gerry Fawkes, Aline Saulnier, Roger Hoar, Art Wilson, Hugh Akagi, and Natalie Hamilton. We also thank the following individuals for building and designing flow simulators at SABS: Joe Johnson, Frank Langley, Hill Brownrigg, Ron Greenlaw, Clyde Tucker, Herb Small, John Carrothers, Tim Foulkes, Sam Polar, Art Carson, Brian Kohler, and Richard Carson. Thanks also to George Steeves and his staff at the Mechanical Engineering department at the Bedford Institute of Oceanography, Dartmouth, Nova Scotia, for flow simulators built and designed after 1993.

1 H. Schlichting, *Boundary-Layer Theory* (New York: McGraw-Hill, 1955), 1– 535; S. Vogel, *Life in Moving Fluids: The Physical Biology of Flow* (Boston: Willard Grant Press, 1981, 1996), 1–484; Mark W. Denny, *Air and Water: The Biology and Physics of Life's Media* (Princeton, NJ: Princeton University Press, 1993), 1–360; K.H. Mann and J.R.N. Lazier, *Dynamics of Marine Ecosystems: Biological–Physical Interactions in the Oceans* (Boston: Blackwell, 1991), 1–496; D.J. Wildish and D.D. Kristmanson, *Benthic Suspension Feeders and Flow* (New York: Cambridge University Press, 1997), 1–409; C.B. Jorgensen, *Biology of Suspension Feeding* (New York: Pergamon Press, 1966), 1–357.

2 C.J. Kerswill, "Some Environmental Factors Limiting Growth and Distribution of the Quahaug, *Venus mercenaria L.*," Fisheries Research Board of Canada Manuscript Report 187 (1941): 1–104; and University of Toronto Ph.D. dissertation, 1941; C.J. Kerswill, "Effects of Water Circulation on the Growth of Quahaugs and Oysters," *Journal of the Fisheries Research Board of Canada* 7 (1949): 545–51.

3 The reader should consult the original publication to check which velocity measure was used.

4 R.L. Saunders, "Adjustment of Buoyancy in Young Atlantic Salmon and Brook Trout by Changes in Swim Bladder Volume," *Journal of the Fisheries Research Board of Canada* 22 (1965): 335–52.

5 J.B. Sprague, "Avoidance of Copper-Zinc Solutions by Young Salmon in the Laboratory," *Journal of the Water Pollution Control Federation* 36 (1964): 990–1004.

6 P.E.K. Symons, "Behaviour of Young Atlantic Salmon (*Salmo salar*) Exposed to or Force-fed Fenitrothion, an Organophosphate Insecticide," *Journal of the Fisheries Research Board of Canada* 30 (1973): 651–5.

7 W.W. Kirby-Smith, "Growth of the Bay Scallop: The Influence of Experimental Water Currents," *Journal of Experimental Marine Biology and Ecology* 8 (1972): 7–18.

8 H.E. Saunders and C.W. Hubbard, "The Circulating Water Channel of the David W. Taylor Model Basin," *Society of Naval Architects and Marine Engineers, Transactions* 52 (1944): 325–64.

9 Wildish and Kristmanson, *Benthic Suspension Feeders and Flow.*

10 D.C. Rhoads and D.K. Young, "The Influence of Deposit-feeding Organisms on Sediment Stability and Community Trophic Structure," *Journal of Marine Research* 28 (1970): 150–78.

11 C.J. Kerswill, "Effects of Water Circulation on the Growth of Quahaugs and Oysters," *Journal of the Fisheries Research Board of Canada* 7 (1949): 545–51.
12 Wildish and Kristmanson, *Benthic Suspension Feeders and Flow*
13 D.J. Wildish and D.D. Kristmanson, "Control of Suspension-feeding Bivalve Production by Current Speed," *Helgoländer Wissenschaftliche Meeresuntersuchungen* 39 (1985): 237–43.
14 D.J. Wildish and D.D. Kristmanson, "Growth Response of Giant Scallops to Periodicity of Flow," *Marine Ecology Progress Series* 42 (1988): 163–9.
15 D.J. Wildish, D.D. Kristmanson, and S.M.C. Robinson, "Does Skimming Flow Reduce Growth in Horse Mussels?" *Journal of Experimental Marine Biology and Ecology* 358 (2008): 33–8.
16 D.J. Wildish, D.D. Kristmanson, R.L. Hoar, A.M. DeCoste, S.D. McCormick, and A.W. White, "Giant Scallop Feeding and Growth Responses to Flow," *Journal of Experimental Marine Biology and Ecology* 113 (1987): 207–20; and W.W. Kirby-Smith, "Growth of the Bay Scallop: The Influence of Experimental Water Currents," *Journal of Experimental Marine Biology and Ecology* 8 (1972): 7–18.
17 Wildish and Kristmanson, *Benthic Suspension Feeders and Flow.*
18 D.J. Wildish, D.D. Kristmanson, and A.M. Saulnier, "Interactive Effect of Velocity and Seston Concentration on Giant Scallop Feeding Inhibition," *Journal of Experimental Marine Biology and Ecology* 155 (1992): 161–8.
19 D.J. Wildish, and M.P. Miyares, "Filtration Rate of Blue Mussels as a Function of Flow Velocity: Preliminary Experiments," *Journal of Experimental Marine Biology and Ecology* 142 (1990): 213–19.
20 C.R. Newell, D.J. Wildish, and B.A. MacDonald, "The Effects of Velocity and Seston Concentration on the Exhalant Siphon Area and Valve Gape in the Mussel, *Mytilus edulis*," *Journal of Experimental Marine Biology and Ecology* 262 (2001): 91–111.
21 D.J. Wildish, H.M. Akagi, and N. Hamilton, "Effect of Velocity on Horse Mussel Initial Feeding Behaviour," *Canadian Technical Report of Fisheries and Aquatic Science* 2325 (2000): 1–34.
22 See figure 7.12 in Wildish and Kristmanson, *Benthic Suspension Feeders and Flow.*
23 D.J. Wildish, P. Lassus, J.L. Martin, A.M. Saulnier, and M. Bardouil, "Effect of the PSP-causing Dinoflagellate, *Alexandrium sp.*, on the Initial Feeding Response of *Crassostrea gigas*," *Aquatic Living Resources* 11 (1998): 35–43.
24 R.L. Saunders, "Adjustment of Buoyancy in Young Atlantic Salmon and Brook Trout by Changes in Swim Bladder Volume," *Journal of the Fisheries Research Board of Canada* 22 (1964): 335–52.

25 N.M. Neave, C. L. Dilworth, J. G. Eales, and R. L. Saunders, "Adjustment of Buoyancy in Atlantic Salmon Parr in Relation to Changing Water Velocity," *Journal of the Fisheries Research Board of Canada* 23 (1966):1617–20.
26 J.M. Byrne, F.W.H. Beamish, and R.L. Saunders. "Influence of Salinity, Temperature, and Exercise on Plasma Osmolality and Ionic Concentration in Atlantic salmon (*Salmo star*)," *Journal of the Fisheries Research Board of Canada* 29 (1972):1217–20.
27 R.H. Peterson and P. Harmon, "Swimming Ability of Pre-feeding Striped Bass Larvae," *Aquaculture International* 9 (2001): 361–6.
28 R.L. Saunders and M.N. Kutty, "Swimming Performance of Young Atlantic Salmon (*Salmo salar*) as Affected by Ambient Oxygen Concentration," *Journal of the Fisheries Research Board of Canada* 30 (1973): 223–7.
29 Symons, "Behaviour of Young Atlantic Salmon."
30 These estimates are calculated from velocities given in the paper as body lengths per time and assume a standard body length of 10 centimetres. See R.H. Peterson, "Influence of Fenitrothion on Swimming Velocities of Brook Trout (*Salvelinus fontinalis*)," *Journal of the Fisheries Research Board of Canada* 31 (1974): 1757–62.
31 B. Holmberg and R.L. Saunders, "The Effects of Pentachlorophenol on Swimming Performance and Oxygen Consumption in the American Eel (*Anguilla rostrata*)," *Rapports et Procès-verbaux des Réunions du Conseil International pour l'Exploration de la Mer* 174 (1979): 144–9.
32 Sprague, "Avoidance of Copper-Zinc Solutions by Young Salmon in the laboratory"; and J.B. Sprague, P.F. Elson, and R.L. Saunders, "Sublethal Copper-Zinc Pollution in a Salmon River: A Field and Laboratory Study," *International Journal of Air and Water Pollution* 9 (1965): 531–43.
33 Saunders, "Adjustment of Buoyancy in Young Atlantic Salmon."
34 D.J. Wildish and A.M. Saulnier, "Hydrodynamic Control of Filtration in *Placopecten magellanicus*," *Journal of Experimental Marine Biology and Ecology* 174 (1993): 65–82.
35 For example, see the proceedings of an Advance NATO Research Workshop in R.F. Dame, ed., *Bivalve Filter Feeders. NATO ASI Series G. Ecological Sciences* (Berlin: Springer-Verlag, 1992), 579.
36 See C.R. Newell, D.E. Campbell, and S.M. Gallagher, "Development of the Mussel Aquaculture Lease Site Model MUSMODC: A Field Program to Calibrate Model Formulations," *Journal of Experimental Marine Biology and Ecology* 219 (1998): 143–69.
37 Themes of the benthic workshops were as follows: "Biology of the Sediment-Water Interface" (1983); "Fluxes of Particulate Matter across Benthic Boundaries" (1985); "Advective Transport and Benthic

Productivity" (1988); "Mariculture Impacts on Coastal Systems" (1990); "Hydrodynamics and Marine Benthos" (1992); "Bivalve Culture" (1994); "Modelling the Benthos" (1996); 'Novel Benthic Methods' (1998); "Physiological versus Ecological Themes in Benthic Studies' (2000); "Pelagic-Benthic Coupling" (2002); "Biology, Geology and Mapping of Pockmarks" (2004); "Practical Benthic Ecology" (2006); "It's a Question of Scale" (2008).

38 These were Marcel Frechette, whose doctoral thesis at the Université Laval in 1985 was entitled "Interactions pelago-benthiques et flux d'énergie dans une population de moules bleues, *Mytilus edulis L.*, de l'estuaire du Saint-Laurent"; and Jennifer A. Cahalan, who studied at the Marine Sciences Research Center, State University of New York at Stony Brook. Her 1988 thesis was titled "The Effects of Flow Velocity, Food Concentration and Particle Fluxes on Growth Rates of Juvenile Bay Scallops, *Argopecten irradians*."

39 These graduates included C.W. Emerson of the University of Guelph zoology department; L. Roseberry of the Université du Québec à Rimouski; A. Chevrier of the Université de Montréal; and C.E. Newell of the University of New Brunswick biology department.

40 D.A. Greenberg, J.A. Shore, F.H. Page, and M. Dowd, "A Finite Element Circulation Model for Embayments with Drying Intertidal Areas and Its Application to the Quoddy Region of the Bay of Fundy," *Ocean Modelling* 10 (2005): 211–31.

41 Recent developments in flow-related measuring technology include three-dimensional internally recording current meters, flow cytometry, multi-spectral fluorometry, automatic plankton counters, floc cameras, image analysis, and remote nutrient analysis.

42 For example, Wildish attended twelve international meetings including four European Marine Biology Symposium meetings over the years and made many contacts and a few lifelong friends, including Prof. John Gray of Oslo University and Dr Vibeke Brock of Arhaus University.

# 8 A Personal Perspective on the Historical Role of the St Andrews Biological Station in Investigations of Canadian Scallop Fisheries

JOHN F. CADDY

*This paper is dedicated to the memory of John (Jack) Stevenson, the first scallop investigator at St Andrews, who in 1936 drowned while sampling scallop larvae from a rowing boat in Annapolis Basin, and to my colleague Bill McMullon, who was always ready to help.*

This chapter describes St Andrews Biological Station's research contributions during a century of investigations on the biology and fisheries of the Atlantic sea scallop (*Placopecten magellanicus*). St Andrews scientists pioneered exploratory surveys early in the twentieth century that identified inshore and offshore beds. They investigated the biology of oysters and their environment, gear performance and the spatial distribution of resources, dredge performance, and the impacts of dragging on the grounds. Observations in the 1960s and 1970s used SCUBA and submersibles to supplement scallop dredge surveys. Their research and contributions have resulted in a sophisticated resource management scheme, which with a few improvements suggested towards the end of this chapter, could lay the basis for successful global management of scallops.

Indeed, much of the science was directed at conserving the resource. Before the extension of national jurisdiction to 200 nautical miles, the former International Commission for Northwest Atlantic Fisheries (ICNAF) was responsible for managing all fisheries resources beyond twelve nautical miles. ICNAF introduced an innovative regulation limiting scallop landed meat sizes for offshore scallops; further adjustments by Canadian fisheries officials helped to reduce the Canadian harvesting of small scallops. When jurisdiction was extended to

200 nautical miles, Canada negotiated with the United States to define the international boundary across Georges Bank; scallops were seen as the most valuable renewable resource. In Washington, DC, during negotiations, publications of the St Andrews scallop program demonstrated Canada's duty of care for Georges Bank renewable resources. St Andrews scientists continued to contribute after scallop research was later coordinated by CAFSAC – the Canadian Atlantic Fisheries Scientific Advisory Committee (1977–94). Later still, the scallop industry became more involved in the scientific work, and cooperated with the St Andrews scientific research program.

The onset of regular offshore scallop assessments demonstrated that irregular recruitment had a major influence on the effects of exploitation, and showed that the dynamic pool assumption, the theory that individuals in exploited populations will re-distribute themselves when a fishery removes individuals in one area, was invalid for semi-sedentary shellfish populations. St Andrews scientists developed strategies for countering fishing effort concentration onto patches of new recruits. They achieved effective access control by assigning enterprise allocations. These successes and ongoing research strongly suggested that future scallop investigations could usefully aim at enhancing stock productivity by applying spatial strategies. This could include rotating harvesting schemes and reducing recruitment irregularities in these sedentary populations through local closures to take advantage of "source/sink phenomena" – in other words, local areas of regeneration that will allow the dispersion of scallops in the pelagic larval phase into depleted areas.

This short historical review discusses the history of Canadian scallop research, in which the St Andrews Biological Station played a key role, and includes my own experience as a research scientist studying *Placopecten magellanicus* from 1966 to 1979. It supplements the major overview of scallop biology and the early history of the Maritimes scallop fishery,[1] and includes limited references to US studies and those elsewhere, if they provide a key to understanding the flow of events. An unbroken research history of almost a century of investigations on *Placopecten magellanicus* was mainly focused at St Andrews, but later also included Nova Scotia; the program on Georges Bank scallop assessment and management was transferred to the Halifax Fisheries Research Laboratory in the mid-1980s. In 1989, a refocused program returned to St Andrews, and this research program has continued to provide a service to the fishing industry of the Maritime provinces.

The coverage of developments after 1979 when I left St Andrews is less comprehensive, but I do include a discussion of the implications of some recent research results, and how these can be adapted to future scallop research and management.

## Biological Studies on Scallops at St Andrews

According to local tradition, the scallop fishery began in Passamaquoddy Bay. Professor A.G. Huntsman recounted that the fishery for scallops began as early as 1895 in the Bay of Fundy, and exploded in 1920 with the discovery of major beds in Digby. The biological station at St Andrews became associated with scallop investigations when Joseph Stafford worked in St Andrews from 1905 to 1910, even before the inauguration of the permanent laboratory. He made extensive plankton tows between St Andrews and Seven Islands (Sept Isles), Quebec, and described the larval stages of the scallop.[2] As Atlantic Biological Station director, Huntsman hired the first full-time scallop investigator, a twenty-one-year-old Queen's University undergraduate student, Jack Stevenson, in 1931, and a slow rise in staff size ensued. Over the next five years, during which time he completed his MA at the University of Western Ontario and entered the zoology department at the University of Toronto as a Ph.D. student, Stevenson established the basis for a descriptive biology of the sea scallop, determined the scallop spawning season in the Bay of Fundy,[3] and published a paper on the growth rate from shell ring reading with a (then) summer student, Lloyd Dickie.[4] This continued the tradition of introducing summer students to future careers in Canadian fisheries. Stevenson was the first to suggest that recruitment success in the Bay of Fundy depended on whether or not larvae were flushed out of the bay by tidal currents. In parallel with Frederick Baird Jr. in the United States, he provided the biological basis for "scallop science" until his untimely death by drowning in 1936 while conducting field studies of scallops near Digby, Nova Scotia.[5]

The early 1960s investigations of J. Carl Medcof and Neil Bourne[6] on catastrophic mortalities of scallops in the Northumberland Strait focused attention on environmental aspects of scallop biology. In 1947 Medcof, Alex H. Leim, Alfred W.H. Needler, and several other colleagues had documented the risk of paralytic shellfish poisoning by consuming scallop rims from the Bay of Fundy,[7] and in 1963 A. Prakash furthered these investigations.[8] Outbreaks of paralytic shellfish poisoning (PSP) have been the subject of many studies, and the full history of this phenomenon is addressed elsewhere in this volume by Jennifer Martin. Studies over the years – by Medcof et al. in 1947, Alfreda B. Needler in 1949, Prakash in 1965, Bourne in 1956, Caddy and R. A. Chandler in 1968, and G.S. Jamieson and Chandler in 1983[9] – showed that scallops can hold onto the toxin for long periods of time, posing a risk if Bay of Fundy body parts other than the PSP-free adductor muscle are consumed. The scallops themselves were also potentially affected, and Medcof and Bourne in 1964 considered that occasional catastrophic mortalities of scallops in the Northumberland Strait were due to paralytic shellfish poisoning or starfish predation,[10] though my own studies in 1968 suggested that scallop gills might be being clogged by fines lifted by intensive dredging on muddy sand.[11]

In 1949 Medcof also began studies on meat yields of scallops,[12] and basic biological studies have continued at St Andrews until today. These have included collaborative investigations of the relationship between scallop growth and the speed and direction of water currents, the history of which is discussed elsewhere in this volume by Wildish and Robinson. During the 1970s my own studies mainly aimed at documenting the role of swimming in gear avoidance;[13] my work also built on earlier work by Maine biologist F.T. Baird Jr. on the importance of byssal attachment for fixing young scallops to nursery grounds, and I investigated the role of bryozoans – tiny communal animals that form crusts or mats on the sea floor or in some cases coral-like formations – as settlement substrates for larval scallops.[14] I was able to collaborate closely with scientists at the Bureau of Commercial Fisheries laboratory at Boothbay Harbor, Maine, and the Woods Hole laboratories, and in February 1966 collected post-settlement stages of the scallop on a bryozoan, *Gemellaria loricata*[15] in Penobscot Bay, Maine. Later studies from the 1990s using suspended culture showed that scallop larvae settle on a variety of substrates resembling fine epifaunal organisms (i.e., organisms that settle on rocks, pilings, marine vegetation, and the like).[16] Later they pass through a semi-cryptic stage, often attached under adult shells. This supports the practice of returning shells with their epifaunal cover to the grounds after shucking.

The fact that both swimming activity and byssal attachment to the bottom drop off for scallops with shell diameters over 100 mm suggests that attachment is a means of maintaining scallops over gravel bottom as opposed to sand – where swimming reactions are more readily induced by predators.[17] Melanie Bourgeois, Jean-Claude Brêthes, and Madeleine Nadeau of the Institut des sciences de la mer de Rimouski[18] found dispersal rates and predation significantly higher on sand than on gravel or dead shell. From underwater studies off the St Andrews wharf, high densities of scallops, especially if damaged by fishing, attract accumulations of starfish leading to high rates of predation.[19] This raises a question, not yet resolved, as to the relative importance of dispersal as opposed to predation in rapidly reducing the dense populations of young scallops, observed, for example, on the northern edge of the Georges Bank in 1971–2.[20]

## Surveying the Maritime Shellfish Resources

In the post-war years, the priority in Canada was to establish what fishery resources were available within national waters, given that before the declaration of a limit of 200 nautical miles, the resources of the continental shelf were under international jurisdiction. The management priority was mainly to ensure fair access to national resources by Canadians after payment of a nominal licence fee, and scallop research was largely focused on mapping beds. A series of informal laboratory leaflets provided surveys and mappings of beds to industry until the early 1960s, especially in the Gulf of St Lawrence.[21] Much more recently, 3-D seabed mapping of offshore scallop fishing areas, using high-resolution acoustics, has helped the scallop industry target fishing on areas with higher scallop densities.[22] Fishing in lower scallop density areas requires a greater area to be fished to achieve the same catch, thus increasing bottom disturbance and incidental mortalities to scallops and epibenthos in the bigger area fished.

Locally important fisheries have developed in the Gulf since the 1950s, following these exploratory surveys. The discovery of offshore scallop beds on St Pierre Bank,[23] including concentrations of the smaller Icelandic scallop *Chlamys islandica*, was recognized as of potential interest in 1965, but only small sporadic fisheries for this latter species have occurred, and especially when Georges Bank scallop abundance was low.[24] Thus, the importance of scallop fishing on the offshore Scotian Shelf banks has generally been inversely proportional to the level of recruitment for the more productive Georges Bank stock.

This early phase of exploration led to the flourishing inshore Maritimes scallop industry. Surveys by the biological station's scientists helped fill the blanks in the Maritime map of marine resources. The production of "industry leaflets" describing these surveys perhaps fitted better within the aims of the then recently founded Resource

Development Branch as the investigative arm of the Department of Fisheries. Certainly, exploratory work occurred at the expense of a lower priority given to other research directions – a situation perhaps paralleled today where regular repeated surveys inevitably reduce the time, manpower, and funds available for innovative research and the development of new research and management tools.

In the early 1970s, alternative shellfish resources were as yet hardly touched – e.g., *Arctica islandica* (ocean quahog), *Chlamys islandica* (Icelandic scallop), and *Spisula polynima* (Stimpson's surfclam)[25] – but even in the late 1960s, it was evident that more was needed if scallop resources were to continue to be viable. In the 1980s, to ensure that scientific assessments of fishery productivity underlay the limited number of licences issued for fishing these resources, the Invertebrates Committee of CAFSAC was created to provide the context for reporting on East Coast scallop evaluations.

By the early 1960s, exploration for new beds had largely come to an end, and researchers switched to providing a scientific basis for managing known resources. Lloyd M. Dickie's early work on scallop population dynamics off Digby[26] led the way here. St Andrews' links with the fisheries department in Halifax became closer once the Fisheries Research Board ceased to exist and it became part of the Department of Fisheries and Oceans. The subsequent merger with the Resource Development Branch inevitably made the station a satellite institute of the fishery management group in Halifax. Given the closer linkage nowadays between surveys, assessments, and licensing, one can understand that as it became more important to optimize fleet economic performance, an independent research arm would be seen as somewhat of a luxury. However, institutionalizing the function of research within an operational context should not hinder the institution's ability to respond to new research challenges, as will be seen in this chapter.

## The Start of the Offshore Fishery

Inshore scallopers made occasional summer trips to Georges Bank as early as 1945, and Canadian research interest was coordinated with the United States National Marine Fisheries Laboratory of Woods Hole, Massachussetts, through the International Commission for Northwest Atlantic Fisheries (ICNAF). In 1959, the first full-time offshore scallop biologist, Neil Bourne, was employed at St Andrews. Staff expanded significantly post–Second World War, but the station's main research vessel, the *Harengus*, a wooden side trawler, was inadequate for offshore scallop surveys, and before 1972, most offshore studies were experimental. However, in the mid-1960s, surveys of the Digby beds were extended to Georges Bank because Canada's national interest and priorities had shifted towards development of a much larger offshore fishery from Nova Scotian ports. ICNAF was then the body with management responsibilities for fisheries outside the 12 nautical miles limit. The onset of the offshore fishery had notable economic impacts. Before extensive offshore harvests began, Lloyd Dickie observed that scallop prices tended to fluctuate seasonally and inversely with landings.[27] Prices, however, tended to remain constant once the Digby fishery became a "price taker" to the much larger volume of meats from Georges Bank which now dominated the market.

In 1986, the minister of fisheries announced the permanent separation of the inshore and offshore fleets, and this led to separate management strategies and fleets, offshore and inshore. Given limited staff, and without an offshore research platform, the research priority for Georges Bank fishery investigations early in the 1960s was not aimed at offshore surveys and assessment. Offshore fishing data, rather, was gathered through analysis of log books volunteered by scallop boat skippers since 1945. These logs gathered spatial information on catch and effort, and were the means of monitoring offshore operations. This new source of information resulted from cooperation between the Fisheries Research Board of Canada and the offshore scallop industry. Since the start of the Canadian Georges Bank fishery, volunteer skippers had passed a spare copy of their log books to the Fisheries Research Board to aid scientific research; their tow locations were recorded by Decca and Loran cross-coordinates. Eventually, regulations required all scallopers over twenty-five metres in length to submit copies of logs to the fisheries department. A. (Sree) Sreedharan's efforts on the first St Andrews computer transformed these log records into a valuable research resource showing how spatial allocation of fishing effort differed from the random distribution of resources assumed by many finfish models. Before annual research vessel surveys, these log-book records provided a window into events in the offshore fishery, and formed the basis for the first fishery spatial model[28] and the computer program YAREA[29] derived from it. This spatial dynamic model projected the effects of fishing on scallops using various fishery scenarios.

The arrival of the stern trawler RV *E.E. Prince* in 1966 enabled St Andrews scientists to begin regular offshore scallop surveys. The extension of Canadian jurisdiction to 200 nautical miles in 1977 meant that Canada needed additional resources and manpower for fisheries research, but this had little impact on work on scallops in what was now NAFO Subarea 5 (Georges Bank).

**Scallop Dredge Performance**

In 1932 Jack Stevenson[30] and Carl Medcof started gear studies. Medcof first described scallop size selection by dredge rings, by making paired tows of gear with different ring sizes in 1952.[31] Field trials by Neil Bourne in 1964, on the performance of the offshore dredge with different ring sizes, effectively sought to implement yield per recruit (Y/R) studies,[32] which had shown that at current effort levels, achieving a maximum Y/R would require postponing harvesting to approximately 5–6 years of age.[33] Unfortunately, a mechanical solution to size selection is complicated by the wide range of sizes partially retained, due to escapement through the larger inter-ring spaces, as well as through the rings. This results in an extended selection curve, meaning that scallops of many sizes are partially retained.[34] This persuaded some fishermen to add multiple links between adjacent rings, making the dredge much heavier and presumably adding to the probability of injuring escapees. Although the rate of "hidden" mortality on escapees is still not quantified in the Canadian fishery, in Australia, R.J. McLoughlin, P.C. Young, R.B. Martin, and J. Parslow, in a 1991 study,[35] showed that long-term mortalities on fished beds are nine to fifteen times higher than on unfished beds due to multiple recapture with incidental damage, and consequent predation on escapees. Multiple escapements lead to multiple check marks,[36] and this reflects a still unknown incidental mortality rate of damaged escapees. Any mortality due to repeated fishing and escapement in the several years needed to grow to full retention should be taken into account in deciding to move to a larger dredge ring. A routine estimation of shock mark frequency to evaluate indirect mortality at size and the intensity of fishing would also be worthwhile,[37] with comparison between areas receiving different fishing intensities.

Early scallop research had focused on studies of gear effectiveness: the 3- inch rings used in the dredges retain a significant number of small scallops, and experiments with 4-inch rings showed a slight increase in efficiency for commercial sizes.[38] The larger rings significantly reduced the trash landed with the dredge. ICNAF considered the issue of ring size in 1961 and 1962, and though scientific opinion favoured the larger ring, the lack of conclusive evidence (and perhaps industry opinion) led to legislation of this approach being dropped on the Canadian side of the Georges Bank, though it was later adopted by the US industry.

In the 1960s, I resumed studies of scallop dredge performance and efficiency in the southern Gulf of St Lawrence.[39] My studies showed that inshore Digby-style gang dredges are very inefficient, and on a rough bottom may dump their catch at intervals along the tow track. The offshore dredge was somewhat more efficient, but on a sand bottom in the Northumberland Strait, scallops under a 100-mm shell diameter showed avoidance reactions, often swimming over the pressure plate or through the rope back to escape.[40] Although some small scallops did escape through the back of the dredge, three to four times as many passed out through the belly of the dredge, where there is a greater potential for damage, than through the top of the dredge. The study confirmed the relatively low efficiency of offshore dredges for small scallops (from 9 to 20% between 50 and 100 mm shell height).

Cameras mounted on the pressure plate of a modified dredge enabled estimates of high densities in recruitment patches (see figure 8.1), and over successive years revealed high apparent local mortalities (instantaneous total mortality coefficient, Z, as high as 1.18), though to what extent this estimate also reflected dispersion was not established.

I can only conclude that further refinements on direct observation methods would seem desirable, and there is considerable scope for studies on the relative importance of dispersal and predation in the first two years of life, as well as for technological advances in the fishing gear – not just to increase efficiency, but also to reduce incidental breakages on the bottom. In other words, investing in an improved method of harvesting scallops which does not damage uncaught individuals, or the epifauna needed for scallop larval settlement, would seem long overdue.

These direct observation studies were paralleled on Georges Bank by camera surveys in which the camera was mounted ahead of the dredge, and photos were compared with the dredge catch to obtain an estimate of capture efficiency. Work with tagged scallops introduced into the dredge, and a fine mesh cover over the back of the dredge confirmed that some small scallops escape through the rope and chain back of the dredge, but about two-thirds pass out through the rings and inter-ring spaces in the chain belly.[41] This helped explain why healed breakage marks peak on shells at, or around, the internal diameter of 3-inch rings (figure 8.2).

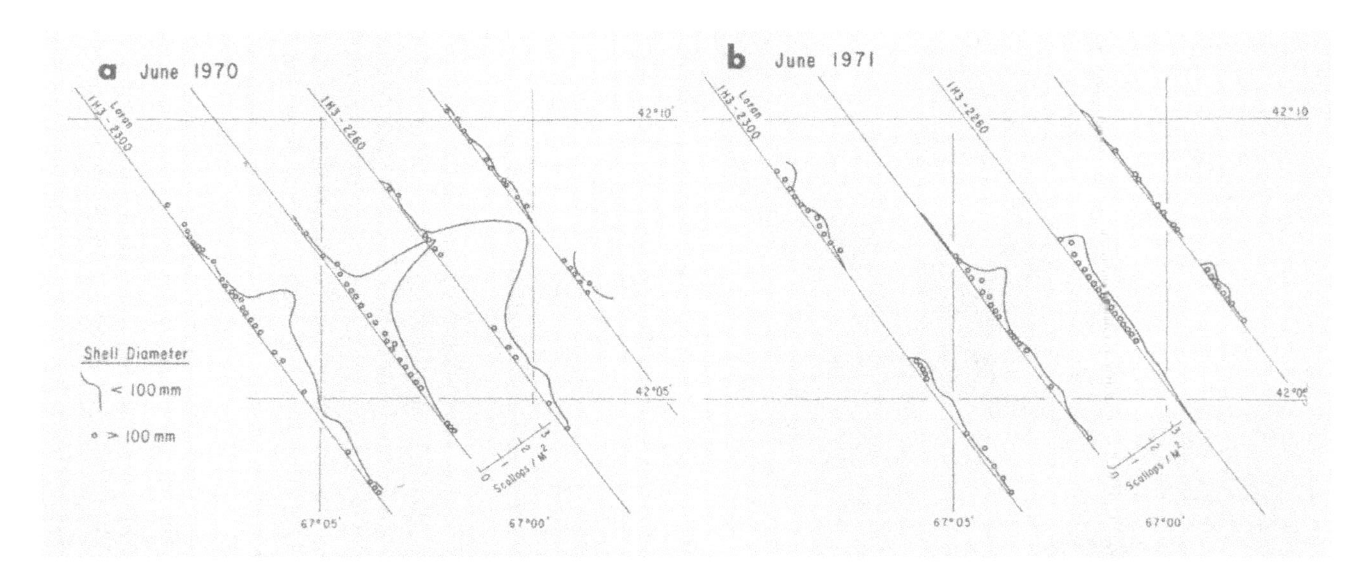


8.1 Camera surveys of high-density patch of recruits (newly settled young spat and juvenile scallops)

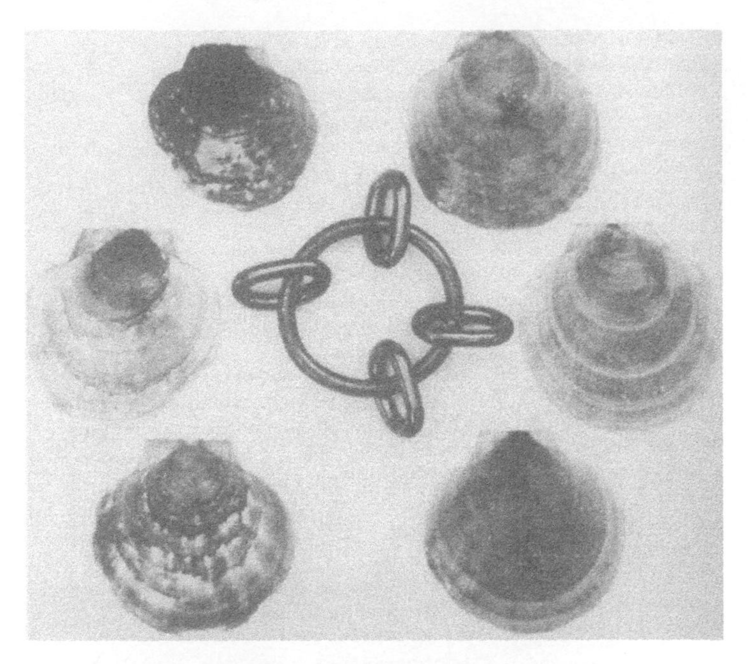

8.2 Breakage marks on scallops. From Caddy, "Perspective on the Population Dynamics and Assessment of Scallop Fisheries" (1989)

## Underwater Science Applied to Scallop Studies

Jacques-Yves Cousteau's television programs in the 1960s promoted the idea that marine scientists can use direct observation methods on marine benthos and fish populations. SCUBA and underwater photography were adopted in St Andrews to overcome the obvious deficiencies of towed bottom gear for sampling. Scientists at St. Andrews carried out pioneering field studies using diving. D.J. Scarratt led the way by using underwater photography to study lobster biology in1968.[42] G.S. Jamieson, M. Etter, and R.A. Chandler in 1981 undertook similar studies on scallop populations and on the impacts of scallop fishing on lobster populations.[43] In the late 1960s and 1970s, the spectacular success of the US lunar landing program led to an air of optimism that underwater technology could similarly solve all problems of fisheries research. These were also the golden years of large government research budgets

– used at St Andrews to apply underwater science and technology to scallop studies.

We mounted Edgerton cameras on the dredge to survey abundance and determine dredge efficiency,[44] and Tim Foulkes and I devised methods of surveying sedentary populations along a transect using specifically developed towed camera platforms.[45] The design and adaptation of vehicles for remote photo surveys, described by Tim Foulkes in chapter 5, benefited from the competence and innovation of station staff such as himself, John Carrothers, and Sam Polar. Photo and video scallop surveys are currently used on the US side of Georges Bank,[46] and there have been experiments in Canada with towed camera vehicles for scallop surveys from 2002 onward,[47] though assessments still largely depend on dredge surveys. This militates against an improved understanding of environmental impacts of fishing grounds – an area of current international concern.[48]

With John Anderson as director, the station extended SCUBA and underwater camera studies to manned vehicle surveys. The *Cubmarine*, *Shelf Diver*, and *Pisces* submersible were hired for underwater research,[49] and a popular film of work during this period – *Down to the Sea* by Giles Walker – played for years at the Huntsman Aquarium. The *Cubmarine* and *Shelf Diver* (figure 8.3) were fitted with a towed odometer to quantify transects across scallop beds, and St Andrews technicians designed a recording system that was then "state-of-the-art," and registered visual observations, photos, height off bottom, and other performance variables[50] that could serve as a model for future quantitative benthic estimation. We used submersibles in the Gulf of St Lawrence and off Digby to map benthos, study gear marks on the sea floor, and demonstrate the contagious nature of scallop populations and other resources.[51] The *Shelf Diver* also documented the relationship of scallops to other benthic epifauna for the Digby beds.[52] Thus the submersible study off Digby in 1969 noted 51 species statistically associated with scallop beds, and developed a run-like statistical approach to analysing faunal adjacencies on a tape recording of observations along a dive track.[53] We discovered non-random zonation by depth and sediment type, and faunal distribution, perhaps reflecting shelter or commensalism (two species sharing and deriving benefit from a common source without adversely affecting each other). Although the continued use of submarines for underwater surveys and the confirmation of remote sampling from the surface offered a more accurate ecological perspective, this method's expense curtailed the use of manned submersibles in future years.[54]

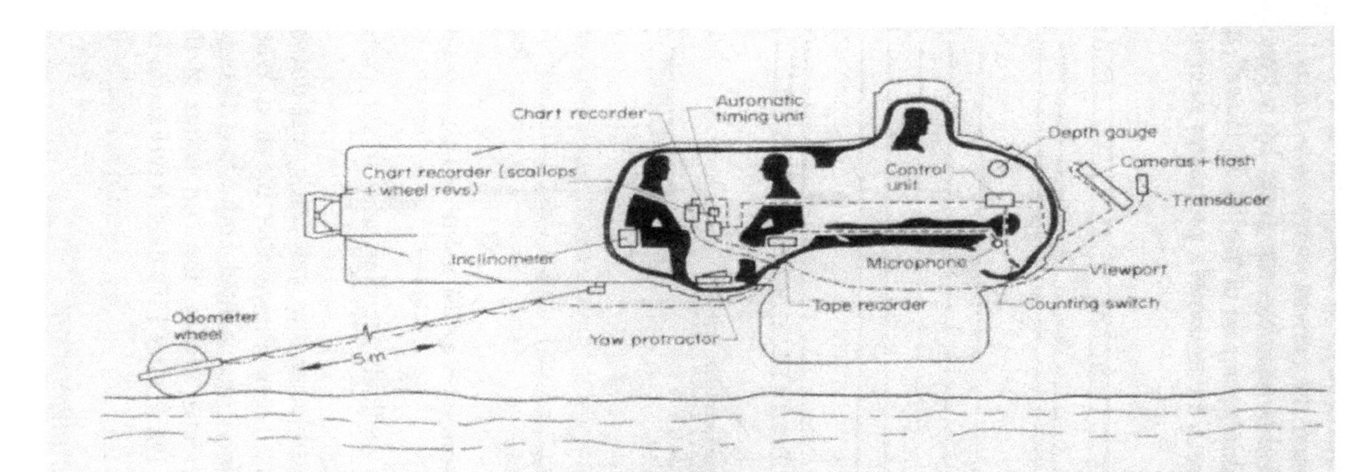


8.3 Technical configuration of shelf diver. From Caddy, "Practical Considerations for Quantitative Estimations" (1976)

**Stock Assessments and Scallop Surveys**

Lloyd Dickie's University of Toronto Ph.D. thesis in 1953[55] included an early application of stock assessment methods already developed for finfish to an invertebrate resource. Dickie applied Leslie/Delury depletion methods to assess the rate of removal of scallops from the Digby beds, calculated yield to recruit isopleths (i.e., it mapped the distribution contours of yield per recruit), and determined, from the abundance of paired dead shells ("cluckers") and their rate of separation, the natural mortality rate of Digby scallops at around 10 per cent annually. This early quantitative work on population dynamics was done at a time when similar studies on shellfish elsewhere in the world were in their infancy. Dickie's work also led to the realization that environmental conditions play a large part in determining recruitment success, partly supporting the earlier views of Jack Stevenson.

Julius Posgay from Woods Hole showed how yield is a function of mortality and growth; he applied tagging to growth and mortality estimates, and used underwater photography to estimate densities.[56] His work established the need to postpone scallop capture to ages 5–6 years old in order to maximize yield per recruit. Glen S. Jamieson, N.B. Witherspoon, and R.A. Chandler revisited a formal yield per recruit analysis of Bay of Fundy stocks in 1980; their analysis, based on age frequencies, log books, and landings, suggested the fishery was then close to giving the maximum yield per recruit. Jamieson, Chandler, and M. Etter also undertook stock assessments of Digby scallops in 1981; more recent stock assessments can be found in a 1995 study by Ellen Kenchington, Dale L. Roddick, and Mark J. Lundy, and for Georges Bank scallops in the 2006 Scientific Advisory Report of the Department of Fisheries and Oceans.[57]

In a 1971 report to ICNAF, I noted that 90–120 meats per pound (each "meat" being the adductor muscle from one scallop),[58] were being reported from some Canadian landings, while catch per hour had dropped to one-fifth of the 1961 level;[59] and so the next year, I proposed different options to control the excessive focus on small scallops, including restrictions on the numbers of licences issued, limits on crew size for the time-consuming shucking of meats, and an increased ring size.[60] The proposal for an increased ring size had already been suggested by the Pacific Biological Station scallop expert Neil Bourne several years earlier.[61] I noted that though setting an absolute minimum size of adductor muscle landed was impractical due to local variations in growth rate, "the most feasible approach would seem to be to specify an upper limit for the mean number of meats per pound in the catch."[62] Dr A.W.H. Needler, then head of the Canadian delegation to ICNAF, did not negate the value of indirect control measures on small scallop capture, such as limits on crew size to discourage shucking of small scallops, but felt that a fisheries regulation should not achieve its results indirectly! In 1994, on the US Georges Bank, an increase in dredge ring size from 3 to 3½ inches, and a further half-inch increase in 2004 to a 4-inch ring, was mandated.[63]

John Gulland, then technical secretary of ICNAF, agreed to the 1972 Canadian proposal of a limit on mean meat size landed. This unusual regulation did not limit the absolute size of meats landed, but set a minimum for the mean count of landed scallop meats landed in bags, to not exceed 40 meats per pound. Ralph Halliday and co-authors F.D. McCracken, A.W.H. Needler, and Ron W. Trites, in their history of Georges Banks Canadian fisheries research, noted that this led to the fishery department monitoring average meat counts landed by the offshore fishery.[64] However, Canada soon had to lodge an objection to its own proposal since a significant proportion of the Canadian fleet was still determined to harvest small scallops! The Canadian government-industry offshore scallop committee resorted to a prolonged campaign that finally reduced mean meat counts over an extended period from about 70 per pound[65] to average meat counts successively required to be less than or equal to 60 per pound in 1974; 50 per pound in 1975; 45 per pound in 1976; 40 per pound by 1983 – until 35 per pound (or 39 per 500 g) was reached in 1986, a meat count still valid up to the present.[66]

A lack of ship time constrained early surveys of Georges Bank scallops to focus on the main productive areas on the Northern Edge and Peak, and log book data was used to stratify statistical areas. US surveys in the 1980s were stratified by depth and latitude, while the Canadian survey design started with stratifying by past commercial catches and depth,[67] but the research survey design has been a subject of controversy and modified several times. Robert K. Mohn, Ginette Robert, and Dale L. Roddick compared research designs and found advantages to each method,[68] the choice being largely determined by the research objective.[69] The bed shape and local density gradients of scallop concentrations, as seen by underwater technologies, contrasts with the substantial area, with variable densities, swept by a dredge in a ten- to twenty-minute tow, and hence with much lower spatial resolution.

When the Canadian government made its declaration of extended jurisdiction to 200 nautical miles in 1977, the problem of how to implement stronger scientific backstopping became an urgent one, and St Andrews' assessment work intensified.[70] At the same time, a more direct responsibility by industry for the state of the scallop resource was also inevitable. In 1986–9, enterprise allocations were granted to the principal companies, and this led to a rationalization and downsizing of the offshore fleet from 68 active vessels in 1986 to 28 in 1999. The institution of total allowable catches (TACs) in 1990 reduced fishing pressure and inevitably led to an increase in the individual sizes harvested, reducing dependence on blending (fishing smaller scallops and mixing with larger ones) to meet the meat count regulation. However, though TACs have significantly reduced fishing effort, the stock was still being fished in 1986 at a level over Fmax (the fishing rate maximizing yield); that is, growth overfishing was occurring.[71] Since 1987, according to DFO sources, effort and exploitation rates have dropped and mean sizes and catch rates have risen.

There have been criticisms of the mean meat count method,[72] but given that there was no effective effort control in the earlier years, it was partially successful in persuading fishers to spend more time fishing outside dense patches of juveniles. Subsequent simulations by Mohn and Robert in 1984[73] showed that a blending strategy resulted in a reduction in yield variation over time. This was more successfully achieved by this strategy than with an absolute minimum size limit, which would have entailed extensive discarding and consequent future loss of yield. Indeed, blending tends to stabilize year-to-year variations in total catch despite marked irregularities in annual recruitment. Perhaps the way to visualize the blending strategy is as a means of spreading the yield from irregular good recruitments over a longer period of years so that landings from a good cohort are spread out, and peaks and troughs in landings minimized. While somewhat wasteful in fuel and fishing time, blending was perhaps the least of all possible evils without an effective control of fishing effort. Its alternative, an absolute size limit, would have been problematic, given variation in growth of meats in different areas,[74] and would likely have increased discarding.

Steven J. Smith and Paul Rago in 2004 also suggested concentrating effort in low productive areas with the overall intention of protecting new recruits.[75] Closed areas and rotating harvest schemes would require new technology to be applied for day-to-day monitoring of fishing locations, and this has now been introduced in the offshore fishery. All licence holders in the Canadian fleet have opted, and paid, for electronic monitoring devices, and industry is now voluntarily protecting a number of "seed box" closures to avoid aggregations of juvenile scallops. Nonetheless, as earlier noted by Robert, Lundy, and Jamieson,[76] a fishery dependent on single year-classes will inevitably suffer drastic reductions of catches following frequent poor years of recruitment unless recruitment is maintained or enhanced.

The "dynamic pool" approach commonly adopted when assessing finfish species is that variations in abundance, spatially and over time, caused by fishing can be ignored, and that by swimming, survivor biomass is effectively remixed between successive fishing trips. This must be discarded for sedentary species where abundance is very patchy, and in consequence, effort is focused onto high-density patches. Shellfish studies have shown that the common assessment assumption – that a dynamic pool can be assumed for yield analysis – can lead to dramatically inaccurate conclusions for sedentary shellfish resources.[77] Some groundfish studies are also adopting methods that account for the spatial discontinuities in fisheries stocks, although the effects are much less evident for mobile fish stocks.

Since 1990 scientists and managers have experienced a growing realization that closing part of the scallop grounds will considerably increase the spawning potential of the whole stock. In 1994, areas of the bank on the US side of the boundary were closed, with limited reopenings in 1999–2001, and in 1998, a "days-at-sea" limit and cap on crew size were declared. Rotating harvest schemes[78] also have considerable potential for scallop management. This system ideally depends on the existence of one or more closed spawning areas for its success. From US experience, the six-year closure to scalloping of part of the US side of the bank led to an almost ten-fold increase in stock biomass in the closed area. In 2003 Deborah R. Hart at the National Marine Fisheries Service at Woods Hole, Massachusetts, also noted that conserving areas where no fishing is allowed may improve recruitment to the whole area[79] – the key question is which areas are the sources of recruits to the whole metapopulation?

## Negotiating the Boundary Line on Georges Bank

During the prolonged negotiations leading up to the international agreement on Georges Bank, it became evident that scallops were the most valuable fishery resource. This is still the case today, especially following the earlier depletion of Georges Bank finfish resources (some of which are now recovering).[80] The focus for Canadian investigations on groundfish and pelagic fish at the time of the boundary dispute was inevitably on the finfish resources of the Scotian Shelf and northern areas, so that most research papers describing research vessel surveys on Georges Bank finfish resources in the 1960s to 1970s were by American scientists. Although there were a slightly higher number of Canadian papers on finfish than scallops for Georges Bank, this is in part an artefact of wider stock areas for many finfish species which extended north to the Scotian Shelf or the Bay of Fundy, and into Canadian waters, where much of the field work was then carried out.

For this reason, regular publications by Canadian scallop investigators were of particular value to Canadian negotiators in showing the "duty of care" exerted by Canada over its southern fishery resources, and surely helped Canada's case during the boundary negotiations.[81] It has probably been forgotten that early on in these negotiations on the fisheries resources of Georges Bank, a firm boundary line was still a mirage, and so a separate agreement on scallops being negotiated by both parties never reached public attention. As conceived in draft form, this would have allocated the Northern Edge beds exclusively to Canada, beds in the Great South Channel exclusively to the United States, while the remaining sparsely populated grounds would remain a common fishing ground. The higher priority given to oil or mineral resources put that solution aside, and a firm line was eventually agreed that split the scallop resource into two mutually exclusive fishing zones, when, in 1984, the International Court of Justice (ICJ) established an international boundary across the Gulf of Maine and Georges Bank, known as the "ICJ line" or "Hague line."

## A Few Ideas for the Future

I would like to end this chapter by offering a few ideas to improve future scallop conservation. These suggestions include improved dredge design to reduce the impacts of fishing on escapees; the use of spatial modelling and source and sink phenomena; and the use of stock enhancement measures including scallop culture.

First, I would like to see studies done to improve dredge design to reduce fishing impacts on escapees. In the 1960s and early 1970s, my SCUBA studies with Carl Medcof on the performance and efficiency of clam and scallop dredges[82] were among the first direct-observation studies to identify gear impacts on the benthos as significant. Robert W. Elner and Glen S. Jamieson also demonstrated that predation on damaged shellfish in the dredge track may be a significant component of mortality.[83] The disturbance of critical benthic habitat by dredging is an aspect of scallop biology which I believe merits further investigation.

Escapement of undamaged smaller scallops is not helped by the narrowing of the rope back as the dredge fills. Improvements to a dredge design, largely unchanged for sixty years, are called for. These include using a more open rope back to promote escapement of small scallops (figure 8.4a), and either a rigid basket held off bottom as used in scallop dredges elsewhere in the world, or incorporating a lifting bar at the tail of the dredge slightly off bottom to reduce damage to escaping scallops (figure 8.4b).

Second, good spatial modelling and diagnostics of source and sink phenomena are needed. A broader outcome of shellfish work at St Andrews in the 1970s was the changeover from applying standard finfish methods to invertebrates, to developing specific models that take into account the particular characteristics of shellfish resources, developments I reviewed in 1989, and re-examined with J.-C. Seijo in 2008.[84] An example was my 1975 model of the Georges Bank scallop fishery[85] which, according to Hilborn and Walters,[86] was the first spatial model for a fishery resource. This model showed that the combination of patches of high-density juveniles, with a wide range of partial retention at size by the gear, led to peak mortality rates occurring on smaller scallops only partially retained by the dredge. Hence, without controls, high density patches of small scallops were attractive to the fleet and soon depleted. This spatial modelling work, initiated on Georges Bank, much later resulted in a spatial software (YAREA) for semi-sedentary resources.[87] In later work on GIS applications to fisheries, F. Carocci and I developed a model for an inshore scallop fishery that was applied to the Digby fishery in the late 1990s,[88] which treated travel time when fishing different distances from port as a key variable.

For bivalve molluscs and other sedentary species there often seem to be "source" areas producing larvae, which due to the local hydrographic regime, are returned to the source beds. Larvae may also colonize "sink" areas, in which, for hydrographic reasons, sub-populations do not replenish themselves. Georges Bank scallop studies had led to a focus on the potential importance of source-sink phenomena in marine resource recruitment. Even the main population components may

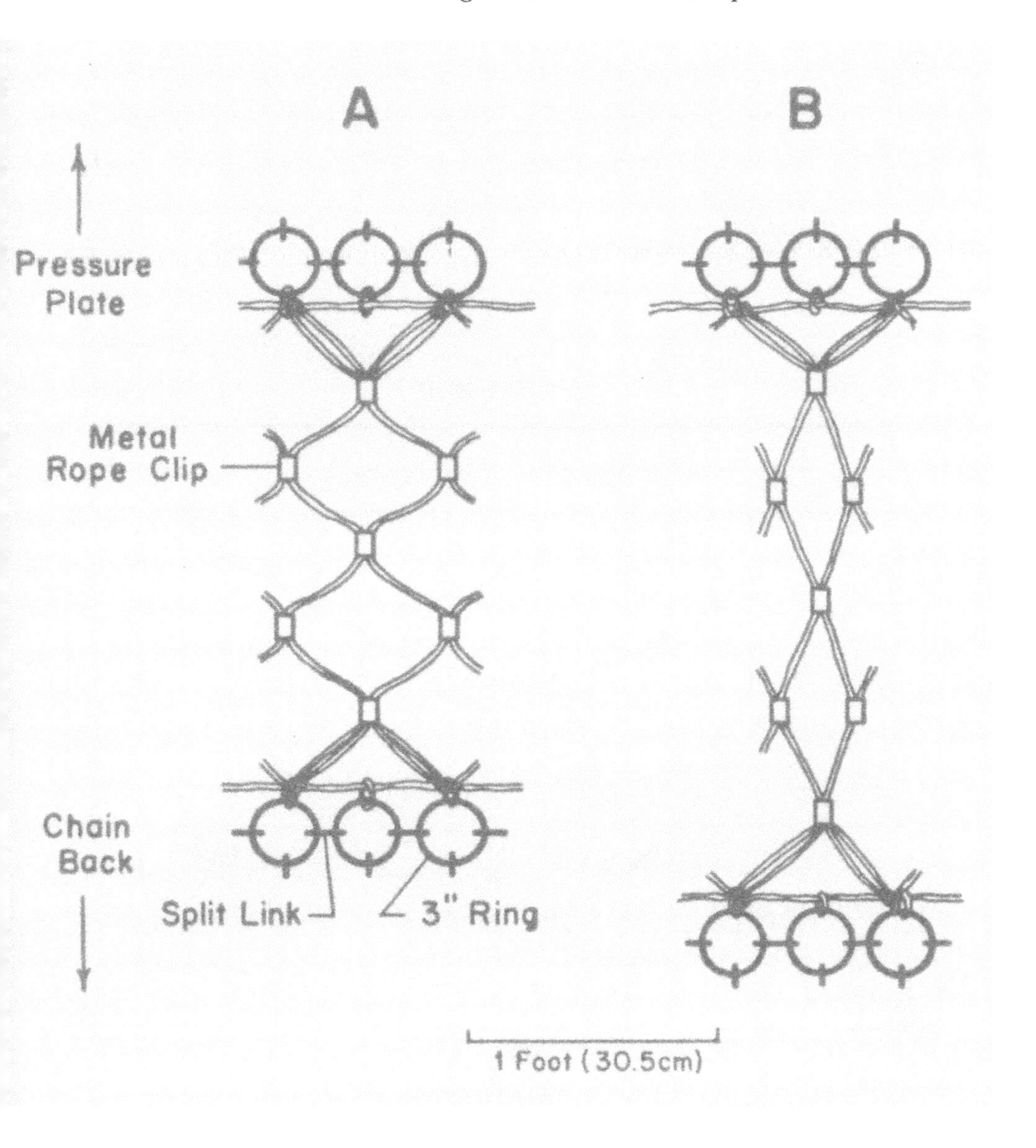


8.4a Stretching of rope back with a full dredge. From Bourne, "Relative Fishing Efficiency of Three Types of Scallop Drags" (1966).

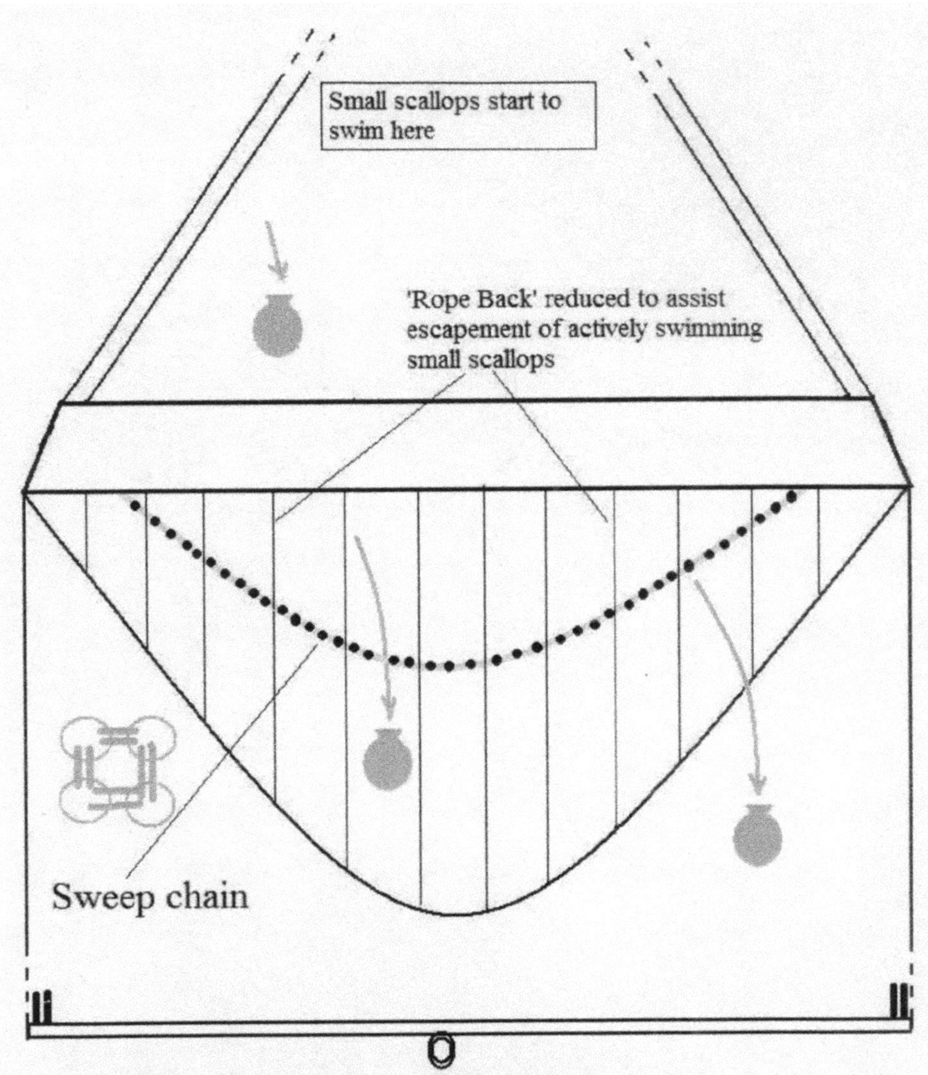


8.4b Dumping bar slightly off bottom to reduce breakages

achieve only poor recruitment in most years; self-recruitment is only good under unusually favourable hydrographic conditions.[89] Helping to diagnose "source" areas would require shell aging studies which identified that most age groups are present, while sink areas are typically dominated by only a few year-classes.

Third, I would like to suggest that the scallop fishery be improved through incorporating stock enhancement measures, including scallop culture. Various researchers, including K.S. Naidu in 1970, Naidu and F.M. Cahill in 1986, and M.J. Dadswell and M.S. Sinclair in 1989,[90] showed that cultivating scallops in suspended culture is feasible in eastern Canada, and this technology is currently being applied in the Magdalen Islands. I must mention a less obvious enhancement measure, however. Many commercially exploited scallop populations around the world are characterized by irregular or pulsed recruitment, which is the extreme case, for example, as was seen off Peru by M. Wolff in 1987.[91] The disadvantage of this is intermittent fisheries separated by long intervals when the resource is only present at lower abundance.[92]

The occurrence of peak spawnings and recruitment pulses at decadal or longer intervals seems characteristic of sedentary molluscs, and two schools of thought have emerged, which are not necessarily incompatible. The first school of thought argues that an underlying stock recruit relationship (SRR) such as the Ricker curve may produce oscillations in recruitment success, but the local density of spawners, and not the total spawning biomass, could be the key to reproductive success in this case. Given epidemic spawning at high densities, only adult populations in patches are likely to achieve a high fertilization rate. Scallops also spawn in conjunction with tidal cycles. The patchiness of scallop populations may, in turn, represent the settlement of a "cloud" of larvae in a local area, but subsequent redistribution to areas of gravel or shell bottom then seems to occur before depletion by fishing. Fishing may disperse survivors further. Local depletion of densities by fishing may, in turn, affect spawning and fertilization success, which typically requires synchronized or "epidemic" spawning to ensure mass fertilization, and scallops also spawn in conjunction with tidal cycles.[93] All these points seem relevant to enhancement. In 1989 S. Aoyama observed that the natural recruitment of Mutsu Bay scallops was enhanced by scallops in suspended culture,[94] and in 1994 G. Jay Parsons, Shawn M.C. Robinson, and James D. Martin also commented on this phenomenon adjacent to a Canadian scallop culture site.[95] This suggests that adding more spawning stock may reduce variance in production cycles. Having the spawning stock in the correct location to ensure a recruitment pulse could be the key to more regular recruitment – but perhaps the ideal spawning location changes from year to year. Is one strategy worth considering, then, to ensure more than one focus of high density spawners is maintained at critical points in the stock range, perhaps including animals in suspended culture to take advantage of the varying hydrographic regime from year to year? As noted by Robert Repetto, it is particularly fascinating that different strategies to improve stock management are being pursued for the same species on two sides of the same bank.[96]

A second school of thought believes that scallop populations are often situated within a gyre that brings larvae back to the adult beds. However, with unstable oceanographic conditions or oscillations in some years, larvae are carried offshore or lost elsewhere. H.B. Hachey's studies suggested such gyres existed in the Bay of Fundy,[97] and later studies, based on "survival at stage" and modelling studies and further studies that combined hydrography with larval survey data, also suggested gyres existed on Georges Bank.[98] The basis for practical prediction of the degree of larval retention in the gyre over the Bank now exists; but what are we to do with this information? That oscillations in production occur over long periodicities was suggested both by the Georges data, and by the Bay of Fundy production series, and are common in world fisheries.[99] I hypothesized in 1979 that this may be linked to long-term tidal oscillations of a high-energy tidal regime, with a gyre which only brought many larvae back to the adult beds at roughly decadal intervals.[100]

The possible migration of scallops has been an issue over the years and is still not fully resolved. In 1953 Frederick T. Baird, of the Maine Department of Sea and Shore Fisheries, described scallop migrations,[101] but in same year Julian Posgay of the Bureau of Marine Fisheries at Woods Hole opined that scallops do not migrate, but almost certainly do disperse in the first few years on bottom.[102] Especially on sand bottom where byssal attachment is impossible, they may disperse in relation to dredges or predators. That both swimming and byssal attachment frequency drop off with age seems to show that evolution reduces dispersal of the more active younger age groups, keeping them on shell and gravel bottom. This perhaps was to ensure that broadcast spawning (only efficient at high densities) is successful. It seems reasonable to suppose, in addition to high incidental mortalities due to fishing on high-density patches of young scallops, that predators target such high densities (reaching six per square metre in some cases) so their natural mortality is higher than the usually assumed $M = 0.1$. Having a better idea of what happens to high-density patches of juveniles would seem, therefore, to merit careful study. Wildish and Robinson elsewhere in this volume outline an alternative hypothesis linked to flow and the trade-off between increased suspension feeding opportunities at higher flows and the increased danger of being swept away.

**A Continuity of Research Was Achieved!**

In conclusion, it is fair to summarize that there have been five focal points for Canadian research on scallops, especially where the St Andrews Biological Station has played a key role. In the earliest phase the focus was descriptive biology of the scallop, a phase that continued from 1905 to the 1970s. Coinciding with the descriptive phase was the exploratory phase, which continued up to the early 1960s. The third focal area was investigating fishing impacts on scallop resources and their environment, the assessment of which began in the 1950s and continues to the present. By the 1970s, the fourth research focus was emerging: the need for controls, and for allocations based on surveys and assessment methods. Again, this area of research is continuing. While I participated in the latter two phases, another focus emerged after I disappeared from the scene, which must be mentioned: this was the development of enterprise allocations which began in 1986 on the Canadian side of Georges Bank. This sets the stage for potential enhancement methodologies to be applied there and elsewhere. The lessons to be learned from two different management regimes being applied on Georges Bank in US and Canadian waters has been a source of interest and deserves further comparisons such as those by Brander and Burke (1995).[103]

At the start of this story I described some early biological investigations of an unexploited resource. At the end, about a century later, we have a national resource under a sophisticated management regime based on the scientific and practical inputs of many researchers, managers, and fishermen, where user rights have apparently begun to be allocated successfully.

NOTES

Maritimes scallop research over the period discussed benefited from technical expertise that converted ideas into reality: Stuart MacPhail, Ross Chandler,

Esther Lord, and a long series of summer students, some of whom later moved on to become investigators themselves, including Alan McIver, Terry Rowell, Chris Radley-Walters, and Michael Sinclair. All these deserve credit for the work reported here. I thank especially Stephen Smith and Shawn Robinson for help in locating source material.

1 N. Bourne, "Scallops and the Offshore Fishery of the Maritimes," *Bulletin of the Fisheries Research Board of Canada* 145 (1964).
2 Rigby and Huntsman, "History of the Fisheries Research Board of Canada," 54.
3 John A. Stevenson, "Growth of the Giant Scallop (*Placopecten grandis*)," Fisheries Research Board of Canada, St Andrews Biological Station, Manuscript 420-S00092 (1932); J.A. Stevenson, "Note on the Advantages and Disadvantages of a 4-inch Mesh in the Fundy Scallop Drags," Fisheries Research Board of Canada Original Manuscript 823; and J.A. Stevenson "Growth in the Giant Scallop (*Placopecten grandis*)," 1936, St Andrews library reprint.
4 John A. Stevenson and Lloyd M. Dickie, "Annual Growth Rings and Rate of Growth of the Giant Scallop, *Placopecten magellanicus* (Gmelin) in the Digby Area of the Bay of Fundy," *Journal of the Fisheries Research Board of Canada* 11 (1954): 660–71.
5 Anonymous, "John Alexander Stevenson, M.A. 1910–1936," *Journal of the Biological Board of Canada* 3:3 (1936–7): 188.
6 J.C. Medcof and N. Bourne, "Causes of Mortality of the Sea Scallop, *Placopecten magellanicus*," *Proceedings of the National Shellfisheries Association* 53 (1964): 33–50.
7 J.C. Medcof, A.H. Leim, A.B. Needler, A.W.H. Needler, J. Gibbard, and J. Naubert, "Paralytic Shellfish Poisoning on the Canadian Atlantic Coast," Atlantic Biological Station Circular no. 9, St Andrews, NB, May 1947.
8 A. Prakash, "Source of Paralytic Shellfish Toxin in the Bay of Fundy," *Journal of the Fisheries Research Board of Canada* 20 (1963): 983–96.
9 See Medcof et al., "Paralytic Shellfish Poisoning"; A.B. Needler, "Paralytic Shellfish Poisoning and *Gonyaulax tamarensis*," *Journal of the Fisheries Research Board of Canada* 7 (1949): 490–504; A. Prakash, "Source of Paralytic Shellfish Poisoning"; N. Bourne, "Paralytic Shellfish Poison in Sea Scallops (*Placopecten magellanicus*, Gmelin) ," *Journal of the Fisheries Research Board of Canada* 22 (1965): 1137–49; John Caddy and R.A. Chandler, "Accumulation of Paralytic Shellfish Poison by the Rough Whelk (*Buccinum undatum L.*) ," *Proceedings of the National Shellfisheries Association* 58 (1968): 46–50; and G.S. Jamieson and R.A. Chandler, "Paralytic Shellfish Poison in Sea Scallops

(*Placopecten magellanicus*) in the West Atlantic," *Canadian Journal of Fisheries and Aquatic Science* 40 (1983): 313–18.

10 J.C. Medcof and N. Bourne, "Causes of Mortality of the Sea Scallop, *Placopecten magellanicus*," *Proceedings of the National Shellfisheries Association* 53 (1964): 33–50.

11 J.F. Caddy, "Underwater Observations on Scallop (*Placopecten magellanicus*) Behaviour and Drag Efficiency," *Journal of the Fisheries Research Board of Canada* 25 (1968): 2123–41.

12 J.C. Medcof, "Meat Yield from Digby Scallops of Different Sizes," *Fisheries Research Board Progress Report* 44 (1949): 6–9.

13 Caddy, "Underwater Observations on Scallop."

14 F.T. Baird, Jr., "Observations on the Early Life History of the Giant Sea Scallop (*Pecten magellanicus*)," *Maine Department Sea Shore Fisheries Research Bulletin* 14 (1953): 7; F.T. Baird, "Migration of the Deep Sea Scallop (*Pecten magellanicus*), ibid.: 1–7; J.F. Caddy, "Progressive Loss of Byssal Attachment with Size in the Sea Scallop, *Placopecten magellanicus* (Gmelin)," *Journal of Experimental Marine Biology and Ecology* 9 (1972): 179–90.

15 Caddy, "Progressive Loss of Byssal Attachment."

16 M. Harvey, E. Bourget, and N. Gagné, "Spat Settlement of Giant Scallop, *Placopecten magellanicus* (Gmelin 1791), and Other Bivalve Species on Artificial Filamentous Collectors Coated with Chitinous Material," *Aquaculture* 148 (1997): 277–98.

17 R.W. Elner and G.S. Jamieson, "Predation of Sea Scallops, *Placopecten magellanicus*, by the Rock Crab, *Cancer irroratus*, and the American Lobster, *Homarus americanus*," *Journal of the Fisheries Research Board of Canada* 36 (1979): 537–43.

18 M. Bourgeois, J.C. Brêthes, and M. Nadeau, "Substrate Effects on Survival, Growth and Dispersal of Juvenile Sea Scallops, *Placopecten magellanicus* (Gmelin 1791)," *Journal of Shellfish Research* 25 (2006): 43–9.

19 J.F. Caddy, ed., "A Perspective on the Population Dynamics and Assessment of Scallop Fisheries, with Special Reference to the Sea Scallop, *Pacopecten magellanicus Gmelin*," in *Marine Invertebrate Fisheries: Their Assessment and Management* (Toronto: John Wiley, 1989), 559–90.

20 J.F. Caddy, "Size Selectivity of the Georges Bank Offshore Dredge and Mortality Estimate for Scallops from the Northern Edge of Georges in the Period June 1970 to 1971," *International Commission for the Northwest Atlantic Fisheries Redbook*, 1972: 79–85, figure 1.

21 J.S. MacPhail, "The Inshore Scallop Fishery of the Maritime Provinces," Fisheries Research Board of Canada, Atlantic Biological Station, St Andrews, NB, General series, 22 (1954); L.M. Dickie and C.D. MacInnes, "Gulf of St

Lawrence Scallop Explorations – 1957," Fisheries Research Board of Canada Manuscript Report Series 650; and N. Bourne and A. McIver, "Gulf of St Lawrence Scallop Explorations – 1961," *Fisheries Research Board of Canada, Biological Station, St Andrews, NB, General Series Circular* 35 (March 1962).

22 V.E. Kostylev, R.C. Courtney, G. Robert, and B.J. Todd. "Stock Evaluation of Giant Scallop (*Placopecten magellanicus*) Using High-resolution Acoustics for Seabed Mapping," *Fisheries Research* 60 (2003): 479–92.

23 L.M. Dickie and L.P. Chiasson, "Offshore and Newfoundland Scallop Explorations," *Fisheries Research Board of Canada General Series Circular* 25 (1955): 1–4.

24 G. Robert, M.J. Lundy, and M.A.E. Butler-Connolly, "Scallop Fishing Grounds on the Scotian Shelf," Canadian Atlantic Fisheries Science Advisory Committee Research doc. 85/28 (1985).

25 J.F. Caddy, R.A. Chandler, and D.G. Wilder, "Biology and Commercial Potential of Several Unexploited Molluscs and Crustaceans on the Atlantic Coast of Canada," in *Proceedings of Government-Industry Meeting on the Utilization of Atlantic Marine Resources*, Montreal, 5–7 February 1974, 57–106.

26 L.M. Dickie, "Fluctuations in Abundance of the Giant Scallop, *Placopecten magellanicus*, (Gmelin) in the Digby Area of the Bay of Fundy," *Journal of the Fisheries Research Board of Canada* 12 (1955): 797–857.

27 L.M. Dickie, "Fluctuations of the Giant Scallop, *Placopecten magellanicus* (Gmelin), in the Digby Area of the Bay of Fundy," Ph.D. thesis, University of Toronto, 1953.

28 J.F. Caddy, "Spatial Model for an Exploited Population, and Its Application to the Georges Bank Scallop Fishery," *Journal of the Fisheries Research Board of Canada* 32 (1975): 1305–28.

29 J.-C. Seijo, J.F. Caddy, and J. Euan, "SPATIAL: Space-time Dynamics in Marine Fisheries. A Bioeconomic Software Package for Sedentary Species," FAO Computerized Information Series (Fisheries), 6 (1994): 116 (plus discs).

30 J.A. Stevenson, "Note on the Advantages and Disadvantages of a 4-inch Mesh in the Fundy Scallop Drags," Fisheries Research Board of Canada Original Manuscript 823.

31 J.C. Medcof, "Modifications of Drags to Protect Small Scallops," *Journal of the Fisheries Research Board of Canada Atlantic Progress Report* 52 (1952): 9–14.

32 Neil Bourne, "Scallops and the Offshore Fishery of the Maritimes," *Bulletin of the Fisheries Research Board of Canada* 145 (1964): 1–61.

33 J.A. Posgay, "Population Assessment of the Georges Bank Sea Scallop Stocks," *International Council for the Exploration of the Sea Rapports et Procès-verbaux* 175 (1979): 109–13.

34 J.F. Caddy, "Efficiency and Selectivity of the Canadian Offshore Scallop Dredge," *International Council for the Exploration of the Sea C.M. 1971/K:25*

(1971): 1–8; and J.F. Caddy, "Size Selectivity of the Georges Bank Offshore Dredge and Mortality Estimate for Scallops from the Northern Edge of Georges in the Period June 1970 to 1971," *International Commission for the Northwest Atlantic Fisheries Redbook* (1972): 79–85.

35 R.J. McLoughlin, P.C. Young, R.B. Martin, and J. Parslow, "The Australian Scallop Dredge: Estimates of Catching Efficiency and Associated Indirect Fish Mortality," *Fisheries Research* 11 (1991): 1–24.

36 See Caddy, "Perspective on the Population Dynamics," figure 2.

37 Ibid.

38 N. Bourne, "Relative Efficiency of Catches by 4- and 5-inch Rings on Offshore Scallop Drags," *Journal of the Fisheries Research Board of Canada* 22 (1965): 313–33; and N. Bourne, "Relative Fishing Efficiency of Three Types of Scallop Drags," *International Commission for the Northwest Atlantic Fisheries Bulletin* 3 (1966): 15–25.

39 J.F. Caddy, "Underwater Observations on Tracks of Dredges and Trawls and Some Effects of Dredging on a Scallop Ground," *Journal of the Fisheries Research Board of Canada* 30 (1973): 173–80.

40 Ibid.

41 Caddy, "Size Selectivity of the Georges Bank Offshore Dredge."

42 D.J. Scarratt, "An Artificial Reef for Lobsters, *Homarus americanus*," *Journal of the Fisheries Research Board of Canada* 24 (1968): 2683–90.

43 G.S. Jamieson, M. Etter, and R.A. Chandler, "The Effect of Scallop Fishing on Lobsters in the Western Northumberland Strait," Canadian Atlantic Fisheries Scientific Advisory Committee Research Document 81/71 (1981): 1–19.

44 Caddy, "Efficiency and Selectivity of the Canadian Offshore Scallop Dredge" and "Size Selectivity of the Georges Bank Offshore Dredge."

45 T.J. Foulkes and J.F. Caddy, "Towed Underwater Camera Vehicles for Fishery Resource Assessment," *Underwater Journal* 6 (1973): 110–16.

46 K.D.E. Stokesbury, B.P. Harris, M.C. Marino II, and J.I. Nogueira, "Estimation of Sea Scallop Abundance Using a Video Survey in Off-shore US Waters," *Journal of Shellfish Research* 23 (2004): 33–40.

47 S.J. Smith, D. McKeown, M. Lundy, D. Gordon, J. Anderson, M. Strong, and M. Power, "TOWCAM – Towed Camera Array for Video/Still Benthic Surveys," *Journal of Shellfish Research* 25 (2006): 308–9.

48 See review in J.F. Caddy, *Marine Habitat and Cover: Their Importance for Productive Coastal Fishery Resources* (Paris: UNESCO Publishing, *Oceanographic Methodology Series*, 2007).

49 J.F. Caddy, "Use of Manned Underwater Vehicles for Fisheries Research," Oceanology International Conference, 1972 (conference papers), 278–81; J.F. Caddy, "Practical Considerations for Quantitative Estimations of

Benthos from a Submersible," in *Underwater Research*, ed. E.A. Drew, J.N. Lythgoe, and J.D. Woods (New York: Academic Press, 1976), 285–98; and J.F. Caddy and J. Watson, "Submersibles for Fisheries Research," *Hydrospace* 2 (1969): 12–16.

50 Caddy, "Quantitative Estimations of Benthos from a Submersible."

51 J.F. Caddy, "A Method of Surveying Scallop Populations from a Submersible," *Journal of the Fisheries Research Board of Canada* 27 (1970): 535–49; and Caddy and Watson, "Submersibles for Fisheries Research."

52 Caddy, "Perspective on the Population Dynamics."

53 J.F. Caddy and J.A. Carter, "Macro-epifauna of the Lower Bay of Fundy – Observations from a Submersible and Analysis of Faunal Adjacencies," *Canadian Technical Reports of Fisheries and Aquatic Science* 1254 (1984): 1–35, http://www.dfo-mpo.gc.ca/Library/13922.pdf.

54 Caddy and Watson, "Submersibles for Fisheries Research."

55 Dickie, "Fluctuations of the Giant Scallop."

56 J.A. Posgay, "Sea Scallop Investigations," in *Sixth Report on Investigations of Methods of Improving the Shellfish Resources of Massachusetts*, Commonwealth of Massachusetts, Dept. of Natural Resources, Division of Marine Fisheries (1953): 9–24.

57 G.S. Jamieson, M. Etter, and R.A. Chandler, "The Effect of Scallop Fishing on Lobsters in the Western Northumberland Strait," Canadian Atlantic Fisheries Scientific Advisory Committee Research Document 81/71 (1981): 1–19; G.S. Jamieson, N.B. Witherspoon, and R.A. Chandler, "Bay of Fundy Scallop Stock Assessment – 1980," Canadian Atlantic Fisheries Scientific Advisory Committee Research Document 81/27: 1–25; and E. Kenchington, D.L. Roddick, and M.J. Lundy, "Bay of Fundy Scallop Analytical Stock Assessment and Data Review 1981–1994: Digby Grounds," DFO Atlantic Fisheries Research Document 95/10 (1995): 1–70; and Department of Fisheries and Oceans, "Assessment of Georges Bank Scallops (*Placopecten magellanicus*)," DFO Canadian Scientific Advisory Secretariat Scientific Advisory Report 2006/032 (2006): 1–11. www.dfo-mpo.gc.ca/Library/323220.pdf.

58 Since metrification, the meat counts are per 500 g.

59 J.F. Caddy, "Recent Scallop Recruitment and Apparent Reduction in Cull Size by the Canadian Fleet on Georges Bank," *International Commission for the Northwest Atlantic Fisheries Redbook, Part III* (1971): 147–55.

60 J.F. Caddy, "Some Recommendations for Conservation of Georges Bank Scallop Stocks," International Commission for the Northwest Atlantic. Fisheries Research Document 72/6 (1972): 1–4.

61 N. Bourne, "Relative Efficiency of Catches by 4- and 5-inch Rings on Offshore Scallop Drags," *Journal of the Fisheries Research Board of Canada* 22 (1965): 313–33; N. Bourne, "Relative Fishing Efficiency of Three Types of Scallop Drags," *International Commission for the Northwest Atlantic Fisheries Bulletin* 3 (1966): 15–25.
62 Caddy, "Some Recommendations for Conservation of Georges Bank Scallop Stocks."
63 See Dvora Hart, "Management and Recovery of U.S. Federal Sea Scallop Fishery" (Woods Hole, MA: NEFSC, 2012), available at www.maine.gov/dmr/rm/scallops/hartforum2012.pdf.
64 R.G. Halliday, F.D. McCracken, A.W.H. Needler, and R.W. Trites, "A History of Canadian Fisheries Research in the Georges Bank Area of the Northwestern Atlantic," *Canadian Technical Reports of Fisheries and Aquatic Science* 1550 (1987): 1–37; and see S.J. Smith and G. Robert, "Scallops, Sampling and the Law," *American Statistical Association, 1991 Proceedings of the Section on Statistics and the Environment* (1992): 102–9.
65 J.F. Caddy and A. Sreedharan, "The Effect of Recent Recruitment to the Georges Bank Scallop Fishery on Meat Sizes Landed by the Offshore Fleet in the Summer of 1970," *Fisheries Research Board Technical Report* 256 (1971): 1–10.
66 Smith and Robert, "Scallops, Sampling and the Law," 102–9.
67 J.F. Caddy and R.A. Chandler, "Accumulation of Paralytic Shellfish Poison by the Rough Whelk (*Buccinum undatum L.*)," *Proceedings of the National Shellfisheries Association* 58 (1969): 46–50.
68 R.K. Mohn, G. Robert, and D.L. Roddick, "Research Sampling and Survey Design for Sea Scallops (*Placopecten magellanicus*) on Georges Bank," *Journal of Northwest Atlantic Fisheries Science* 7 (1987): 117–21; S.J. Smith and G. Robert, "Getting More out of Your Survey Information: An Application to Georges Bank Scallops (*Placopecten magellanicus*)," in *North Pacific Symposium on Invertebrate Stock Assessment and Management*, ed. G.S. Jamieson and A. Campbell, special publication of *Fisheries and Aquatic Science* 125 (1998): 3–13; and R.K. Mohn, G. Robert, and G.A.P. Black, "Georges Bank Scallop Stock Assessment – 1988," Canadian Atlantic Fisheries Science Advisory Committee Research Document 89/21 (1989): 1–21.
69 R.K. Mohn, G. Robert, and D.L. Roddick, "Georges Bank Scallop Stock Assessment – 1986," Canadian Atlantic Fisheries Science Advisory Committee Research Document 87/9 (1987): 1–23.
70 See, for example, Mohn, Robert, and Black, "Georges Bank Scallop Stock Assessment – 1988"; and G. Robert, G.A.P. Black, M.A.E. Butler, and S.J.

Smith, "Georges Bank Scallop Stock Assessment – 1999," Department of Fisheries and Oceans Canada, Stock Assessment Secretariat Research Document 2000/016 (2000): 1–11, www.dfo-mpo.gc.ca/Library/248548.pdf.

71 Mohn, Robert, and Black, "Georges Bank Scallop Stock Assessment – 1988."

72 K.S. Naidu, "An Analysis of the Meat Count Regulation," Canadian Atlantic Fisheries Science Advisory Committee Research Document 84/73 (1984): 1–18.

73 R.K. Mohn and G. Robert, "Comparison of Two Harvesting Strategies for the Georges Bank Scallop Stock," Canadian Atlantic Fisheries Scientific Advisory Committee Research Document 84/10 (1984): 1–35.

74 S.J. Smith, E.L. Kenchington, M.J. Lundy, G. Robert, and D. Roddick, "Spatially Specific Growth Rates for Sea Scallops (*Placopecten magellanicus*)," in *Spatial Processes and Management of Marine Populations*, Eds. G.H. Kruse, N. Bez, A. Booth, M. Dorn, S. Hills, R. Lipcius, D. Pelletier, C. Roy, S.J. Smith, and D. Witherell (Fairbanks: University of Alaska Sea Grant College Program, 2001), 211–31.

75 S.J. Smith and P. Rago, "Biological Reference Points for Sea Scallops (*Placopecten magellanicus*): The Benefits and Costs of Being Nearly Sessile," *Canadian Journal of Fisheries and Aquatic Science* 61 (2004): 1338–54.

76 G. Robert, G.S. Jamieson, M.J. Lundy, "Profile of the Canadian Offshore Scallop Fishery on Georges Bank, 1978–1981," Canadian Atlantic Fisheries Scientific Advisory Committee Research Document 82/15 (1982): 1–33.

77 Caddy, "Spatial Model for an Exploited Population"; Smith and Rago, "Biological Reference Points for Sea Scallops"; and J.M. Orensanz and G.S. Jamieson, "The Assessment and Management of Spatially Structured Stocks: An Overview of the North Pacific Symposium on Invertebrate Stock Assessment and Management," in *Proceedings of the North Pacific Symposium on Invertebrate Stock Assessment and Management*, ed. G.S. Jamieson and A. Campbell, *Canadian Special Publication of Fisheries and Aquatic Science* 125 (1998): 441–60.

78 J.F. Caddy and J.-C. Seijo, "Application of a Spatial Model to Explore Rotating Harvest Strategies for Sedentary Species," *Special Publication of Fisheries and Aquatic Science* 125 (1999): 359–65; D.R. Hart, "Yield- and Biomass-per-recruit Analysis for Rotational Fisheries, with an Application to the Atlantic Sea Scallop (*Placopecten magellanicus*)," *Fisheries Bulletin* 101 (2003): 44–57; and Smith and Rago, "Biological Reference Points for Sea Scallops."

79 D.R. Hart, "Yield- and Biomass-per-recruit Analysis for Rotational Fisheries."

80 J.F. Caddy and D.J. Agnew, "An Overview of Global Experience to Date with Recovery Plans for Depleted Marine Resources and Suggested Guidelines for Recovery Planning," *Fish and Fisheries* 14 (2004): 43–112.

81 Halliday, McCracken, Needler, and Trites, " History of Canadian Fisheries Research in the Georges Bank Area," figure 50.

82 J.C. Medcof and J.F. Caddy, "Underwater Observations on Performance of Clam Dredges of Three Types," Fisheries Research Board Manuscript Report Series 1313 (1974): 1–9; Caddy, "Underwater Observations on Scallop Behaviour and Drag Efficiency"; and Caddy, "Underwater Observations on Tracks of Dredges and Trawls."

83 R.W. Elner and G.S. Jamieson, "Predation of Sea Scallops, *Placopecten magellanicus*, by the Rock Crab, *Cancer irroratus*, and the American Lobster, *Homarus americanus*," *Journal of the Fisheries Research Board of Canada* 36 (1979): 537–43.

84 See Caddy, " Perspective on Population Dynamics"; and J.F. Caddy and J.-C. Seijo, "Application of a Spatial Model to Explore Rotating Harvest Strategies for Sedentary Species," *Special Publication of Fisheries and Aquatic Science* 125 (2008): 359–65.

85 Caddy, "Spatial Model for an Exploited Population."

86 R. Hilborn and C.J. Walters, *Quantitative Fisheries Stock Assessment* (London: Chapman and Hall, 1992).

87 Seijo, Caddy, and Euan, "SPATIAL."

88 J.F. Caddy and F. Carocci, "The Spatial Allocation of Fishing Intensity by Port-based Inshore Fleets: A GIS Application," *International Council for the Exploration of the Sea Journal of Marine Science* 56 (1999): 388–403.

89 See J.F. Caddy and O. Defeo, "Enhancing or Restoring the Productivity of Natural Populations of Shellfish and Other Marine Invertebrate Resources," *FAO Fisheries Technical Paper 448* (Rome: FAO, 2003), 1–159.

90 K.S. Naidu, "Reproduction and Breeding Cycle of the Giant Scallop *Placopecten magellanicus* Gmelin in Port au Port Bay, Newfoundland," *Canadian Journal of Zoology* 48 (1970): 1003–12; K.S. Naidu and F.M. Cahill, "Culturing Giant Scallops in Newfoundland Waters," Canadian Manuscript Report of Fisheries and Aquatic Science 1876 (1986): 1–23; and M.J. Dadswell and M.S. Sinclair, "Aquaculture of Giant Scallop *Placopecten magellanicus* in the Canadian Maritimes: Its Use as an Experimental Tool to Investigate Recruitment and Growth," *International Council for the Exploration of the Sea ENEM/63* (1989): 1–17.

91 M. Wolff, "Population Dynamics of the Peruvian Scallop *Argopecten purpuratus* during the El Nino Phenomena 1983," *Canadian Journal of Fisheries and Aquatic Science* 44 (1987): 1684–91.

92 J.F. Caddy and J.A. Gulland, "Historical Patterns of Fish Stocks," *Marine Policy* 7:4 (1983): 267–78.

93 G.J. Parsons, S.M.C. Robinson, R.A. Chandler, L.A. Davidson, M. Lanteigne, and M.J. Dadswell, "Intra-annual and Long-term Patterns in the

Reproductive Cycle of Giant Scallops *Placopecten magellanicus* (*Bivalvia: Pectinidae*) from Passamaquoddy Bay, New Brunswick, Canada," *Marine Ecology Progress Series* 80 (1992): 203–14.

94 S. Aoyama, "The Mutsu Bay Scallop Fisheries: Scallop Culture, Stock Enhancement, and Resource Management," in *Marine Invertebrates Fisheries: Their Assessment and Management*, ed. J.F. Caddy (Toronto: John Wiley and Sons, 1989), 525–39.

95 G.J. Parsons, S.M.C. Robinson, and J.D. Martin, "Enhancement of a Giant Scallop Bed by Spat Naturally Released from a Scallop Aquaculture Site," *Bulletin of the Aquaculture Association of Canada* 94:2 (1994): 21–3.

96 R. Repetto, "A Natural Experiment in Fisheries Management," *Marine Policy* 25 (2001): 251–64.

97 H.B. Hachey and H.J. McLellan, "Trends and Cycles in Surface Temperature of the Canadian Atlantic," *Journal of the Fisheries Research Board of Canada* 7 (1948): 355–62.

98 See, for example, R. McGarvey, F.M. Serchuk, and I.A. McLaren, "Statistics of Reproduction and Early Life History Survival of the Georges Bank Sea Scallop (*Placopecten magellanicus*) Population," *Journal of Northwest Atlantic Fisheries Science* 13 (1992): 1–17. See also M.J. Tremblay and M.M. Sinclair, "Planktonic Sea Scallop Larvae (*Placopecten magellanicus*) in the Georges Bank Region: Broadscale Distribution in Relation to Physical Oceanography," *Canadian Journal of Fisheries and Aquatic Science* 49 (1992): 1597–615; and M.J. Tremblay, W. Loder, F.E. Werner, C.E. Naimie, F.H. Page, and M.M. Sinclair, "Drift of Sea Scallop Larvae *Placopecten magellanicus* on Georges Bank: A Model Study of Roles of Mean Advection, Larval Behaviour and Larval Origin," *Deep-Sea Research, Part II, Topical Studies in Oceanography* 41 (1994): 7–49.

99 Caddy and Gulland, "Historical Patterns of Fish Stocks."

100 J.F. Caddy, "Long-term Trends and Evidence for Production Cycles in the Bay of Fundy Scallop Fishery," *Rapports et Procès-verbaux des Réunions du Conseil International pour l'Exploration de la Mer* 175 (1979): 97–108.

101 F.T. Baird, "Migration of the Deep Sea Scallop (*Pecten magellanicus*)," *Maine Department of Sea and Shore Fish Research Bulletin* 14 (1953): 1–7.

102 Posgay, "Sea Scallop Investigations."

103 L. Brander and D.L. Burke, "Rights-Based vs. Competitive Fishing of Sea Scallops *Placopecten magellanicus* in Nova Scotia," *Aquatic Living Resources* 8 (1995): 279–88.

# 9 Fifty Years of Atlantic Salmon Field Studies at St Andrews Biological Station

RICHARD H. PETERSON

This chapter reviews the history of research done from the St Andrews Biological Station on the population biology of Atlantic salmon from its beginnings in the early 1920s, primarily due to the interests of individual researchers rather than from an overwhelming need for intervention or worry about the fishery. It follows the expansion of the research program in the 1940s to early 1960s, when the fishery began to become heavily overused; then its contraction and eventual cessation in the early 1970s, when the commercial salmon fishery had to be halted owing to the almost complete disappearance of the salmon due to overfishing. It ends when the Department of Fisheries and Oceans transferred research support elsewhere, and devolved responsibility for salmon research in the most productive streams to other research centres. This review is based primarily on the station's annual reports for the period 1913–74,[1] in addition to several publications by the scientists themselves, which were sometimes included in the annual reports.

In the 1920s, when the biological station was open only during the summer months, some of the visiting scientists' research concerned responses to requests for help from salmon hatcheries. Dr Archibald Gowanlock Huntsman and Dr R.H. McGonigle, a pathologist, were called to assist when pathological epidemics occurred.[2] Dr Edward Ernest Prince designed and supervised construction of the first fishway on the Magaguadavic River at this time.[3] Mr Harley C. White, later a full-time researcher at St Andrews, studied early salmon development at St Andrews during the summer, while spending the winters at Cornell University.

Research by Drs R.B. Kerr and A.A. Blair, who were based at the University of Toronto in the early 1930s, formed the groundwork for subsequent studies on the population biology of salmon in Maritime rivers. They collected basic data such as length, weight, sex, and scale samples, from several hundred salmon caught in the commercial and angling fisheries from the Margaree, Saint John, and Miramichi Rivers. Their interpretation of scale circuli (rings) permitted them to determine freshwater and stream ages, ages at sea, and whether the fish had previously spawned. In the Margaree and Saint John systems most smolts – fish at the juvenile stage at which salmon are covered in silvery scales and first migrate from the river to the sea – were two years of age, with perhaps a third of Saint John smolts being three years old. Most Miramichi smolts were three years of age, with some two- and four-year-olds. As late as the 1940s, Dr Blair was called upon to instruct field personnel on how to read scales. Their studies were the first to indicate the existence of early and late runs of salmon in the Miramichi River.

In 1931, H.C. White investigated the migratory behaviour of salmon in the Apple River, a river system in Cumberland County, Nova Scotia, that empties into Chignecto Bay. The east branch had apparently lost its run of salmon, though salmon still ran up the western branches. In 1933, he stocked fry from the Restigouche River into the east branch, and subsequently marked trapped smolts. In 1935, about 5 per cent of the marked fish returned as late-run grilse – that is, Atlantic salmon first returning after one sea year and entering brackish waters leading to upriver migrations – although they were a stock that returned as early-run grilse in the Restigouche River. He suggested that the returns data indicated that after the smolt migration, the fish remained within a few miles of the river mouth – a prevailing opinion of salmon researchers of the period. There is no further mention of the Apple River, so the subsequent fate of these salmon is unknown. Huntsman in 1931 showed that inner Bay of Fundy rivers – termed the "Chignecto" fishery – in general had their salmon runs practically eliminated by 1875 through unregulated exploitation, with no evidence of significant recovery for the next fifty years.[4] As an aside, it is interesting to observe that White also noted the inverse correlation between salmon parr swim-bladder inflation and the velocity of their stream habitats,[5] a phenomenon later more extensively studied by R.L. Saunders and his co-workers – see Wildish and Robinson's chapter in this volume.

In the early 1930s Huntsman published a series of reports on the Margaree River and its estuary, dealing primarily with their physical characteristics. These undoubtedly proved useful to subsequent work on bird predation that Dr Paul F. Elson carried out on the Margaree from the 1950s to 1974. Dr Huntsman was assisted in some of the Margaree field studies by William S. Hoar, who later distinguished himself as an endocrinologist and co-editor of the *Physiology of Fishes* series.[6] Huntsman also documented an unusual run of salmon entering the Saint John River in November – obviously very fat fish that were not in spawning condition. These fish, the so-called Serpentine run, would spend an entire year in fresh water prior to spawning.

Although Huntsman became a renowned (or infamous) salmon expert, he had his failings.[7] In 1983 I was in the process of doing lake surveys in connection with acid rain research, one lake being Chisholm Lake – a private lake owned by a consortium of Saint John doctors and lawyers. The gentleman who arranged access to the lake for the author was a Mr Harris Hoar, brother of Dr Hoar. During a conversation he recalled going angling on the Margaree with Dr Huntsman in the 1930s. He related that Huntsman became very excited at a large run of "salmon" passing by their fishing spot. Mr. Hoar said he did not have the heart to tell Dr Huntsman that the fish were, in fact, gaspereaux (*Alosa pseudoharengus*).

The commercial salmon harvests in the Maritimes were accurately documented from the early 1870s in federal fisheries statistics. In 1931, Huntsman published an analysis of these harvests from the 1870s to the late 1920s. His primary objective was to "discover the factors that determine fluctuations in the fisheries, or that limit the abundance of the fish" – a tall order.[8]

As late as the early 1970s, Dr Huntsman was circulating memos lamenting the lack of a "science of where organisms are and when." He was preoccupied for over a forty-year period with the questions he first posed in his 1931 bulletin "The Maritime Salmon of Canada." He asked, among other things: "What are the characters and habits of the fish outside the estuary? What are the factors that cause them to enter the estuary? What are the characters and habits of the fish in the estuary throughout the season? What are the factors that cause them to enter the river proper? [and] What are the characters of the fish in the river proper throughout the season?" He stated: "Our knowledge of the salmon in the sea is almost a complete blank."[9] However, he was convinced that they remained within the influence of the home estuary, in spite of the fact that over 9000 kelts – salmon that had already spawned – were tagged before release from 1913 to 1929, and some of these tagged fish were taken by fisheries in Labrador and Newfoundland.[10]

Dr Huntsman's analysis of yearly fluctuations in commercial catches was of necessity descriptive, considering the absence of mathematical methods of waveform and trend analyses in early population models. Commercial catches were divided into four areas: western gulf (including the Miramichi and Restigouche Rivers), eastern gulf (including the Margaree River), the Atlantic (including such rivers as the LaHave), and the Fundy area (mainly the Saint John River).

Although the landings showed no obvious downward trend from 1870 to 1930, unregulated exploitation for the previous three decades markedly reduced adult returns. Also, catch per effort would probably show a different picture, since the number of drift nets increased from 49 to 104 between 1922 and 1930, and the set nets from 187 to 215. The commercial landings from the Saint John River in 1881 had declined by about 90 per cent from 700,000 pounds in 1875 to 80,000 pounds.[11] During this period drift nets extended all along the coast from Saint John Harbour to Point Lepreau. The Miramichi system provided about 20 per cent of the total annual Maritime catch from 1910 to 1930 – around 14,000,000 out of 65,000,000 pounds.[12]

The landings from 1870 to 1930 were characterized by two large spikes at each end, with lesser fluctuations in between. Huntsman postulated three cycles of periodicity for this period, with the most apparent being a 9.6-year period that he compared with similar cycles in mammals and birds. This periodicity seemed to persist when the landings were broken down into the four areas he had designated. He also postulated a three-year cycle in the Miramichi landings that he thought reflected the preponderance of three-year smolts in this system. It should be mentioned at this point that these landings were primarily for multi-sea-year fish, as grilse were not harvested to a great extent in the commercial fishery. Since, in all probability, grilse were mating with multi-sea-year fish, this would no doubt smear any three-year cycle, if such existed, grilse being progeny of a different year-class. The most adventurous proposed cycle was a 48-year cycle based upon the spikes in landings at the extremes of the data set. Dr Huntsman considered this to be related to the 9.6-year cycle, as five times 9.6 is 48.[13] At any rate, the validity of a 48-year cycle would never be tested because the commercial fishery was banned in the mid-1970s. By then there were also other factors at play in diminishing the returns of salmon, such as the Greenland fishery.

Although Dr Huntsman was no longer the director at the Atlantic Biological Station after 1935, his final assessment of the salmon fishery in the 1931 paper may have influenced the expanded research programs of the 1940s and 1950s. As he argued, "Full solution of the problem or problems of the Maritime salmon requires far more information than is available at the present time, and much of it can be obtained only by extensive investigation."[14]

**Pollet River Studies**

We now jump ahead to the mid-1940s when Dr Alfred W. Needler was the director, and Dr Elson, assisted by White, was hired to investigate factors affecting smolt production.[15] The selection of a section of the Pollett River was an excellent choice to study the relationship between yearling stocking densities and smolt production, as this section was bounded by an impassable dam downstream and an impassable falls upstream. After eight years of research, Elson concluded that five to six smolts could be produced per one hundred square yards of stream, requiring ten to twelve large parr for this, and that around two hundred eggs per one hundred square yards were needed to produce these smolts (a survival of 2 to 3 per cent from egg to smolt). These values were later corroborated by a similar study on the Dungarvon River.

Much effort during these years was directed to predator control, so that maximum smolt production could be realized. The main predators were deemed to be mergansers, which Elson concluded should be limited to less than one bird per fifty acres of stream. This aspect commanded much of Elson's attention for the next 25–30 years. Trained hunters were required for effective control; young ducklings were the most difficult to destroy, as they would leave the water and had to be chased through streamside vegetation. Kingfishers were controlled initially; researchers found they were killed most effectively by erecting poles in the stream which the kingfishers used as perches during hunting. A leg-hold trap, set on the top of the poles, proved effective in reducing kingfisher numbers – at a cost to other species perching on the poles. In one year a cowbird, two flickers, three spotted sandpipers, three sharp-shinned hawks, and five barred owls were trapped.

White exerted much effort to determine the food habits of kingfishers, which regurgitate pellets of undigestible fragments of prey. He collected bone samples from kingfisher pellets of several watersheds, and acquired the ability to identify fish species eaten from his bone reference collection. Since only about 15 per cent of the kingfisher diet was salmon – as compared to 65 per cent minnows and suckers – and these were smaller salmon than those eaten by mergansers, Elson concluded that kingfishers were not important enough salmon predators to warrant continued trapping.[16]

Merganser removal resulted in a 250 per cent increase in fish populations, including a great increase in minnows and suckers. Elson did not determine the influence of this increase in fish densities on juvenile salmon growth rates, but decided that the Margaree River was better for future studies of the effects of merganser control as this watershed contained no cyprinids or suckers.

Eel predation of salmon fry presented a more intractable problem. In one study, examination of the stomach contents of six eels one day after stocking salmon fry yielded ninety-one salmon fry.[17] Eel removal would have been very disruptive of salmon occupying the same habitat, however; also, electrofishing techniques were still fairly primitive in the 1940s.

The predicted beneficial effect of merganser control was tested in 1949–50. Before predator control, a 3 per cent yield of smolts was the rule, resulting in one to four thousand smolts. In 1950, over 17,000 smolts were produced from 235,000 yearlings stocked (or around a 7 to 8 per cent yield).[18]

The 1949–50 merganser control program essentially ended the Pollett River predator control investigations, which would continue on the Margaree River for the next twenty years. Dr Elson spent about twenty-five years in trying to improve salmon returns by controlling merganser populations on the Miramichi and Margaree systems,[19] but was confounded by external factors, including insecticide side-effects on fish populations, overfishing elsewhere, and mining pollution. The influence of predator control on adult salmon returns remained inconclusive. The returns were assessed primarily from angling catches that were greatly influenced by water levels and temperatures. I recall Dr Elson remarking ruefully that when he expected to demonstrate the beneficial effects of predator control on adult returns, the Greenland fishery reared its ugly head.

Whether such programs would be possible now, with increased environmental concern for ecosystems, is questionable. The primary goal of freshwater studies at St Andrews in those days was to increase yields, and predator control was an important aspect. In Dr Smith's studies on lake fertilization, a caretaker was hired to kill all fish-eating birds and mammals around Creasey Lake. It should be noted that the Wildlife Service assisted the control programs on the Pollett River, and possibly elsewhere. In retrospect, the program took a simplistic approach to salmon stream ecosystems, in effect, trying to transform salmon rivers into hatchery systems. Many prey-predator studies have since assigned a function to the predator of culling "weaker" members of the prey population. Resultant lower smolt survival at sea may have reduced some of the effects of merganser control.

Elson was later recruited to compare this study with a similar one on the Foyle River in Ireland, resulting in a 1975 monograph he co-authored with Dr A. Tuomi, *The Foyle Fisheries: New Basis for Rational Management*.[20] One of the study's products was a mathematical stock-recruitment model that indicated an escapement of 12,000 large salmon was required to fully stock the Miramichi system.[21] The model indicates that too many spawners can result in lower recruitment, an aspect that received some criticism. Like many models, perhaps it is unwise to extend it much beyond the extent of the database. Considering the incredible numbers of spawners entering streams in the 1800s, reduced recruitment seems unlikely, but back then the stream environment was probably quite different. Dr Elson was justifiably proud of the Foyle monograph, telling J.W. (Wilf) Saunders, a scientist who investigated anadromous fish that "the Foyle Publication is required reading for all (N.B.) Natural Resources personnel." Wilf enjoyed puncturing balloons: his rejoinder was "Oh yeah? Well so are the yellow pages."

## Miramichi River Program

From 1950 onward, the salmon research program emphasis shifted to the Miramichi system, by far the most productive in the Maritimes. Counting fences were installed on two branches, the Northwest Miramichi and the Dungarvon Rivers, with a fishway installed on the latter tributary. From 1950 to 1963 a counting fence was located at Curventon, about seven miles above tidehead.[22] Adult and smolt traps were installed in the estuary. Dr C. James (Jim) Kerswill was hired as senior scientist in charge, with Elson continuing as senior scientist and White as associate scientist. The Miramichi program objectives were: to determine annual production of young salmon over a long period; to determine the whereabouts of salmon after leaving the river as smolts; to determine the proportions of Miramichi stock taken by various sport and commercial fisheries; to determine what improvement of production resulted from bird control; and to obtain information on times of various runs of fish into the river, the size and age composition of the runs, plus "other matters" relating to proper salmon fishery regulation. The Pollett River program was designed to isolate freshwater salmon production from all external perturbation, while the Miramichi program would jump with both feet into the much more complex world of human–fishery interactions.

The demonstration of beneficial effects of predatory bird control on the Miramichi system would prove impossible due to the negative impacts of several human perturbations: the spraying of watershed forests with DDT, beginning in the mid-1950s; the spillage, in the early 1960s, of toxic mine effluent into the Tomogonops River – a tributary of the Northwest Miramichi upstream of the Curventon fence;[23] the decimation of two-sea-year salmon by the estuarine fisheries, resulting in increasing proportions of grilse in the escapement; the increasing exploitation of Maritime salmon stocks by the Greenland fishery from the early to the mid-1960s; and the interference with salmon migratory behaviour by industrial pollution flowing through the estuary.

It was estimated that 1,650,000 smolts were produced in the Miramichi system in 1951. Elson and his team arrived at this estimate by marking smolts on the Dungarvon and Northwest Miramichi tributaries, and recapturing smolts which were sampled in the estuary, using the formula of multiplying total smolts caught in estuary by the total marked smolts at tributary traps, and dividing the product by the total marked smolts caught in the estuary. This assumes no differential loss of marked smolts between the tributary traps and the estuarine trap, and complete mixing of smolts above the estuarine trap.[24]

By the mid-1950s, however, the commercial landings were, at most, only half that of the average for the previous twenty-five years. The commercial catch in 1956 was 988,500 pounds – 12 per cent higher than 1955 – and the angling catch was 47,312 fish.[25] However, the numbers of grilse and large salmon that passed through the Curventon fence were down in 1956. Forest spraying with DDT had begun in 1954, and it was considered probable that resultant mortality of juvenile salmon was beginning to be reflected in lower adult returns by 1956.

Research that continued on the Pollett River in the mid- to late 1950s yielded basic information on the survival rates from egg to smolt, as adults were allowed into the experimental segment. St Andrews scientists determined that two to three hundred eggs per one hundred square yards were required to produce five to six smolts. The average survival with bird control from egg to under-yearling was 5 per cent, from under-yearling to parr 60 per cent, and 50 per cent from parr to smolt.[26]

Growing concern for the future of salmon fisheries had led in 1949 to the formation of a coordinating committee, comprising representatives of the federal and provincial government programs on Atlantic salmon.[27] The lack of knowledge on salmon movements at sea, and the consequent contributions of different stocks to the various fisheries, led to expanded smolt marking programs in the Maritimes, Newfoundland, and Quebec, and monitored recovery of marked fish in the various fisheries. St Andrews Biological Station was tasked with collecting and analysing statistics on commercial and recreational landings; determining the size of spawning runs, and subsequent success from egg to smolt; and determining the contributions of "typical" rivers to each fishery. The station's scientists were also tasked with discovering effective ways of increasing smolt production through hatchery programs; assessing the value of controlling predatory birds; and studying the effects of sea lampreys. This was obviously an overly ambitious program for the manpower available for salmon research in the mid-1950s. Over the next ten years, several more scientists were hired to work on different aspects of salmon biology, namely, Drs Miles H.A. Keenleyside and John B. Sprague in the late 1950s, Dr Richard L. Saunders in 1960, and Drs Philip E.K. Symons and Aivars B. Stasko in the mid-1960s.

The Dungarvon counting fence was transferred to "Camp Adams" on the Northwest Miramichi, some thirty-three miles upstream of the Curventon fence, at the end of 1956. This was of significance in distinguishing spawning areas of late-run and early-run salmon stocks. The late-run fish did not make it up to the Camp Adams fence. I was a part of the counting fence program as a young researcher. One special memory is of an evening of mid-September 1959. I was driving down from Camp Adams to Curventon to spend the last two days of summer employment on the counting fence there. The other students had left, and I would have the fence to myself those two days. Possibly there would be some large late-run fish going through the fence. There was frost in the air and a heavy mist was rising off the river which coated the windshield. The beauty of the river that evening has stayed with me for the past fifty years. It had been a good summer. Dr Kerswill had picked me up at my house on his way to the Miramichi. There I spent the first part of the summer on the budworm spray monitoring, working with Dr F.P. Ide, an authority on mayfly taxonomy. I spent the rest of the year at Camp Adams under the supervision of the very capable E.J. Schofield.

By 1957, the practice of spraying DDT at half a pound per acre was beginning to have disastrous consequences on Miramichi, Restigouche, and Saint John salmon stocks. At this rate of application, direct toxic effects on young salmon and indirect effects on growth through mortality of stream insects – stream insect work was primarily done by Dr Frederick P. Ide under contract from University of Toronto – resulted in reduced smolt production.[28] It also resulted in a decline of mergansers, due to lack of insect food for the early young. Possibly eggshell thinning also affected merganser reproduction, as would be demonstrated for other birds of prey. Merganser control on the Miramichi was stopped in 1961 because DDT effects obscured the influence of bird control. The study of bird control was shifted solely to the Margaree watershed in 1962. By the late 1950s, the rate of DDT application was reduced to one quarter of a pound per acre, and over the next ten years, was replaced by less toxic pesticides, such as fenitrothion, the effects of which were monitored by St Andrews scientists.[29] In the early 1960s, the program to monitor effects of forest spraying followed the spray program around the province: the Keswick watershed in 1959, the Boisetown area in 1960, and the Molus River in Kent County in 1961. Salmon fry and yearlings were held in cages in sprayed streams, and drift and bottom samples of stream insects collected.[30] Truckloads of stream substrate from unsprayed tributaries were distributed in the Northwest Miramichi to attempt to re-establish insects lost to the spray. The late 1950s and early 1960s probably marked the peak effort on the Miramichi program, with five scientists, seven full-time technicians, and a dozen part-time employees and students involved.

In 1960, spillage from settling ponds of the Heath-Steele Mine to the Tomogonops River, a tributary of the Northwest Miramichi, resulted in avoidance by migrating adult salmon, many of which were processed through the downstream traps at Curventon a few days after they went through the upstream trap. Field observations, combined with laboratory observations, on avoidance of copper and zinc by juvenile salmon by Sprague confirmed the avoidance of mine effluent by salmon.[31] It also marked the beginning of toxicology studies at St Andrews Biological Station, the history of which is the focus of a chapter by Peter Wells in this volume. The mine spillage had several effects on salmon in the Northwest Miramichi River. The late-run fish spawned in some unpolluted tributaries such as the Portage River. Many fish that avoided the polluted Northwest Miramichi were diverted into unpolluted areas such as the Sevogle River. A decade later, heavy metal pollution was still a problem. In 1969, adult salmon mortality occurred due to torrential rainfall on the mine property.

Things began heating up politically on salmon exploitation in the late 1940s. The Atlantic Salmon Association was founded in 1948, and began to push for more restrictions on the commercial fishery. A Federal-Provincial Coordinating Committee on Atlantic salmon was organized in 1949 to shape research and management. Projects were reviewed and redirected in 1950, with responsibilities divided between the Fisheries Research Board and the Conservation and Development Service. The Research Board, which ran the Pacific and Atlantic Biological Stations at that time, was given responsibility to review commercial and angling statistics, conduct "natural history" studies, and determine important physical and chemical factors responsible for survival and growth of all life-cycle stages. The board was charged with determining the effects of various pollutants, studying the reactions of migrating fish to barriers, and determining the relation of spawners to fry, parr, and smolt production. The Fisheries Research Board also began to do population estimates on all freshwater stages, assessed effects of water level fluctuations, and continued the studies on predator control. Small wonder that the manpower for salmon investigations was increased in the late 1950s; in fact, the employment of additional scientists and full-time technicians was recommended in the early 1960s. A fair number of these tasks already had received considerable attention by Elson's studies of the Pollett and Dungarvon Rivers, and the Miramichi River program of the previous half-dozen years.

Those concerned with commercial exploitation of salmon in the sea certainly had reason for concern. The New Brunswick commercial salmon fishery (the bulk of it Miramichi fish) rose from 500,000 to 600,000 pounds in the early 1960s to one million pounds by 1965.[32] Over the same period the Greenland salmon catch, one quarter of which was made up of salmon originating in Atlantic Canada, increased from about 200,000 pounds to between one and a half and three million pounds. By 1966, Dr Kenneth R. Allen had replaced M.W. Smith as head of the anadromous fish program;[33] he and Richard Saunders were primarily concerned with the Greenland fishery at this time. The Miramichi angling catch was certainly of concern at this time as well, as it totalled only around 300,000 pounds, and consisted mostly of grilse.[34] Contrast these figures with the forty million pounds taken in the western gulf annually around 1930, of which a large percentage, approximately 30 to 40 per cent, were of Miramichi origin.

By the mid-1960s, the impact of the commercial fishery on salmon production was well understood. The preponderance of grilse at the

Curventon fence, and in the angling catch, indicated a probable insufficiency of egg deposition in the Miramichi system, particularly since about 66 per cent of Miramichi grilse are males.[35] In 1964–5, fewer than 150 large (multi-sea-year) salmon passed the Curventon fence as opposed to an average of 3000 grilse. It was feared that the "late run" in the Northwest Miramichi, consisting mostly of large salmon, might be eradicated. The percentage of grilse angled in the Miramichi system in 1965 was ninety-two. The abolition of the Maritime commercial salmon fishery – mostly multi-sea-year fish – in the early 1970s was a foregone conclusion.

The runs continued to decline in the early 1970s. In 1971, the commercial catch reached an all-time low of 340,000 pounds and the angling catch was only six thousand fish – mostly grilse.[36] With commercial fishing banned in 1972, the angling catch increased to 21,000 fish, still a small amount compared with six years previously. Anglers were estimated to take 30 per cent of all salmon that passed through the Curventon fence. The total Curventon count was 450 multi-sea-year salmon and 2300 grilse. On a bright note, the late run amounted to two hundred, the best in ten years. Juvenile populations that year were the lowest in twenty-two years – since the early DDT spraying program. Elson estimated the run provided only enough for 50 to 75 per cent of required spawning. The low densities of juveniles resulted in faster growth rates, with 80 per cent being two-year-old smolts, as compared to 20 per cent at higher densities.

The final year of a management-oriented program at Curventon was not optimistic, with only 116 large salmon passing through the fence. Anglers caught 116, so the escapement of large salmon was probably negligible. The late run totalled 54 large salmon and 22 grilse. The counted smolt run was about five thousand – probably about half the total run – due to the poor spawning run in 1970. Elson estimated that the large salmon run would probably not improve until 1978–9.[37]

## Final Years

The late 1960s and early 1970s were periods of instability in the salmon research program at the St Andrews Biological Station. Four different program heads followed after Dr Kerswill transferred elsewhere in 1963, with Dr M.W. Smith being head of anadromous investigations in 1964, Dr Kenneth R. Allen in 1965, and Dr Elson from 1966 until the program ceased to exist in the mid-1970s. The dissolution of the Fisheries Research Board in the early 1970s, in retrospect, was probably not a positive factor for salmon fisheries research. Also, about that time, the Federal Department of Fisheries was merged with the Department of Environment, which also may have diluted the importance of this type of program on the Miramichi River: it was thought by some managers to have regressed to being a monitoring program. In fact, there is a thin line between monitoring and research; the advent of any stream perturbation converts a program from monitoring to research. Consequently, funding and manpower were transferred from the Miramichi program to other, more environmentally oriented projects. In fairness, the division of the Maritimes into Gulf and Scotia-Fundy regions within the next few years would have removed the Miramichi drainage from the St Andrews sphere of concern.

One of the last projects of importance for the Miramichi program was the study of estuarine pollution from various estuarine industries, conducted by Dr Elson and Dr Aivars B. Stasko, a specialist in the sonic tracking of fish. The tracking study, of a somewhat preliminary nature, indicated that industrial effluents were interfering with salmon passage through the estuary. As a result, salmon resided in the estuary longer, increasing the time of exposure to commercial nets.[38] The study was perceived to be too expensive to continue past the early 1970s.

The period from 1930 to 1970 was marked by a general decrease of adult escapement. If Dr Elson's figures were correct, then the survival from egg to smolt is in the order of 2 per cent on average. The nearly 100-year policy of catching only multi-sea-year salmon was a major factor in the long-term decline of the salmon runs. Salmon as a species are stated to produce around 800 eggs per pound of fish, so a typical female grilse (only constituting 30 per cent of the total grilse run) would spawn about 2400 eggs, which would result in about fifty smolts or so. Whether a salmon run of over 90 per cent grilse is sustainable depends upon survival from smolt to adult return. These estimates varied from 1.5 per cent (estimated on per cent return of marked grilse) to about 5 per cent. This amounts to two to four adults per female grilse spawned. About 30 per cent of the grilse run was angled in the late 1960s and 1970s, so the escapement would amount to 1.4 to three adults per spawning female. At the lower end, this escapement production would not be sustainable, since the male contribution has to be taken into account. The abolition of the estuarine fishery in the early 1970s should have resulted in an increase in escapement of a few tens of thousands of adults. Yet the Atlantic Salmon Federation had a news item in 2009 urging anglers to release any grilse caught (large salmon cannot be retained by law), due to anticipated inadequate egg deposition that fall.[39] However, the last thirty years is not part of the St Andrews story, and is beyond the scope of this chapter.

The Miramichi has been fortunate in some ways; it has not been modified by power dams as has the Saint John. The influences of mining, pesticide sprays, and estuarine pollution were documented by programs out of St Andrews, but possible effects of wide-spread logging on thermal regimes, siltation, and ground-water discharge may have had more subtle effects, examined in a cursory fashion by Dr Elson.[40] If my memory is accurate, the thermal maxima at Camp Adams never got much above 18 degrees Celsius in the late 1950s, and brook trout could be found in the main Northwest Miramichi there throughout the summer. Temperatures that are too warm hurt salmon survival. Possibly due to loss of tree cover,[41] temperatures in the range of 26 degrees Celsius were recorded in 2010 in some parts of the river system. In retrospect, the commercial fishery probably had more long-term negative effects on salmon runs than DDT or heavy metal spillage.

In spite of circumstances beyond the control of scientists at the St Andrews Biological Station that resulted in the loss of a great commercial fishery for salmon, and the near loss of the salmon sport fishery, St Andrews scientists made important contributions to our understanding of the life history, population dynamics, and migratory behaviour of Atlantic salmon as well as external physical, chemical, and predation factors affecting Atlantic salmon populations on the Miramichi and elsewhere. It is possible that the merganser control practised on the Miramichi benefited the grilse runs, for example. The focus on this important fish in addition led to studies on toxic mine effluents, and an early awareness of the unanticipated side-effects of DDT on the environment, and the toxic side-effects of organophosphate insecticides, which replaced the DDT used in forest pest-control programs. The St Andrews Biological Station was also very much involved, as Dr Bob Cook documents elsewhere in this volume, with the establishment and growth of salmon aquaculture in the Bay of Fundy (where millions of salmon now swim in endless circles in net pens, rather than making incredible oceanic migrations). So the St Andrews research reflects, in retrospect, a microcosm of the evolution of human endeavour from hunter-gatherer to farmer.

NOTES

I would like to thank R.L. Saunders, W. Taylor, and D.J. Wildish for providing useful comments and editing.

1 Dr Elson provided a more detailed review than is given here of the northwest Miramichi program. See P.F. Elson, "Impact of Recent Economic Growth and Industrial Development on the Ecology of Northwest Miramichi Atlantic salmon (*Salmo salar*)," *Journal of the Fisheries Research Board of Canada* 31:5 (1974): 521–4. Much of my information was obtained from St Andrews Biological Station annual reports, which were extremely detailed until the early 1970s and provided more continuity than would be possible by resorting to primary publications.
2 See, e.g., A.G. Huntsman, "North American Atlantic Salmon," *Rapports et Procès-verbaux, Conseil Permanent International pour l'Exploration de la Mer* 101 (1938): 11–15; A.G. Huntsman, "Cyclical Abundance and Birds versus Salmon," *Journal of the Fisheries Research Board of Canada* 24 (1941): 21–32; R.H. McGonigle, "Some Effects of Temperature and Their Relation to Fish Culture," *Transactions of the American Fisheries Society* 62 (1932): 119–25; R.H. McGonigle, "Acute Catarrhal Enteritis of Salmonid Fingerlings," *Transactions of the American Fisheries Society* 70 (1940): 297–303.
3 Pronounced by locals as "Mackadavic" River.
4 A.G. Huntsman, "The Maritime Salmon of Canada," *Biological Board of Canada, Bulletin 21* (1931): 15–17.
5 H.C. White, "Some Facts and Theories concerning the Atlantic Salmon," *Transactions of the American Fisheries Society* 64 (1934): 360–2.
6 W.S. Hoar and D.J. Randall, eds., *Fish Physiology*, vol. 1, *Excretion, Ionic Regulation, and Metabolism* (New York: Academic Press, 1969); vol. 2, *The Endocrine System* (New York: Academic Press, 1969); vol. 3, *Reproduction and Growth, Bioluminescence, Pigments and Poisons* (New York: Academic Press, 1969); vol. 4, *The Nervous System, Circulation and Respiration* (New York: Academic Press, 1970; vol. 5, *Sensory Systems and Electric Organs* (New York: Academic Press, 1971); vol. 6, *Environmental Relations and Behavior* (New York: Academic Press, 1971); vol. 7, *Locomotion* (New York: Academic Press, 1969).
7 For a view of the mixed legacy of Huntsman's salmon work, please see J. Hubbard, "Home, Home, Sweet Home? A.G. Huntsman and the Homing Behaviour of Canadian Atlantic Salmon, *Acadiensis* 19:2 (1990): 40–71, available as a PDF at http://www.google.ca/#hl= en&source=hp&biw=

1568&bih=735&q=A.G.+Huntsman+Atlantic+salmon&aq=f&aqi=&aql=&oq=&fp=3e41beee0f6b4c75.

8 See Huntsman, "The Maritime Salmon of Canada."

9 Ibid, 5–7.

10 Ibid, 90–1.

11 Ibid., 30.

12 Ibid., 43.

13 Ibid, 9–14, 96–8.

14 Ibid, 96.

15 P.F. Elson, "Increasing Salmon Stocks by Control of Mergansers and Kingfishers," Fisheries Research Board of Canada, Progress Reports of the Atlantic Coast Stations 51 (1950): 12–15. Harley White first published on this issue in 1937. See H.C. White, "Local Feeding of Kingfishers and Mergansers," *Journal of the Biological Board of Canada* 3 (1937): 323–38.

16 H.C. White, "The Food of Kingfishers and Mergansers on the Margaree River, Nova Scotia," *Journal of the Biological Board of Canada* 2 (1936): 299–309; H.C. White, "The Eastern Belted Kingfisher in the Maritime Provinces," *Bulletin of the Fisheries Research Board of Canada* 97 (1953): 1–44.

17 P.F. Elson, "Eels as a Limiting Factor in Salmon Production," Fisheries Research Board of Canada Manuscript Report of the Biological Station 213 (1941).

18 Elson, "Control of Mergansers and Kingfishers," 2–15.

19 P.F. Elson, "The Fundamentals of Merganser Control for Increasing Production of Salmon Smolts from Streams," Fisheries Research Board of Canada Manuscript Report 694 (1960): 1–10.

20 P.F. Elson and A. Tuomi, *The Foyle Fisheries: New Basis for Rational Management* (Lurgan, Ireland: LM Press, 1975).

21 Elson, "Impact of Recent Economic Growth."

22 R.L. Saunders and J.B. Sprague, "Copper-Zinc Mining Pollution on a Spawning Migration of Atlantic Salmon," *Water Research* 1 (1967): 431.

23 Ibid., 419.

24 P.F. Elson, "Magnitude of Smolt Runs Measured by Sampling," Fisheries Research Board of Canada, Report of the Atlantic Biological Station for 1951 (1951): 95–7.

25 K.R. Allen and J.K. Lindsey, "Commercial Catches of Atlantic Salmon in the Maritimes Area, 1949–1965," *Fisheries Research Board of Canada Technical Report* 29 (1967): 1–160.

26 P.F. Elson, "Adult Salmon Required to Maintain Stocks," Fisheries Research Board of Canada, Biological Station, St Andrews, NB,

Investigators' Summaries no. 66 (1955–6): 173–6; P.F. Elson, "Number of Salmon Needed to Maintain Stocks," *Canadian Fish Culturist* 21 (1957): 19–23.

27 B. Wilks, *Browsing Science Research at the Federal Level in Canada: History, Research Activities, and Publications* (Toronto: University of Toronto Press, 2004), 276.

28 F.P. Ide, "Effects of Forest Spraying with DDT on Aquatic Insects of Salmon Streams in New Brunswick," *Journal of the Fisheries Research Board of Canada* 24 (1967): 769–805; P.F. Elson, "Effects on Wild Young Salmon of Spraying DDT over New Brunswick Forests," *Journal of the Fisheries Research Board of Canada* 24 (1967): 731–67; M.H.A. Keenleyside, "Effects of Forest Spraying with DDT in New Brunswick on Food of Young Atlantic Salmon," *Journal of the Fisheries Research Board of Canada* 24 (1967): 807–22.

29 M.H.A. Keenleyside, "Effects of Spruce Budworm Control on Salmon and Other Fishes in New Brunswick," *Canadian Fish Culturist* 24:2 (1959): 17–22.

30 P.F. Elson and C.J. Kerswill, "Impact on Salmon of Spraying Insecticide over Forests," *Advances in Water Pollution Research* 1 (1966): 521–44.

31 J.B. Sprague, "Lethal Concentrations of Copper and Zinc for Young Atlantic Salmon," *Journal of the Fisheries Research Board of Canada* 21 (1964): 17–26.

32 Allen and Lindsey, "Commercial Catches of Atlantic Salmon."

33 Saunders and Sprague, "Copper-Zinc Mining Pollution," 431.

34 R.L. Saunders, "Seasonal Pattern of Return of Atlantic Salmon in the Northwest Miramichi River, New Brunswick," *Journal of the Fisheries Research Board of Canada* 24:1 (1967): 21–32.

35 Elson, "Recent Economic Growth and Industrial Development."

36 C.J. Kerswill, "Relative Rates of Utilization by Commercial and Sport Fisheries of Atlantic Salmon (*Salmo salar*) from the Miramichi River, New Brunswick," *Journal of the Fisheries Research Board of Canada* 28:3 (1971): 351–63; S.J. Smith, "Atlantic Salmon Sport Catch and Effort Data, Maritimes Region, 1951–79," Canadian Data Report of Fisheries and Aquatic Science 258 (1981): 1–267.

37 Elson, "Recent Economic Growth and Industrial Development."

38 A.B. Stasko "Progress of Migrating Atlantic Salmon (*Salmo salar*) along an Estuary, Observed by Ultrasonic Tracking," *Journal of Fish Biology* 7:3 (1975): 329–38.

39 Atlantic Salmon Federation, "Serious Decline in Small Salmon Necessitates Live Release," press release, 15 September 2009.

40 Elson, "Recent Economic Growth and Industrial Development."

41 R.A. Cunjack, "Addressing Forestry Impacts in the Catamaran Brook Basin: An Overview of the Pre-logging Phase," in *Water, Science and the Public: The Miramichi Ecosystem*, Special publication of the *Canadian Journal of Fisheries and Aquatic Science* 123 (1995): 191–210.

# 10 Paralytic Shellfish Poisoning (PSP) Research – Seventy Years in Retrospect

JENNIFER L. MARTIN

Historical evidence indicates that paralytic shellfish poisoning (PSP), a syndrome derived from eating shellfish containing toxins produced by a phytoplankton species, has been present in the Bay of Fundy for over a century. Shellfish, such as soft-shell clams (*Mya arenaria*) and blue mussels (*Mytilus edulis*), feed through filter feeding and can accumulate, acquire, and store toxins in their tissues following consumption of phytoplankton known to produce paralytic shellfish poisoning toxins. Illness can result when people (or other vertebrates) consume shellfish with toxins in their tissues. A person who has eaten shellfish with unsafe levels of toxins can suffer from PSP and may exhibit the following symptoms: a tingling sensation, headache, dizziness, nausea, incoherent speech, stiffness, non-coordination of limbs, muscular paralysis, respiratory difficulty, choking, and, in severe cases, death. To this day, there are no antidotes.

Although early records are not detailed, historical records from natives and native lore indicate that PSP toxins have been associated with shellfish in the Bay of Fundy for many years. The first documentation was by Marc Lescarbot in 1609, who wrote the first history of New France. He wrote that the Indians at Port-Royal, Annapolis Basin, NS, did not eat mussels. He documented anecdotal evidence of illnesses associated with toxins and the consumption of shellfish.[1] In earlier years, they chose not to eat shellfish during months that did not have an "r" in their spelling.

As a result of the dangers posed by PSP if consumed by humans, there has been an ongoing research program studying PSP toxins at the St Andrews Biological Station for several decades.[2] However, in the last couple of decades, paralytic shellfish poisoning has been documented in other areas of the world including Alaska and tropical waters. Scientists in other institutions are now also engaged in such investigations, demonstrating the great importance of this research program pioneered by St Andrews Biological Station scientists.

Prominent past researchers from the St Andrews Biological Station include Dr Alfreda B. Needler, who was engaged in bivalve research from 1930 to 1951. She established the link between the organism *Gonyaulax tamarensis* (now called *Alexandrium fundyense* – both names are used interchangeably here) and the production of PSP toxins and accumulation in shellfish. Carl Medcof, a researcher at St Andrews from 1937 to 1973, investigated cases of PSP, and was involved in a number of research studies including initiating a monitoring program for toxins in shellfish; this determined uptake and depuration of toxins (i.e., cleansing through elimination of body fluids) in shellfish and the level of risk to the human consumer. He implemented a shellfish survey along the Atlantic coast, established zones for PSP toxicity and investigated a "human" bioassay – that is, a method to determine if a person has PSP. He was honoured by the international community for his contributions to studies of harmful marine algae in 1985 when the Third International Conference on Toxic Dinoflagellates was held at St Andrews in his honour. Annand Prakash, who arrived in 1960 and remained at St Andrews for around five years, investigated annual cycles of toxicity, isolating and culturing *Gonyaulax*, and linked growth to salinity. Alan White, a researcher at St Andrews from 1973 to 1986, linked herring mortalities to the uptake of PSP toxins through the food web and suggested that PSP levels in shellfish had a great degree of interannual variability. Lucie Maranda worked on growth studies and food web dynamics between 1974 and 1977.

I personally joined Alan White in the PSP research program in 1977 and have been active in this field ever since: I have collaborated with Murielle LeGresley in this research since 1987. We have focused on remote sensing, prediction and hindcasting, modelling, linkages between phytoplankton and shellfish toxicity, and toxin uptake and depuration. We have also investigated spatial and temporal trend in numbers of *Alexandrium* in relation to numbers of the whole phytoplankton community, the relationship between overwintering cysts and "vegetative" cells, determining sources of salmon mortalities and domoic acid–related studies.

Through the years other researchers have undertaken related resereach, including Neil Bourne, John Caddy, Glen Jamieson, Ross Chandler, Kats Haya, and Murielle LeGresley. More recently, Shawn

Robinson, Les Burridge, and Dawn Sephton have studied toxicity and biotransformation of toxins in bivalves.

**History of PSP Research**

As the organism responsible for producing PSP toxins is a member of the phytoplankton community, it is important to acknowledge the early work on the phytoplankton community in the Bay of Fundy, initiated in the early years at the Atlantic Biological Station. Clara Fritz was the first researcher to document phytoplankton in the area from samples collected in 1916 and 1917,[3] but unfortunately her work was restricted to the diatoms, and the organism that was later determined to be responsible for producing PSP toxins is a member of the dinoflagellates. Viola Davidson[4] (figure 10.1) did another study of the diatoms over seven years from 1924–31, and determined conditions that might be favourable for some species.[5]

The earliest recorded work on the dinoflagellates of the Bay of Fundy was initiated in 1931 when Norwegian scientist Haaken H. Gran was appointed by the International Passamaquoddy Fisheries Commission as an expert on phytoplankton production to study the possible effects of the projected Cooper Dam.[6] In 1932, Trygve Braarud, also from Norway, was appointed to the position of assistant expert of this commission. The commission collected phytoplankton samples in 1931, initially intending to begin sampling for phytoplankton, hydrographic data, and chemical parameters early in 1932. Unfortunately, a fire at the Atlantic Biological Station in March destroyed much of the equipment that was to be used for the work, and so sampling was delayed until April and the chemical work until May. In addition, cuts to funding, due to the deepening Great Depression, forced the commission's researchers to scale down their work, so that it was stopped in October 1932. This made the work less complete than originally intended. However, their analyses indicated that the organism later linked to the production of PSP toxins was a major component of the summer dinoflagellate community.

In 1937 Carl Medcof, a summer student in Prince Edward Island at the Ellerslie substation, did oyster research and worked with the oyster industry on the Canadian East Coast. He became interested in PSP when he heard that poisonings and a death had occurred the previous year in the Bay of Fundy. He became a full-time researcher with the Fisheries Research Board of Canada in 1938, and remained with the department for forty-four years, retiring in 1973. He moved to the biological station

10.1 Dr Viola Davidson

in St Andrews in 1940, although he continued to return with his family to Ellerslie for the summers until 1944.[7] During his career he investigated cases of PSP in humans and animals, and found that young children, women, and non-locals were more susceptible to the toxins. He also determined that Sundays were worse in Bay of Fundy for poisonings because many went for Sunday drives (presumably gathering shellfish along the way, or stopping at fish shacks and purchasing marine produce), while Fridays were worse along the Gulf of St Lawrence, when most of the population ate seafood for religious reasons. He initiated the current shellfish monitoring program and suggested the temporary closure of the mussel fishery. The shellfish monitoring program he began surveys the Atlantic coast for PSP toxins in shellfish, establishes shellfish zones for PSP toxicity, and looks at toxin uptake and depuration in relation to commercial processing. Medcof also worked with R.M. "Bush" Bond on the human bioassay.[8]

Carl Medcof is recognized as a pioneer for his work in the field of marine toxin research. He documented much of the history of PSP in the Canadian Maritimes and established the early records of toxicity. He is still remembered today for his meticulous notes and attention to detail. He was instrumental in developing in 1943 the ongoing shellfish toxicity surveillance program in Eastern Canada that serves as a model for PSP monitoring programs internationally. In 1943, in order to meet the war-time need, commercial shellfish processors asked permission to process mussels in cans. Because blue mussels rapidly accumulate toxins, and do so at higher levels than clams, it was suggested that the commercial fishing of mussels be prohibited in the Bay of Fundy while researchers studied the problem. Investigations in the winters of 1943–4 and 1944–5 showed that mussel toxicities, even in the worst areas, were low, which Medcof felt was justifiable evidence for treating them in a similar manner to the soft-shell clams.[9] However, the closure of wild mussel beds to harvesting remains in effect today – even though the years with the greatest toxins measured in mussels were in the mid-1940s.[10] In 1985, thirteen years after his retirement, at the Third International Conference on Toxic Dinoflagellates held at the Algonquin Hotel in St Andrews, Medcof gave the dinner speech,[11] which is still referenced by attendees today.[12] The case history information he collected over the years helped define the symptomatology of PSP and the approximate human dose–response to the toxins.

Carl Medcof also made invaluable contributions to understanding the unique oceanographic events known as "red tides." During the

1940s, Medcof worked with Alfreda B. Needler on red tides together with Charlotte Sullivan (a fellow bivalve researcher),[13] who also studied oyster larval development in eastern Canada. Sullivan did a 24-hour series for *Gonyaulax* abundance in 1946 that indicated that the species left the surface during the night and rose again in the morning. Needler (figure 10.2) studied phytoplankton and did a four-year field survey of phytoplankton populations during the years 1944–7. She worked as a volunteer at the biological station, as her husband, Alfred Needler, was employed with the Department of Fisheries as the station's director, and spouses were not permitted to be employed at the same institute. This regulation was not changed until the early 1970s. Her findings were published in 1949;[14] her career is discussed in chapter 2, on women marine scientists, written by her daughter, Mary Needler Arai.

In her 1949 paper Needler presented data and reasons why *Goniaulax tamarensis* – also called *Gonyaulax, Protogonyaux,* and *Gessnarium,* and finally renamed *Alexandrium fundyense* in 1985 by Enrique Balech[15] – was the principal cause of the toxicity in shellfish. Needler compared plankton tows with toxicities of shellfish from nearby sites and found a close relationship, whereas when she compared toxicities with other phytoplankton counts, there was no close relationship to toxicity in shellfish. Outside Passamaquoddy Bay, where *Alexandrium* occurred, the shellfish became toxic, and inside, where *Alexandrium* was absent or less abundant, shellfish were seldom toxic and never toxic enough to be harmful. Her research found three principal factors governing *Alexandrium* abundance: water temperature; the presence of zooplankton, especially the predator *Favella,* which appeared to act also as a competitor; and the fact that diatoms can out-compete *Alexandrium* when diatom abundance is high. In 1945, Medcof and Needler suggested looking at phytoplankton in order to link cell concentrations with the PSP toxicity, but they were unable to continue the program "due to practicality." Unfortunately, Alfreda Needler's career ended at a relatively young age when she died from cancer.

During the 1940s and 1950s, D.G. (Dick) Wilder supervised staff who prepared extractions from shellfish for toxicity testing, as a means of determining the species affected and monitoring the incidence of PSP. This was work pioneered at the Atlantic Biological Station. Other bivalve researchers have also contributed to PSP shellfish toxicity research through the years. Neil Bourne studied PSP toxins in sea scallops (*Placopecten magellanicus*) from the major scallop-producing areas on the Canadian Atlantic coast. He found that sea scallop toxicity was

10.2 Dr Alfreda Needler, who worked at St Andrews and at Ellerslie, discovered the link between toxicity in shellfish and *Alexandrium fundeyense*.

greater in the Bay of Fundy than Passamaquoddy Bay; the adductor muscles (the only part of scallops from these areas that is marketed commercially) and gills were poison free, but the digestive glands and remaining tissues contained toxins throughout the year. The majority of the toxins were concentrated in the digestive glands.[16] The study was also expanded to include Georges Bank sea scallops during 1961–3, and scallops from the Gulf of St Lawrence in 1963–4, and from Miscou Island and the Magdalen Islands in 1964. The results indicated that Georges Bank and southern Gulf of St Lawrence scallops contained only small amounts of, or no, PSP toxins.

When a number of mild poisonings occurred after people consumed whelks from the St Lawrence in 1966, John F. Caddy and Ross A. Chandler fed whelks some shellfish containing high PSP toxin concentrations. They determined that high levels of toxins were detected in the digestive glands and were the probable cause of poisonings. Their work determined that when the whelks were starved and then fed a diet without *Gonyaulax* there was a rapid elimination of the poisons.[17] Glen S. Jamieson and Chandler monitored sea scallop tissues between 1977 and 1981 and found that all tissues except the adductor muscle were highly toxic from Bay of Fundy scallops. There was negligible toxicity in scallop tissues from Georges Bank, the outer Scotian Shelf, and Northumberland Strait. They recommended that the scallop roe fishery be allowed from the northern part of Georges Bank and Northumberland Strait.[18] In the late 1960s, Annan Prakash (figure 10.3) continued work on the organism responsible for producing the toxins in poisoning and was able to

10.3 Dr Annan Prakash

grow *Gonyaulax* in the laboratory and determined factors important to its growth, enabling studies on the physiological mechanisms of shellfish toxicity, and toxin accumulation and elimination.[19]

Annan Prakash, Carl Medcof, and A.D. Tennant in 1971 published a compilation of the summary of pertinent knowledge to that time, *Paralytic Shellfish Poisoning in Eastern Canada*.[20] Tennant contributed to the section on measuring shellfish toxicity; Medcof described the features of poisoning, toxicity of shellfish, and control programs; and Prakash did the sections on the causative organism, mechanisms for poison accumulation and elimination, the toxin, and detoxification of shellfish. This reference book is still widely used today.

Alan White (figure 10.5) arrived at the biological station in 1973 and continued growth experiments and *Gonyaulax* studies. His laboratory

10.4 Carrie Lewis and Bill Garnet (from the *J.L. Hart*) sampling for cysts

experiments determined that agitation of *Alexandrium* cells caused death and disintegration of cells, suggesting that small-scale oceanographic turbulence could inhibit growth.[21] In 1976, when herring mortalities occurred off the coast of Grand Manan, Vlado Zitko suggested that White should work on the herring mortalities and possible linkages with PSP. This work resulted in White determining that PSP toxins were able to pass through the food chain and caused the mortalities.[22] Herring mortalities occurred again as a result of PSP toxins in 1979.[23] Lucie Maranda worked with White from 1974 to 1977 and was involved with the early work with toxins and the food chain.[24] In 1977 I (Jennifer Martin) began my career at the biological station and worked with White on his continuing research on toxin uptake through the food chain. Frank Louzon joined the group for a year on a postdoctoral fellowship (1977–8) and was involved in growing mass cultures of *Gonyaulax*, toxin extraction and purification, and determining the susceptibility of other species of fish to PSP toxins. Results indicated that PSP toxins are lethal to herring

(*Clupea harengus harengus*), pollock (*Pollachius virens*), winter flounder (*Pseudopleuronectes americanus*), Atlantic salmon (*Salmo salar*), and cod (*Gadus morhua*).[25]

As researchers in other regions of the world were beginning to unravel the life cycle of *Gonyaulax*, we learned that it had an overwintering dormancy stage. So, in 1980, Carrie Lewis (figure 10.4) joined our lab for a year to work on the overwintering *Gonyaulax* cysts to determine their spatial distribution.[26] During the winter of 1980–1, Lewis and I spent a considerable amount of time aboard the *J.L. Hart* collecting benthic sediment samples. We were actually the first women to participate in a cruise aboard that vessel, which was built in 1967 and commissioned as a research vessel. The delay in women participating in this kind of activity is discussed in Mary Arai's chapter in this volume. During the five years that we conducted sediment surveys, they were also conducted over the five summers (1980–4) to try and establish if there was a link between *Gonyaulax* cell densities in the summer surface waters and the cyst abundance in sediments – and to determine the hotspots for cell concentrations.[27] The major seed beds for *Alexandrium* were found to be in the central Bay of Fundy north and east of Grand Manan in the soft clay muds. The summer "cruise" of 1980 revealed the first recorded instance of actual red water discolouration as a result of more than a million *Gonyaulax* cells per litre in surface waters. Murielle LeGresley joined the group in 1982 and was involved with further field and laboratory experiments. As a number of studies in other regions of the world had suggested that ozone might be used to remove PSP toxins from shellfish,[28] we conducted a study to determine the effectiveness of ozone in depurating PSP toxins from shellfish from the Bay of Fundy.[29] Our results indicated that ozone was not effective in depurating Bay of Fundy soft-shell clams.

White also sought to determine whether there was a relationship between shellfish toxicity and various environmental factors such as wind speed and direction, salinity, and temperature. He found some correlations between environmental factors and shellfish toxicity and noted definite cyclical periods of higher and lower shellfish toxicities; he found that there might be a link to the 18.6-year lunar tidal cycle.[30]

In this period, the Third International Conference on toxic dinoflagellates was held in St Andrews at the Algonquin Hotel; Fisheries and Oceans Canada hosted the 1985 conference, chaired by Alan White and coordinated by Marja White.[31] The conference's objective was to bring together those interested in problems associated with toxic

10.5 Carl Medcof receiving an award from Alan White in 1985

dinoflagellates. Carl Medcof, one of the pioneers in the field of PSP research, was, as mentioned above, recognized at the conference for his contributions (figure 10.5). Enrique Balech, from Argentina, was also recognized for his work in the field. During the conference he formally renamed the Bay of Fundy *Gonyaulax* as *Alexandrium fundyense*. Earlier, he had spent time at the St Andrews Biological Station poring over cultures and samples from the wild.[32]

With the advent of chemical analysis and sophisticated instrumentation for measuring toxins, more than twenty-four toxins were discovered to be involved with PSP. We carried out anatomical distribution studies using high-performance liquid chromatography (HPLC) and the traditional mouse bioassay on soft-shell clams to determine the transformation of toxins between tissues and the retention of toxins.[33] During the late 1980s, deaths of humpback whales occurred off Cape Cod, USA, and Kats Haya led a collaborative effort at the biological station that determined that mackerel from the Bay of Fundy may have been a possible vector or could be a threat to marine mammals.[34] Haya's group, along with Shawn Robinson and colleagues, also looked at PSP toxin uptake and depuration in the sea scallop.[35]

In 1987, SABS established a phytoplankton monitoring program in the south-west New Brunswick portion of the Bay of Fundy in response to a rapidly growing salmonid aquaculture industry and concerns of potential impacts to the environment and industry.[36] This provides a time series for *Alexandrium* concentrations in the region and an opportunity to compare phytoplankton populations from those measured in the 1920s and 1930s as well as an opportunity to compare *Alexandrium* populations with the total phytoplankton community and look at spatial and temporal trends in *Alexandrium* populations.[37]

In 1988, domoic acid was detected in shellfish from the Bay of Fundy and the monitoring program was instrumental in determining the organism responsible, as there was a good documentation of phytoplankton events before and during the event.[38] In the next chapter, Peter Wells discusses domoic acid and the sometimes lethal poisoning related to this toxin, which was first found in Prince Edward Island mussels in a 1987 seafood poisoning outbreak.

David Wildish and Patrick Lassus conducted experiments in the 1990s on how flow and temperature influence toxic and non-toxic strains of *Alexandrium* and their interaction with the initial feeding response of bivalves.[39] It was mentioned above that in 1944 mussel harvesting in the Bay of Fundy was banned due to concerns about PSP toxins.

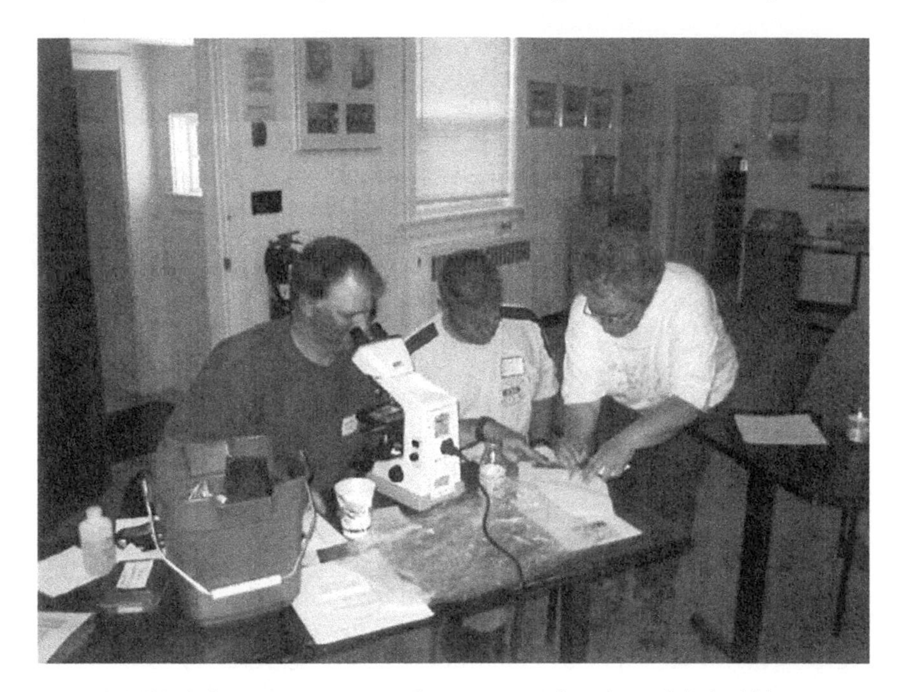

10.6 Workshop to train aquaculturists to analyse harmful algal bloom concentrations

That ban had not been lifted through the years, although subsequent research had suggested that mussels depurate (eliminate) PSP toxins readily compared to other shellfish, and in many parts of the world the mussel culture industry is able to deal with this annual occurrence.

More recent efforts have mitigated this ban. In 2001 a multi-trophic aquaculture study, led by Shawn Robinson at the St Andrews Biological Station and Thierry Chopin at the University of New Brunswick, Saint John, was funded with the intent of integrating different types of aquaculture. Their work aimed to develop responsible practices, optimizing the efficiency of aquaculture operations and diversifying the industry, while maintaining the health of coastal waters.[40] During this study, mussels and seaweeds were cultured to investigate the occurrence of PSP toxins and domoic acid as well as to look at the spatial and temporal distribution of all algal species, with a particular focus on the PSP producing organism. Investigators used this information to design an early warning system for when mussels are accumulating marine toxins, and to regulate the frequency of sampling mussels and seaweeds.[41] The Integrated Multi-Trophic Aquaculture research team's efforts and persistence, with the cooperation of regulatory agencies (Canadian Food Inspection Agency) and the Department of the Environment, led to a commercial mussel aquaculture industry being deemed operational. This was carried out initially on a pilot basis in cooperation with Heritage Salmon and now with Cooke Aquaculture.[42] Robert Cook, in the last chapter, recounts the development of New Brunswick's aquaculture programs.

Work has progressed in recent years in vertical migration studies conducted to determine if, as suspected, the *Alexandrium* in the Bay of Fundy behaves differently from species in other world regions. Indeed, it does not seem to exhibit diel (24-hour) vertical migration.[43] In addition, we continue to work with the Canadian Food Inspection Agency to predict algal blooms and, hence, shellfish toxicity,[44] and to determine the linkages between the bacterial community and *Alexandrium*.[45] We found that elevated right-whale mortalities in 2004 were attributable to PSP toxins.[46]

*Alexandrium* was responsible for salmon mortalities in the aquaculture industry in 2003 and 2004 when *Alexandrium* concentrations were so high that water discolouration was observed.[47] As a result of the 2003 event, a collaborative project was initiated with the industry, and training workshops were held (figure 10.6). Industry partners purchased microscopes and collected samples for analyses of harmful algal bloom concentrations. The training enabled industry personnel to implement techniques to minimize future losses. Further scientific research on the problem entailed culturing large volumes of *Alexandrium* in the laboratory through a collaborative effort with Les Burridge and Monica Lyons; salmon were then exposed to varying concentrations to determine lethal concentrations.[48] Collaborative work has been ongoing for a number of years with the oceanography group under the direction of F.H. Page to explore the hydrographic linkages and transport of *Alexandrium* blooms.

The St Andrews Biological Station became the leading institution in the world for studies of *Alexandrium fundyense* and related species because scientists like Alfreda Needler and Carl Medcof, whose work was based at St Andrews and its substation at Ellerslie, Prince Edward Island, first recognized the link between these dinoflagellate phytoplankton, red tide, and paralytic shellfish poisoning. Having identified an area of investigation with important economic and health implications, Medcof and his successors continued their research program, and in collaboration with Canadian medical and food safety institutions, expanded it to include studies of other toxins. Their work has been emulated around the world as other regions identified problematic concentrations of toxic dinoflagellates and embarked on similar research programs. More recent investigations seek to understand the range of species affected by these toxic organisms, and the conditions that create both high concentrations of toxic dinoflagellates and the conditions that increase toxicity within these organisms. Recent studies of *Alexandrium* as well as historical studies suggest that high concentrations are linked to weather patterns such as fog and associated conditions.[49]

Work on PSP and *Alexandrium* today continues with many collaborative efforts. For example, recent work includes the following projects: remote sensing; prediction and hindcasting; modelling; studying linkages between *Alexandrium* and phytoplankton and shellfish toxicity; toxin uptake and depuration; and spatial and temporal trends. In addition, we are researching the status of *Alexandrium* in relation to the total phytoplankton community, and the relationship between cysts and "vegetative" cells in the life stages of this species. We are also conducting comparisons between ELISA kits and chemical methods for testing toxins, and doing domoic acid–related studies. Aside from the PSP researchers at the St Andrews Biological Station, collaborators include colleagues from the Bedford Institute of Oceanography, the Woods Hole Oceanographic Institute, the University of Maine, Mount Allison University, the University of New Brunswick, the University of Washington, the Huntsman Marine Laboratory, Dalhousie University, the University of Massachusetts, and Rutgers University. The fact that so many scientists in different institutions are engaged in such investigations demonstrates the high importance of this research program begun so many years ago at the St Andrews Biological Station.

## NOTES

Special thanks to Charlotte McAdam, Joanne Cleghorn, and Catie Sahadath from the SABS library for their patience and dedication in finding archived literature and photographs and the many individuals who have contributed to the PSP research program over the years.

1 W.F. Ganong, "The Economic Mollusca of Acadia," *Bulletin of the Natural History Society of New Brunswick* 8 (1889): 1–116.

2 Researchers who have contributed to PSP research over the years at St Andrews include J.A.P. Bastien, R.M. Bond, H.L. Boyd, N. Bourne, L.E. Burridge, T. Braarud, D.F. Bray, J.F. Caddy, the Canadian Food Inspection Agency, R.A. Chandler, F.B. Cunningham, K.R. Freeman, J. Gibbard, R.J. Gibbons, H.H. Gran, K. Haya, M. Hodgson, G. Jamieson, A. Lachance, L.M. Lauzier, A. Lebrie, M.M. LeGresley, A.H. Leim, D.A. Litalien, C.M. Lewis, E. Lord, M. Lyons, L. Maranda, J.D. Martin, P.W.G. McMullin, J.C. Medcof, A.B. Needler, A.W.H. Needler, N. Morin, A. Nadeau, J. Naubert, J. Power, A. Prakash, L. Provasoli, J. Reid, S. Robinson, M.M. Ringuette, D. Sephton, E.S. Silva, A. Sreedharan, C.R. Trask, A.D. Tennant, L.C. Turgeon, B.A. Waiwood, A.W. White, D.G. Wilder, D.J. Wildish, and V. Zitko. I am grateful for all their contributions and extend sincere apologies to anyone who has been omitted.

3 C. Fritz, "Plankton Diatoms, Their Distribution and Bathymetric Range in St Andrews Waters," *Contributions to Canadian Biology 1918–1920* (1921): 49–62.

4 One of the earliest women researchers at St Andrews, Davidson is the first woman scientist to have a Canadian Coast Guard scientific research ship named after her. The CCGS *Viola Davidson* was launched in 2010 and now is based at the St Andrews Biological Station wharf.

5 V.M. Davidson, "Fluctuations in the Abundance of Planktonic Diatoms in the Passamaquoddy Region, New Brunswick, from 1924 to 1931," *Contributions to Canadian Biology and Fisheries*, new series, 8 (1934): 1–36.

6 H.H. Gran and T. Braarud, "A Quantative Study of the Phytoplankton in the Bay of Fundy and the Gulf of Maine (Including Observations on Hydrography, Chemistry, and Turbidity)," *Journal of the Biological Board of Canada* 1 (1935): 279–467.

7 N.F. Bourne and S.M.C. Robinson, "In Memoriam: John Carl Medcof 1911–1997," *Journal of Shellfish Research* 19 (2000): 1–5.

8 R.M. Bond and J.C. Medcof, "Epidemic Shellfish Poisoning in New Brunswick," *Canadian Medical Association Journal* 79 (1957): 19–24.

9 J.C. Medcof, A.H. Leim, A.B. Needler, A.H. Needler, J. Gibbard, and J. Naubert, "Paralytic Shellfish Poisoning on the Canadian Atlantic Coast," *Bulletin of the Fisheries Research Board of Canada* 75 (1947): 1–32.

10 J.L. Martin and D. Richard, "Shellfish Toxicity from the Bay of Fundy, Eastern Canada: 50 Years in Retrospect," in *Harmful and Toxic Algal Blooms. Proceedings of the Seventh International Conference on Toxic Phytoplankton, Sendai, Japan, 12–16, 1995*, ed. T. Yasumoto, Y. Oshima, and Y. Fukuyo (Paris: IOC/UNESCO, 1996), 3–6; and Martin, unpublished manuscript.

11 J.C. Medcof, "Life and Death with *Gonyaulax*: An Historical Perspective," in *Toxic Dinoflagellates*, ed. D.M. Anderson, A.W. White, and D.G. Baden (New York: Elsevier, 1985), 1–8.
12 Personal communications to me over the years.
13 C.M. Sullivan, "Bivalve Larvae of Malpeque Bay, PEI," *Bulletin of the Fisheries Research Board of Canada* 77 (1948): 1–30.
14 A.B. Needler, "Paralytic Shellfish Poisoning and *Goniaulax tamarensis*," *Journal of the Fisheries Research Board of Canada* 7 (1949): 490–504.
15 It is interesting that in early documentation *Gonyaulax* was also spelled *Goniaulax*, although there is no documentation in taxonomy texts or literature as to the reason. Balech, in his description of the Genus *Alexandrium Halim* states: "For a long time scientists expressed doubts about several aspects of the complex genus *Gonyaulax Diesing*. They questioned its real limits, its relationships with other genera, and even the orthography because it is sometimes written as *Goniaulax* (probably etymologically more accurate). The genus evolved rather chaotically." See E. Balech, "The Genus *Alexandrium* or *Gonyaulax* of the *tamarensis* Group," in *Toxic Dinoflagellates*, ed. Anderson, White, and Baden, 33–8.
16 N. Bourne, "Paralytic Shellfish Poison in Sea Scallops (*Placopectin magallanicus Gmelin*)," *Journal of the Fisheries Research Board of Canada* 22 (1965): 1137–49.
17 J.F. Caddy and R.A. Chandler, "Accumulation of Paralytic Shellfish Poison by the Rough Whelk (*Buccinum undatum L.*)," *Proceedings of the National Shellfisheries Association*, June 1968: 46–50.
18 G.S. Jamieson and R.A. Chandler, "Paralytic Shellfish Poisoning in Sea Scallops (*Placopectin magellanicus*) in the West Atlantic," *Canadian Journal of Fisheries and Aquatic Science* 40 (1983): 313–18.
19 A. Prakash, "Source of Paralytic Shellfish Toxin in the Bay of Fundy," *Journal of the Fisheries Research Board of Canada* 20 (1963): 983–96; and A. Prakash, "Growth and Toxicity of a Marine Dinoflagellate, *Gonyaulax tamarensi*," *Journal of the Fisheries Research Board of Canada* 24 (1967): 1589–606.
20 A. Prakash, J.C. Medcof, and A.D. Tennant, "Paralytic Shellfish Poisoning in Eastern Canada," *Bulletin of the Fisheries Research Board of Canada* 177 (1971): 1–87.
21 A.W. White, "Growth Inhibition Caused by Turbulence in the Toxic Marine Dinoflagellate *Gonyaulax excavata*," *Journal of the Fisheries Research Board of Canada* 33 (1976): 2598–602.
22 A.W. White, "Dinoflagellate Toxins as Probable Cause of an Atlantic Herring (*Clupea harangus harengus*) Kill, and Pteropods as Apparent Vector," *Journal of the Fisheries Research Board of Canada* 34 (1977): 2421–4.

23 A.W. White, "Dinoflagellate Toxins in Phytoplankton and Zooplankton Fractions during a Bloom of *Gonyaulax excavata*," in *Toxic Dinoflagellate Blooms*, ed. D.L. Taylor and H. Seliger (New York: Elsevier North-Holland, 1979), 381–4; A.W. White, "Recurrence of Kills of Atlantic Herring (*Clupea harengus harengus*) Caused by Dinoflagellate Toxins Transferred through Herbivorous Zooplankton," *Canadian Journal of Fisheries and Aquatic Science* 37 (1980): 2262–5; A.W. White, "Marine Zooplankton Can Accumulate and Retain Dinoflagellate Toxins and Cause Fish Kills," *Limnology and Oceanography* 26 (1981): 104–10.

24 A.W. White and L. Maranda, "Paralytic Toxins in the Dinoflagellate *Gonyaulax* Excavate and in Shellfish," *Journal of the Fisheries Research Board of Canada* 35 (1978): 397–402.

25 A.W. White, "Sensitivity of Marine Fishes to Toxins from the Red-tide Dinoflagellate *Gonyaulax excavata* and Implications for Fish Kills," *Marine Biology* 3 (1981): 255–60.

26 A.W. White and C.M. Lewis, "Resting Cysts of the Toxic, Red Tide Dinoflagellate *Gonyaulax excavata* in Bay of Fundy Sediments," *Journal of the Fisheries Research Board of Canada* 39 (1982): 1185–94.

27 J.L. Martin and A.W. White, "Distribution and Abundance of the Toxic Dinoflagellate *Gonyaulax* Excavate in the Bay of Fundy," *Canadian Journal of Fisheries and Aquatic Science* 45 (1988): 1968–75; and J.L. Martin and D.J. Wildish, "Temporal and Spatial Dynamics of *Alexandrium fundyense* Cysts during 1981–3 and 1992 in the Bay of Fundy," in *Proceedings of the Fourth Canadian Workshop on Harmful Marine Algae*, ed. J.R. Forbes, Canadian Technical Report of Fisheries and Aquatic Science 2016 (1994): 22–4.

28 W.J. Blogoslawski, M.E. Stewart, J.W. Hurst, and F.G. Kern, "Ozone Detoxification of Paralytic Shellfish Poison in the Softshell Clam (*Mya arenaria*)," *Toxicon* 17 (1979): 650–4.

29 A.W. White, J.L. Martin, M. LeGresley, and W.J. Blogoslawski, "Inability of Ozone to Detoxify Paralytic Shellfish Toxins in Soft-shell Clams," in *Toxic Dinoflagellates*, ed. Anderson, White, and Baden, 473–8.

30 A.W. White, "Relationships of Environmental Factors to Toxic Dinoflagellate Blooms in the Bay of Fundy," in *Rapports et Procès-verbaux des Réunions du Conseil International pour l'Exploration de la Mer* 187 (1987): 38–46.

31 Anderson, White, and Baden, eds., *Toxic Dinoflagellates*.

32 Balech, "The Genus *Alexandrium* or *Gonyaulax*."

33 J.L. Martin, A.W. White, and J.J. Sullivan, "Anatomical Distribution of Paralytic Shellfish Toxins in Soft-shell Clams," in *Toxic Marine Phytoplankton*, ed. E. Graneli, B. Sundstrom, L. Edler, and D.M. Anderson (New York: Elsevier, 1990), 379–84.

34 K. Haya, J.L. Martin, B.A. Waiwood, L.E. Burridge, J.M. Hungerford, and V. Zitko, "Identification of Paralytic Shellfish Toxins in Mackerel from Southwest Bay of Fundy, Canada," in *Toxic Marine Phytoplankton*, ed. Graneli, Sundstrom, Edler, and Anderson, 350–5.

35 K. Haya, J.L. Martin, S.L. Robinson, M.S. Khots, and J.D. Martin, "Does Uptake of *Alexandrium fundyense* Cysts Contribute to the Levels of PSP Toxins Found in the Giant Scallop, *Placopecten magellanicus*?" *Harmful Algae* 2 (2003): 75–81.

36 D.J. Wildish, J.L. Martin, A.J. Wilson, and M. Ringuette, "Environmental Monitoring of the Bay of Fundy Salmonid Mariculture Industry during 1988–89," *Canadian Technical Report of Fisheries and Aquatic Science* 1760 (1990): 1–123; D.J. Wildish and J.L. Martin, "Determining the Potential Harm of Marine Phytoplankton to Finfish Aquaculture Resources of the Bay of Fundy," *Fisken og Harvet* 13 (1994): 115–26; J.L. Martin, D.J. Wildish, M.M. LeGresley, and M.M. Ringuette, "Phytoplankton Monitoring in the Southwestern Bay of Fundy during 1990–1992," Canadian Manuscript Report of Fisheries and Aquatic Science 2277 (1995): 1–154; J.L. Martin, M.M. LeGresley, P.M. Strain, and P. Clement, "Phytoplankton Monitoring in the Southwest Bay of Fundy during 1993–96," Canadian Manuscript Report of Fisheries and Aquatic Science 2265 (1999): 1–132; J.L. Martin, M.M. LeGresley, and P.M. Strain, "Phytoplankton Monitoring in the Western Isles region of the Bay of Fundy during 1997–98," Canadian Manuscript Report of Fisheries and Aquatic Science 2349 (2001): 1–85; and J.L. Martin, M.M. LeGresley, and P.M. Strain, "Plankton Monitoring in the Western Isles Region of the Bay of Fundy during 1999–2000," *Canadian Technical Report of Fisheries and Aquatic Science* 2629 (2006): 1–88.

37 F.H. Page, J.L. Martin, A. Hanke, and M.M. LeGresley, "The Relationship of *Alexandrium fundyense* to the Temporal and Spatial Pattern in Phytoplankton Community Structure within the Bay of Fundy, Eastern Canada," in *Harmful Algae*, ed. K.A. Steidinger, J.H. Landsberg, C.R. Thomas, and G.A. Vargo (St Petersburg: Florida Fish and Wildlife Conservation Commission, Florida Institute of Oceanography, and IOC/UNESCO, 2004), 92–4; F.H. Page, A. Hanke, J.L. Martin, M. LeGresley, B. Chang, and P. McCurdy, "Characteristics of *Alexandrium fundyense* Blooms That Affect Caged Salmon in the Bay of Fundy," *Aquaculture Association of Canada Special Publication* 9 (2005): 27–30; and F.H. Page, J.L. Martin, A. Hanke, and M.M. LeGresley, "Temporal and Spatial Variability in the Characteristics of *Alexandrium fundyense* Blooms in the Coastal Zone of the Bay of Fundy, Eastern Canada," in *Harmful Algae 2004*, ed. G.C. Pitcher, T.A. Probyn, and H.M. Verheye, *African Journal of Marine Science* (2006): 203–8.

38 J.L. Martin, K. Haya, L.E. Burridge, and D.J. Wildish, "*Nitzschia pseudodelicatissima* – A Source of Domoic Acid in the Bay of Fundy, Eastern Canada," *Marine Ecology Progress Series* 67 (1990): 177–82.
39 P.D. Lassus, D.J. Wildish, M. Bardouil, J.L. Martin, M. Bohec, and S. Bougrier "Ecophysiological Study of Toxic *Alexandrium spp.* Effects on the Oyster, *Crassostrea gigas*," in *Harmful and Toxic Algal Blooms, Proceedings of the Seventh International Conference on Toxic Phytoplankton, Sendai, Japan, 12–16, 1995*, ed. T. Yasumoto, Y. Oshima, and Y. Fukuyo (Paris: IOC/UNESCO, 1996), 409–12; and D.J. Wildish, P. Lassus, J.L. Martin, A.M. Saulnier, and M. Bardouil, "Effect of the PSP-Causing Dinoflagellate, *Alexandrium sp.*, on the Initial Feeding Response of *Crassostrea gigas*," *Aquatic Living Resources* 11 (1998): 35–43.
40 T. Chopin, "Integrated Multi-trophic Aquaculture. What It Is, and Why You Should Care ... and Don't Confuse It with Polyculture," *Northern Aquaculture Magazine* 12:4 (2006): 4.
41 K. Haya, D. Sephton, J. Martin, and T. Chopin, "Monitoring of Therapeutants and Phycotoxins in Kelps and Mussels Co-cultured with Atlantic Salmon in an Integrated Multi-trophic Aquaculture System," *Bulletin of the Aquaculture Association of Canada* 104:3 (2004): 29–34.
42 S.M.C. Robinson and T. Chopin, "Defining the Appropriate Regulatory and Policy Framework for the Development of Integrated Multi-trophic Aquaculture Practices: Summary of the Workshop and Issues for the Future," *Bulletin of the Aquaculture Association of Canada* 104:3 (2004): 73–82.
43 J.L. Martin, F.H. Page, A. Hanke, P.M. Strain, and M.M. LeGresley, "*Alexandrium fundyense* Vertical Distribution Patterns during 1982, 2001 and 2002 in the Bay of Fundy, Eastern Canada," *Deep-Sea Research II* 52 (2005): 2569–92.
44 J.L. Martin and D. Richard, "Shellfish Toxicity from the Bay of Fundy, Eastern Canada: 50 Years in Retrospect," in *Harmful and Toxic Algal Blooms*, ed. Yasumoto, Oshima, and Fukuyo, 3–6; and J.L. Martin, F.H. Page, M.M. LeGresley, and D.J.A. Richard, "Phytoplankton Monitoring as a Management Tool: Timing of *Alexandrium* and *Pseudo-nitzschia* Blooms in the Bay of Fundy, Eastern Canada," in *Harmful Algae Management and Mitigation*, ed. S. Hall, S. Etheridge, D. Anderson, J. Kleindinst, M. Zhu, and Y. Zou (Singapore: Asia-Pacific Economic Cooperation Publication, 2004), 136–40.
45 M. Ferrier, J.L. Martin, and J.N. Rooney-Varga, "Stimulation of *Alexandrium fundyense* Growth by Bacterial Assemblages from the Bay of Fundy," *Journal of Applied Microbiology* 92 (2002): 706–16; Y. Hasegawa, J.L. Martin, M.W. Giewat, and J.N. Rooney-Varga, "Microbial Community Diversity

in the Phycosphere of Natural Populations of the Toxic Alga, *Alexandrium fundyense*," *Environmental Microbiology* 9 (2007): 3108–21.

46 G.J. Doucette, A.D. Cembella, J.L. Martin, J. Michaud, T.V.N. Cole, and R.M. Rolland, "Paralytic Shellfish Poisoning (PSP) Toxins in North Atlantic Right Whales *Eubalaena glacialis* and Their Zooplankton Prey in the Bay of Fundy," *Canada Marine Ecology Progress Series* 306 (2006): 303–13.

47 J.L. Martin, M.M. LeGresley, K. Haya, D.H. Sephton, L.E. Burridge, F.H. Page, and B.D. Chang, "Salmon Mortalities Associated with a Bloom of *Alexandrium fundyense* in 2003 and Subsequent Early Warning Approaches for Industry," in *Harmful Algae 2004*, ed. Pitcher, Probyn, and Verheye, 431–4; J.L. Martin, M.M. LeGresley, A. Hanke, and F.H. Page, "*Alexandrium fundyense*: Red Tides, PSP Shellfish Toxicity, Salmon Mortalities and Human Illnesses in 2003–04 – Before and After," in *Proceedings of the 12th International Conference on Harmful Algae*, ed. Ø. Moestrup (Copenhagen: International Society for the Study of Harmful Algae and IOC/UNESCO, 2008), 206–8; and D.H. Sephton, K. Haya, J.L. Martin, M.M. LeGresley, and F.H. Page, "Paralytic Shellfish Toxins in Zooplankton, Mussels, Lobsters and Caged Atlantic Salmon, *Salmo salar*, during a Bloom of *Alexandrium fundyense* off Grand Manan Island, in the Bay of Fundy," *Harmful Algae* 6 (2007): 745–58.

48 Burridge et al., unpublished data; B.D. Chang, F.H. Page, J.L. Martin, G. Harrison, E. Horne, L.E. Burridge, M.M. LeGresley, A. Hanke, and P. McCurdy, "Phytoplankton Early Warning Approaches for Salmon Farms in Southwestern New Brunswick," *Aquaculture Association of Canada Special Publication* 9 (2005): 20–3.

49 Martin, unpublished.

# 11 A History of Research in Environmental Science and Ecotoxicology at the St Andrew's Biological Station

PETER G. WELLS

Environmental science and ecotoxicology have been conducted at the biological station since it was established in St Andrews in 1908. The St Andrews Biological Station was an early player in research on the aquatic environment, chemical oceanography, ecology, physiology, and biochemistry of fish and invertebrates, water pollution, and environmental chemistry. These are all components of what is now called "environmental science." From the start, the biological station's scientists were addressing a wide range of questions, from fisheries–environment interactions to understanding specific threats to the environmental quality of aquatic ecosystems. This chapter briefly describes early work up to the 1950s, but it focuses especially on studies of DDT and Atlantic salmon, the extensive aquatic toxicology studies in the program formally established between late 1950 and 1970, and the many contributions to environmental chemistry, aquatic toxicology, and ecotoxicology since the 1970s in the current environmental science program. The research occurred while the station was part of the Biological Board and the Fisheries Research Board of Canada until the early 1970s, then as Environment Canada and Fisheries and Environment Canada, and finally the Department of Fisheries and Oceans (DFO) from 1979 onwards. The biological station's environmental science contributions occurred in the context of many issues facing marine and aquatic ecosystems and their living resources in the Atlantic provinces and north-west Atlantic in the twentieth century. Although a discussion of these issues in depth is beyond the scope of this chapter,[1] environmental science at St Andrews has been multifaceted and interdisciplinary from its beginning. Research programs were usually linked to questions and concerns about the fisheries, the various species from lobsters to molluscan shellfish to herring, and their habitat requirements, hence showing the many benefits of the interactions of fisheries science and the environmental sciences. It is noteworthy that A.G. Huntsman, one of the St Andrews Biological Station's earliest and most prominent scientists, felt that "how the environment creates important fisheries was perhaps the central question in fisheries biology."[2]

While E.A. Trippel briefly covered the whole history of the St Andrews Biological Station in 1999,[3] this chapter describes in more detail the early years of work in environmental science and ecotoxicology to the 1970s, and research during the most recent years, from 1970 onwards.[4] Scientists at St Andrews during this "early environmental era" contributed to the nascent fields of aquatic toxicology, ecotoxicology, and environmental chemistry. In more recent years, their research has sought solutions to many emerging environmental problems. This chapter gives a summary description of selected studies, their significance, and references to significant publications, and identifies major themes of St Andrews research over the years. Please note that the studies on harmful algal bloom studies, including work on toxins that involved ecotoxicologists (the field of toxinology), are covered by Jennifer Martin elsewhere in this volume.[5] Environmental research over the "early" period of 1960 to 1979 has been described briefly by Zitko[6] and Zitko and Wildish,[7] but as will be seen, environmental studies started at SABS well before 1960.

Federal science policies, fiscal resources, stable organization, and leadership have all influenced marine scientific direction, the recruitment of personnel, research questions and productivity, and advances in knowledge. Commentary on the political, organizational, and fiscal climates of research within Canadian government laboratories is given in Hayes's tome of 1973, *The Chaining of Prometheus*, Johnstone's *The Aquatic Explorers* (1977), Harris's *Lament for an Ocean* (1998), and, most recently, Hubbard's *A Science on the Scales* (2006), as well as numerous recent papers, such as one by L.M. Dickie[8] and Rob Stephenson's chapter in this book. Harris's examination of the Atlantic cod collapse provides the general context for understanding marine environmental conditions critical to a particular fisheries species, and research policies and agendas of the responsible federal department (currently the Department of Fisheries and Oceans). From the Biological Board's beginnings to the era from 1979 onward – in which the science has operated as a branch of the DFO – there have been many highs and lows in conditions for productive and timely marine research, and in the appropriate use of scientific advice for fisheries and environmental management. Science histories such as this volume are important reflections on this record.

## The Period from 1900 to 1950

In the early portable barge laboratory, where studies were conducted before 1908, "chemical studies were commenced on the composition of medusae and the effects of pollution."[9] A.P. Knight (figure 11.1) worked on problems in fisheries and aquatic toxicology (it was not called that then) during the summers of 1900–7. "In New Brunswick, he found that sawdust waste was not lethal to fish, pulp mill effluent was lethal at 10% (a value that held true until the 1980s), gas works effluent at 0.5%, and a nail factory effluent at 0.1%. Professor Knight was a good toxicologist, used a control fish (hopefully more than one!), and anticipated the documents outlining methods for testing."[10] Knight also published "The Effects of Polluted Waters on Fish Life" (1901) and, in 1907, "The Effects of Dynamite Explosions on Fish Life."[11] From 1908 onward, studies at the officially established Atlantic Biological Station "were concerned with the ways in which the physical and chemical conditions of the water and the ocean currents controlled the distribution of various species of fish."[12] Hence, environmental physiology studies, a core component of ecotoxicology, started at the beginning of the St Andrews station.

Huntsman's early observations are both relevant and insightful: it was important to research the "limiting factors for aquatic (fish) life, important for an understanding not only of the conditions in nature but also of those in cultural operations, where unfavourable physical conditions become evident in the appearance of disease or in the summary death of the fish."[13] Huntsman (figure 11.2) was part of life at the biological station from almost the beginning to the early 1970s, and can be considered one of Canada's earliest environmental scientists; anecdotes and short accounts of his and others' early environmental work can be read in Johnstone's 1977 history. The effects of "physical factors on marine animals" were studied in the new bacteriological and biochemical laboratory, built in 1922.[14] From 1925 to 1930, "a number of experimental studies were begun: for example, the effects of different temperatures and salinities on the development of flounder eggs by W.C.M. Scott, and the effects of light on the copepod *Calanus*."[15] In 1934 Morden Whitney

11.1 A.P. Knight

Smith also investigated the occurrences and causes of eutrophication – overgrowth of phytoplankton and other microorganisms in bodies of water, which he found to be due to heavy fertilization.[16]

Early environmental studies also examined potential impacts of physical changes in coastal waters. The International Passamaquoddy Fisheries Commission in 1931 investigated the "probable effects on local fisheries of damming Passamaquoddy and Cobscook bays in a project to develop hydro-electric power from the tides."[17] The commission reported in 1933 that "the herring fishery in Passamaquoddy Bay would probably be reduced to negligible proportions, that herring weirs outside the dams would probably be affected but in unpredictable ways,

11.2 Huntsman stamp

and that there would be little prospect of effects along the coast of Maine or even serious effects at Grand Manan."[18] Historian Jennifer Hubbard described the station's involvement with this tidal power dam commission, and the comprehensive environmental assessment which showed mixed results of little effect or even improved phytoplankton growth outside the inner-Quoddy region, but important changes to the fisheries within.[19] In the end Canada "declined to participate in the project,"[20] and it never proceeded. Hence, from the 1930s onwards, St Andrews' scientists contributed to understanding and warning about the environmental implications of harnessing tidal energy; this work continues as new tidal power technologies are developed and tested, largely in the upper Bay of Fundy.

**The 1950s – "Recognizing" Pollutants from Metals to DDT**

D.G. (Dick) Wilder's study of metal toxicity in traps to lobsters marked a significant early study in water pollution from St Andrews.[21] During the 1950s, St Andrews investigators did detailed investigations of the toxicity of metals to adult lobsters, in relation to the composition of lobster traps, storage containers, and shipping procedures. These complemented other studies that addressed "the effect of various factors on the survival of lobsters to improve methods of holding and shipping live lobsters."[22] The young Don McLeese (see below) assisted with these studies, starting a career that focused on the environmental physiology and ecotoxicology of lobsters.

Major concerns emerged about the effects of dichloro-diphenyl-trichloroethane (DDT) – which was sprayed to control New Brunswick forest insect pests such as the spruce bud-worm – on fishes such as Atlantic salmon and non-target aquatic insects. Many studies and papers from St Andrews appeared from 1955 onwards, based on the work of Paul F. Elson (figure 11.3), Jim (C.J.) Kerswill, Miles H.A. Keenleyside, Fred P. Ide, Don Alderdice, and John Anderson. In 1954 Kerswill and Elson began "a major salmon investigation on the Miramichi River, using fences on three tributaries to obtain exact escapement figures"[23] and initiated various ecological studies on DDT impacts, which are described in this volume by Richard (Dick) H. Peterson. The studies appeared in the annual reports of the board and papers throughout the 1950s,[24] culminating in six articles by C.J. Kerswill and contributing colleagues in the May 1967 issue 24 of the *Journal of the Fisheries Research Board of Canada*.[25] J.L. Hart succinctly reviewed the earlier studies, which uncovered the deadly effects of DDT, in 1958:

> Since 1954, the salmon investigation has given much attention to the effects on salmon and on the aquatic insects on which salmon feed of spraying watersheds with DDT to control spruce budworm outbreaks. A high proportion of young salmon in the rivers are killed and the aquatic insect fauna is first drastically reduced and later much altered, but spraying seemed necessary. Attention in 1957 and 1958 was accordingly being directed to exploring remedial and mitigating action.[26]

Other notable aquatic scientists contributed to the DDT studies. Don Alderdice, who largely worked out of Nanaimo, BC, conducted studies in 1951 in the Maritimes, "doing field work on fisheries and DDT

11.3 Paul Elson

spraying of forests."[27] Frances Premdas and John Anderson studied the uptake and detoxification of DDT by salmon.[28] Especially noteworthy were ecological studies by F.P. Ide of the University of Toronto (who was the thesis supervisor of J.B Sprague, see below) on the effects of DDT on aquatic insects in salmon and trout streams.[29]

Part of these studies' historical importance was that "the work from SABS played a part in the environmental revolution of the 1960s; in the book *Silent Spring* by Rachel Carson ... Chapter 9 "Rivers of Death" was the story of DDT and salmon rivers of northern NB."[30] Carson acknowledged C.J. Kerswill of St Andrews and the Fisheries Research Board for the information he provided for her book; Kerswill's reports are referenced in its "Notes" section. Carson, herself a long-term employee of the US Bureau of Fisheries, as a report writer who specialized in work dedicated to popularizing the work of the bureau, had also worked as an aquatic biologist. *Silent Spring*, considered one of the most influential books of the twentieth century, was the inspiration for the environmental movement of the late twentieth century. St Andrews scientists' DDT-salmon studies made important contributions to Carson's writings and conclusions about the ecological risks of pesticides.

## The 1960s – Establishing the Pollution Program

During the period when F. Ronald Hayes was chairman of the Fisheries Research Board of Canada (1964–9), "some of the changes in the direction of Board activities resulted from the growing awareness of water pollution and lake eutrophication." Research of this era included studies on the "effects of heavy metals and organochlorine pesticides on aquatic organisms."[31] Indeed, a "close liaison between biology and oceanography was required to deal with pollution problems" during this era of "the discovery of the environment, and new work was initiated, especially on freshwater and the Great Lakes, where "the effects of pollution were most serious."[32] At this time the Fisheries Research Board established the Freshwater Institute in Winnipeg, Manitoba; the Canada Centre for Inland Waters in Burlington, Ontario; and the Pacific Environmental Institute in West Vancouver. Sprague's water pollution group at St Andrews was strengthened both in numbers and in its diversity of disciplines. Among their extremely important work in this era were investigations of nitrilotriacetates (NTA), proposed "as substitutes for phosphates in detergents," which "proved non-toxic and much less biogenic than phosphates."[33]

During the ensuing chairmanship of Dr J.R. Weir (1969–72), environmental research was a major thrust of the Fisheries Research Board. Weir's comments on the importance of this research reflect the continuance of the efficiency focus in environmental and resource management, discussed by Jennifer Hubbard elsewhere in this volume:

> Environmental research includes studies on all of the biological aspects of oceanography and limnology as they relate to productivity of marine and inland waters, and to fitness of the aquatic environment: detection, dispersion, and degradation of man-made pollutants in water; understanding and modification of natural ecological changes; and the resistance and tolerance of aquatic organisms to harmful substances ... The desirability of preserving a healthy and viable environment introduces the need for proper overall management of all renewable resources within the ecosystem so

> that its deterioration through such things as pollution can be avoided. This means long-range research and planning to ensure a more efficient, high quality, continuing resource for commercial and other uses that is compatible with demands for human health and enjoyment.[34]

The Fisheries Research Board's key role was providing knowledge needed "to detect deterioration of the environment and to develop remedial measures" and to support government responsibilities in the environmental-quality field.[35] Unfortunately, the early 1970s also witnessed the beginning of the Trudeau government's "Make or Buy" policy, which encouraged "government research units to contract out research and development problems rather than solve them in-house."[36] This began a trend towards less internal research, more contractual research, less continuity of programs, and reduced expertise housed in federal aquatic laboratories in upcoming decades. The short-term goals of the latter conflicted with the long-range research seen by Weir as being so essential for environmental and human health.

**The Sprague Era, 1958–70: Setting up a Water Pollution Research Section at St Andrews**

One of the most important figures in the history of the St Andrews Biological Station's pollution research was John Booty Sprague. Sprague was one of a number of key graduates trained by University of Toronto professors Fred Fry and Fred Ide in the Department of Zoology. Fry's graduate students who eventually joined the St Andrews Biological Station's research program included John Anderson and Don McLeese; Sprague was a student in Ide's lab.[37]

Station director John Hart hired John B. Sprague in 1958 to start a section on water pollution research, then one of two in Canada within the Fisheries Research Board. Hart had also established the other section on the west coast in 1954, when he was director of the Pacific Biological Station, with Michael (Mike) Waldichuk as its first water pollution scientist.[38] Laboratory work in aquatic toxicology began in 1960, when base-metal mining appeared to be affecting salmon in the Miramichi River. As "station staff were doing research work on DDT ... they observed an unusual number of Atlantic salmon turning back downstream at a counting fence in the NW Miramichi River." They began research "on heavy metal toxicity in freshwater streams, work made urgent since a base metal mine – the Heath Steele mine"[39] – "had begun an intermittent operation in the watershed a few years earlier."[40] "The results – high toxicity of zinc and extremely high toxicity of copper in soft water, the introduction of the toxic units (TU) concept, and the observation that salmon avoid metals in water at 0.3TU – formed the basis for future mining pollution control in Canada."[41] The toxic unit concept was much-used in the early days of water pollution control, as it was a simple way to interpret the significance of exposure concentrations and could be applied directly to control discharge levels and total amounts. Several of the studies involved salmon scientists Paul F. Elson and Richard L. Saunders.[42] In the early 1970s, two studies, by Elson, L.M. Lauzier, and Vladimir Zitko (1972), and by Elson, Saunders, A.L. Meister, J.W. Saunders, Sprague, and Zitko (1973), investigated salmon movement in the Miramichi estuary, using sonic tracking techniques;[43] they "demonstrated that salmon move upstream fairly slowly, particularly against a pollution gradient and they tend to seek less polluted water."[44]

John Sprague was a pioneer of aquatic toxicity investigations. His aquatic toxicity studies while at St Andrews were extensive, addressing many water pollution problems facing the Maritimes at the time. Sprague's wide range of field and laboratory research included field surveys on pulp mills and mines in New Brunswick, especially on the Saint John and Miramichi rivers, then experiencing a heavy impact from unregulated industrial pollution; he also investigated mine wastes in Labrador. He studied the effects of food processing plant wastes on the Saint John River's water quality, and especially on its oxygen levels. He was first author of thirty-nine papers and reports in the 1960s and served on many national and international committees, and his contributions drew widespread recognition. He investigated general pollution,[45] factors influencing toxicity,[46] the toxicity of metal and chelating agents,[47] DDT,[48] oil-spill dispersants,[49] sublethal behavioural toxicity,[50] and oxygen levels in rivers and estuaries affected by industrial effluents.[51] He developed a unique apparatus for studying the avoidance of pollutants by fish.[52] While on sabbatical in 1968–9 at the University of Oregon in Corvallis, Sprague wrote three review papers on how to measure pollutant toxicity to fish; the first of these described the principles and concepts of bioassay methods for acute toxicity. The next two papers discussed, respectively, how to utilize and apply bioassay results; and methods to measure and estimate sub-lethal effects and "safe" concentrations.[53] These papers provided a solid foundation establishing aquatic toxicology as a science and are widely quoted citation classics.

Sprague also made many other contributions. He served the American Society for Testing and Materials (ASTM) through a key water pollution committee (E-47) and the American Public Health Association (APHA), where he made an important contribution to the bioassay section in the classic handbook *Standard Methods for the Examination of Water and Wastewater* (1961 edition).[54] He was also involved in one of the St Andrews Biological Station's contributions to UN expert groups. Sprague was an early member of the United Nations Joint Group of Experts on the Scientific Aspects of Marine Pollution (GESAMP). Vladimir Zitko was also an early member of GESAMP Working Group One (Evaluation of Hazardous Substances) of the International Maritime Organization, a UN agency headquartered in London. Sprague refers to the GESAMP report in his paper in *Water Research 3*, which describes the system for classifying different grades of toxicity he helped to develop[55] – a system which has been the basis for classifying industrial chemicals carried by ships under MARPOL since 1974.[56]

In 1970 Sprague left St Andrews for the University of Guelph, where he set up an aquatic toxicology program and laboratory in the Department of Zoology. Many of his first graduate students were still active in the field circa 2008: Peter V. Hodson, Douglas Holdway, George Dixon, Joanna Parrott, William Logan, and Peter G. Wells. Sprague was honoured by the Society for Environmental Toxicology and Chemistry (SETAC) in the 1980s with the SETAC Founders' Award. In a plenary presentation to the 22nd National Aquatic Toxicity Workshop in 1995 in St Andrews, referring to his and other SABS biologists' contributions as pioneers in Canadian aquatic toxicology, he stated that "there has been a strong connection of SABS with early Canadian work on water pollution and toxicology, a connection that continues."[57]

## Environmental Sciences (Ecology/Toxicology) at St Andrews Biological Station – 1968 to Present

Beginning in 1968, St Andrews' environmental science and ecotoxicology programs were staffed by postdoctoral fellows and employees hired in the late 1960s and early 1970s, and included: Vlado Zitko, hired as an expert in chemistry in 1968; Dave Wildish, hired to investigate marine ecology in 1969; and, among the specialists on fish, Phil Symons, an expert in fish behaviour who had been hired in 1965, and who was joined in 1968 by Arne Sutterlin, who investigated fish neurophysiology, and in 1970 by Dick Peterson, who studied fish physiology. Into the 1990s,

Zitko and Kats Haya led the water pollution program as it evolved into a fully integrated program of environmental chemistry and aquatic ecology and toxicology.[58]

The St Andrews program gave ongoing recognition to the importance of interdisciplinary perspectives in environmental toxicology and chemistry. Although interaction between disciplines is a now a fundamental and accepted principle in the field, this was not the case thirty years or more ago. At the St Andrews Biological Station, state-of-the-art chemistry was combined with biochemistry, physiology, behaviour, toxicology, and ecology in studies ranging from the impacts of organic pollutants, such as pulp mill and aquaculture wastes, to the effects of pesticides and aquaculture chemicals. This section looks at the training and work of some important figures in this multidisciplinary field.

Vlado Zitko (figures 11.4a and 11.4b) came to Canada from Czechoslovakia in 1964 as a postdoctoral fellow at the National Research Council, Ottawa. Hired at St Andrews in 1968 in the water pollution section, which needed many studies on industrial pollution and chemicals, Zitko was immediately busy investigating a range of problems: mercury, other metals, marine oil pollution, yellow phosphorus, and pesticide formulations. He and colleagues surveyed methyl mercury in freshwater and marine fishes. Where levels exceeding 0.5 micrograms per gram were measured in some freshwater fish,[59] they suggested that aerial fallout was the source.[60] When the tanker *Arrow* had a large Bunker C (no. 6 fuel) oil spill in Chedabucto Bay, Nova Scotia, in February 1970, the station's staff as a whole were involved in the emergency response and follow-up scientific studies. Zitko developed a spectrofluorometric method for measuring Bunker C in water, sediments, and biota.[61] Sprague and Carson also performed oil and dispersant toxicity studies at the station.[62]

Zitko and staff also investigated the 1970 fish kill in Placentia Bay, Newfoundland, and discovered the "the fish kill was caused by yellow phosphorus from an effluent discharged from the local phosphorus manufacturing plant." They went on to determine that yellow phosphorus was toxic to herring, later proving it by reproducing in the laboratory "the 'red herring' syndrome, observed in the field."[63]

Zitko was the most prominent scientist in the environmental chemistry group. According to his colleague Kats Haya, "Vlado had a knack of predicting what to look at (i.e. which chemicals and where) well before others looked for them, e.g., flame retardants. He was a lead player in Canada in aquatic environmental chemistry."[64] Zitko was active in

11.4a Vlado Zitko

11.4b Vlado Zitko

selecting the priority chemicals list for Canada under the Canadian Environmental Protection Act. He identified nonylphenols in pesticide solvents, which were already suspected from toxicity studies. Zitko also helped identify domoic acid algal toxin during the mussel crisis in Prince Edward Island in 1987. This crisis, in which 107 people across Canada fell ill shortly after eating cultured mussels between 17 and 24 November 1987, resulted in nineteen people being admitted to hospital with neurologic illness, three of whom died. Most people only suffered nausea, cramping, and diarrhea, but some victims suffered disorientation, confusion, and short-term memory loss, while some older men also suffered irreversible short-term memory loss, seizures, and some fell into a coma. Although this complex of symptoms had never before been associated with the consumption of molluscs, the Department of Fisheries and Oceans, National Research Council, and public health officials quickly identified diatoms in the PEI mussels as the source of the mysterious toxins. The mussels were quickly withdrawn from the market.[65] Zitko was on the team of investigators, led by the NRC, who had identified domoic acid as the offending toxin by 18 December 1987. This was a brilliant piece of forensic ecotoxicology, noted globally, as diatom-domoic acid outbreaks have become common in many countries.

Zitko also led investigations on many kinds of industrial chemicals and toxic chemicals in the Bay of Fundy and in the environment in general.[66] In the 1970s, he worked on chlorinated paraffins (CPs) and polybrominated biphenyls (PBBs).[67] He "concluded that long chain CPs are not likely to accumulate in biota to the same extent as PCBs because of CP's high molecular weight, extremely low aqueous solubility, and strong adsorption on suspended matter. The bioaccumulation of PBBs matched that of PCBs but the compounds were more reactive, with different toxicological consequences."[68] Zitko also studied thallium[69] after investigating zinc and copper salts in mining effluents. Thallium in mining effluents had a similar, though slower, toxicity to copper in soft water to fish.[70]

Zitko's many other contributions (he had one of the most extensive publication records of any St Andrews scientist) included studies on flame retardants (HBCD – hexabromocyclododecane) detected in the floats used in salmon aquaculture; other flame retardants (DBDE – decabromodiphenyl) found in plastic composites; and surveyed musks from fragrances, found contaminating the tissues of aquatic fauna, a phenomenon first reported in the early 1980s by Yamagishi et al.[71] in

Japan, but not discovered until the late 1990s in Canadian waters. He did important work on chemometrics as a research tool.

Another innovative and productive scientist, Don McLeese, worked in marine invertebrate physiology and ecotoxicology at the St Andrews Biological Station for almost forty years. He started research on invertebrate physiology and behaviour in the late 1950s, and progressed to studies with lobsters (*Homarus americanus*), looking at chemoreception responses to contaminants, including pulp mill effluents and pesticides.[72] McLeese was internationally known for these studies and contributed to the behaviour panel at a landmark ICES marine biomonitoring workshop in 1979.[73] By the mid-1970s, McLeese was researching the toxicity to lobsters, clams, and other invertebrates of new pesticides being introduced by the forestry industry during this period, research he continued for another fifteen years. He investigated the effects of: phosphamidon;[74] fenitrothion – for which he demonstrated low toxicity to freshwater crayfish and a very high toxicity to lobsters;[75] pyrethroids, then being introduced into agriculture and forestry and proving to be very toxic to marine invertebrates;[76] and Matacil ®, a carbamate insecticide sprayed in a formulation containing nonylphenol.[77] The nonylphenol proved to be the most toxic ingredient, from this and other regional studies.[78] His research contributed to nonylphenol being removed from the Matacil® formulation.[79]

McLeese's work on metal contaminants especially focused on cadmium contamination being discharged in the final aqueous effluents of the Belledune lead smelter at Belledune harbour, New Brunswick. This problem was also extensively studied by the Environmental Protection Service of Environment Canada in the 1970s. Lobsters in the harbour bioaccumulated the cadmium to high levels and for long periods,[80] and cadmium was found to be highly toxic to young salmon.[81] Both McLeese and Peterson collaborated with Sandy Ray, an analytical chemist at St Andrews from the mid-1970s to mid-1980s, who was a recognized expert on the fate and effects of cadmium in the aquatic environment;[82] they conducted various metal toxicity studies.[83]

An expert on benthic ecology joined the St Andrews Biological Station's environmental sciences and ecotoxicology group in 1969. This was Dave Wildish, who came as a postdoctoral fellow from the University of London in the UK. He worked on numerous pollution problems, especially with amphipods (tiny crustaceans), other benthic invertebrates, and sediments. From 1982 to 1997, he led his own group, first called Aquatic Assessments, and after 1987, the Applied Ecology group.

Wildish soon became involved in pesticide studies in New Brunswick. The highly persistent and toxic DDT was being replaced with fenitrothion, an organophosphate pesticide. The station responded by initiating toxicological work on this and other prospective pesticides. Wildish and Lister determined the "no-effect levels" of fenitrothion and concluded that ingestion of insects killed by the pesticide did not affect juvenile Atlantic salmon.[84] However, they noted that detrimental ecological effects could result from the scarcity of insects after forest spraying,[85] an observation similar to Ide's for DDT.

Wildish also studied the fate and effects of pulp mill effluents in the L'Etang Estuary, near St George, New Brunswick. A "new mill (on the L'Etang Inlet) came on line in 1971 and discharged its effluent into a tidal marine inlet. The effluent contained excessive amounts of organic material in the form of wood fibres and dissolved (organic) substances,"[86] the ecological effects of which St Andrews thoroughly documented.[87] Their work contributed to a review of the effects of the pulp and paper industry on aquatic ecosystems.[88] Their work also helped lead to revised federal regulations and guidelines for pulp and paper mill effluent, and resource management recommendations for the upper part of the L'Etang Inlet. The L'Etang Inlet pulp mill study and resulting actions represent "a fine example of the successful application of science to correct a longstanding environmental problem."[89]

Wildish also examined impacts of dredging and dumping on the benthos in and near Saint John harbour,[90] and changes that might result from dam construction in the Minas Basin.[91] In the 1980s and 1990s, he developed biological methods to predict "benthic production and for environmental monitoring in the coastal zone."[92] This led to his involvement with assessing the benthic impacts of salmon aquaculture in and around Passamaquoddy Bay, New Brunswick. Salmon faeces and waste food contributed to organic enrichment, which may show effects in the water column or sediments.[93] Scientists at St Andrews helped to develop cost-effective, electrode-based methods for monitoring redox and sulphides in the sedimentary environment under cages.[94] The electrode methods used for environmental impact assessments of mariculture sites by government agencies[95] are the foundation for the current environmental monitoring program for the Bay of Fundy salmon aquaculture industry.

A number of other St Andrews environmental sciences group scientists made important contributions to the field of aquatic physiology and toxicology, including Richard (Dick) Peterson, a prominent member during the 1970s and 1980s. His many studies included the toxic effects on fish physiology and behaviour of acid rain, pesticides, chlorinated hydrocarbons, metals (copper, zinc, cadmium), and industrial effluents (potash brines).[96] Gilles L. Lacroix also conducted acid rain studies.[97] Dave Aiken, David Scarratt, and Susan Waddy also conducted acid rain studies, investigated various aspects of the growth, reproduction, and ecology of lobsters, and made a series of toxicological studies in support of the environmental and aquaculture programs. They focused on pulp and paper mill effluents, pesticides, and pharmaceuticals; Scarrat also conducted numerous studies of the "Arrow" Bunker C (no. 6 fuel) oil spill and its effects on the benthos.[98] Waddy and her colleagues' most recent work shows potential sublethal effects of sea lice therapeutants placed into salmon feed on lobster molting and reproduction.

One student, Chris Metcalfe, briefly worked at St Andrews during the 1970s with Don McLeese and V. Zitko on pesticides, creosote, and PCBs. His subsequent career shows the importance of SABS' educational opportunities. Metcalfe went on to Ph.D. studies with Ron Sonstegard (Guelph), and to a successful career in aquatic toxicology at Trent University, specializing in the environmental effects of endocrine disruptors and health-care products.

Kats Haya and Les Burridge together developed aquatic and biochemical toxicology at St Andrews. Biochemical toxicology establishes links between chemical exposures and their potential toxic effects. Kats Haya was hired in 1978 under an Environment Canada initiative. Trained in pharmacology and biochemistry, Haya gave the toxicology program a much-needed perspective and capacity to study the uptake, metabolism, and depuration (self-cleansing) of chemicals in various marine species. Haya's research focused on developing biochemical indicators of sublethal effects caused by fish exposure to chemicals to assess the health of wild and cultured finfish populations. He worked on sublethal enzymatic responses, and hence moved "the toxicological research at SABS in the direction of biochemical (biomarker) and physiological bioindicators of exposure and effects."[99] Rather than simply exposing organisms to toxic chemicals to see how many died, or how long it took until they died, the research on "sublethal" responses investigated how organisms exposed to toxic chemicals were somehow affected in terms of some aspect of their bodily functioning – blood enzymes, hormones, and so on. This relatively new way of thinking in ecotoxicology required considerable scientific study to measure and interpret findings. Biochemical markers were macromolecules in the bodies of organisms that could indicate certain changes, and more crucially, were chemicals that would predictably occur or change if there had been stress or exposure. The physiological bioindicators required an understanding of the physiological responses that were indicators of chemical exposure.

Haya initially studied acid rain, examining the biochemical mechanisms of acid rain effects on Atlantic salmon. He demonstrated that decreased hatching success of salmon eggs exposed to low pH (4.5), compared to those held at pH 6.5, might be partially due to the reduced activity of chorionase, an enzyme that acts on the inner layer of the chorion and initiates the hatching process.[100] He also found that chronic exposure of juvenile salmon to low pH (4.5) caused a disruption in energy metabolism and a decreased food intake, resulting in slower growth.[101] Haya produced more than fifty publications in peer-reviewed journals on these and other topics before retiring in 2008.

His colleague Les Burridge, during the 1970s, worked with Don McLeese on pesticides. Much of Burridge's early work involved acute lethal bioassays with fish and lobsters. His later work with Haya focused on a variety of therapeutic chemicals, especially those used to control sea lice on the salmon. They have studied these chemicals' side-effects due to concerns about their effects on non-target species; they also collaborated with Susan Waddy, a lobster growth specialist, in many studies on lobsters at various life stages.[102]

From the 1980s onward, the increasingly important salmon aquaculture industry in south-western New Brunswick and other parts of the Bay of Fundy led Haya and Burridge to examine the biological and environmental interactions of chemical wastes from salmon aquaculture. Their aim was to develop methods and criteria for assessing environmental impacts, identifying hazards, and assessing risks of chemical wastes produced by the industry. Their recent projects have concerned identifying the chemical wastes, their distribution, and environmental fate. They have also investigated how therapeutants and pesticides used in aquaculture interact in the ecosystem and affect commercially important non-target fisheries resources.[103]

Haya and Burridge also took up investigations of the aquatic toxicology of marine phytotoxins, following their involvement in the Department of Fisheries and Ocean's response to the PEI farmed-mussel domoic acid crisis of 1987. Their phycotoxin research, on the impact of toxin-producing marine phytoplankton blooms on wild and cultured finfish and shellfish industries, sought to identify the lethality of marine phytotoxins to fish and determine how phycotoxins move through the marine food web. Their objectives were to develop techniques to detect and quantify these toxins, and to improve predicting and detecting toxic algal blooms and the accumulation of toxins in wild and cultured shellfish. Their recent emphasis, on seafood safety related to the integrated aquaculture of shellfish, salmon, and kelp,[104] is covered in detail by Jennifer Martin in an earlier chapter. The other focus of Haya and Burridge's recent research is to determine the lethal and sublethal biochemical effects on Atlantic salmon and juvenile cod, of pesticides, pharmaceuticals, health-care products, and contaminants from the offshore oil and gas industry (figure 11.5a and figure 11.5b).

## Highlights of the Environmental Science Program

While environmental studies have been conducted at the biological station at St Andrews for the past one hundred years, these have received special emphasis since the late 1950s. This paper, and the SABS website, document numerous contributions, which continued despite increasing funding challenges from the 1970s onward. The trend away from direct internal government funding made St Andrews scientists more dependent on external funding agencies, for which grant applications must be submitted, for both research and salary support, as Rob Stephenson has described in chapter 4, and put St Andrews scientists in competition with university scientists for scarcer and scarcer research funds. This trend diverts time away from research and writing. Despite such challenges, the station's scientists and technical support staff continued to make noteworthy and significant contributions to numerous areas of ecotoxicology.

In environmental chemistry, St Andrews scientists have identified problem chemicals in the marine environment, the seriousness of their status being determined by their properties of persistence, bioaccumulation, and toxicity; and have developed methods to measure these qualities (chemometrics) as a tool in environmental research. In the field of aquatic toxicology, St Andrews Biological Station researchers have investigated how different environmental factors affect the toxicity of identified problem chemicals, and identified lethal thresholds for chemical substances. They have measured the sensitivity of fisheries species such as lobsters to pollutants, and developed a critical area of aquatic toxicology, the study of the toxicity of mixtures such

11.5a Pesticide research

as pesticide formulations. They have developed sublethal molecular biomarkers and investigated behavioural toxicology – the behavioural responses of marine and aquatic animals to toxic chemicals.

St Andrews researchers – scientists and technical experts – have also been heavily involved in field-based biomonitoring. They designed environmental-effects monitoring (EEM) programs to track pollutants' effects on benthic species and communities, applying their knowledge to marine environmental protection. Their chemical and toxicological data on pesticides, metals, industrial chemicals, contaminated dredge spoils, and so on has contributed to the regulatory process at national and international levels.

Staff at the St Andrews Biological Station and its environmental sciences program in recent decades have assisted regional and national programs dealing with water quality and toxic chemicals. Staff scientists, both at St Andrews and the Halifax Department of Fisheries and Oceans Laboratory, in the 1980s and 1990s, served on various advisory teams and task forces.[105] When he was director of the St Andrews Biological Station, Robert (Bob) Cook served on the St Croix International Joint Commission (IJC) Water Quality Committee from 1979 to 1992.[106] In that period this committee sought to control water pollution from the

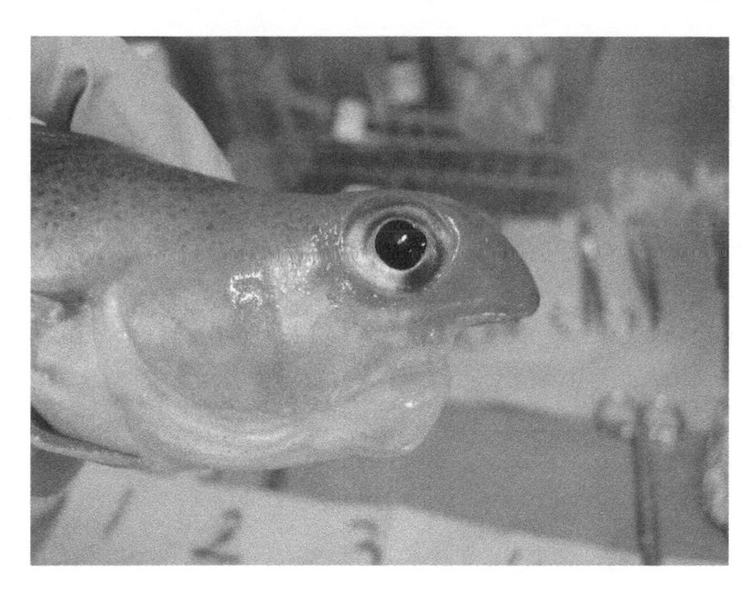

11.5b Pesticide research

Georgia Pacific pulp mill at Woodland, Maine, and to improve salmon passage up and down the St Croix River at Milltown, New Brunswick. In other examples, Kats Haya dealt with the Contaminants Act; Gilles Lacroix and Harry Freeman were involved in the acid rain national program; and from 1989 to 1994, Dave Wildish worked with federal and provincial governments for the New Brunswick Aquaculture Coordination program. On a more local scale, Vlado Zitko worked on the New Brunswick forest spray program; Sandy Ray and Jack Uthe worked as advisers for the Belledune Smelter and the local lobster fishery; and Dick Peterson was involved with the impacts of forestry practice. In addition, Bob Cook worked on the Atlantic Fish Habitat Task Force; and Harris Hord, Donald Riley, and Chris Morry were involved with fisheries environmental coordination. Similar contributions continued into the new millennium.

St Andrews scientists and technical staff have also responded to environmental emergencies: they studied pesticide formulations when awareness sharply grew that these contributed to environmental degradation; they investigated heavy metals in industrial effluents and their threats to the fisheries; and they mobilized efforts to respond to health crises sparked by spikes in algal toxins, such as the PEI mussel domoic acid crisis. The environmental sciences program has always been busy due to the frequency of major marine and freshwater quality issues in the Maritimes.[107] The program was very active on day-to-day issues, due to numerous environmental emergencies in recent decades, from the Chedabucto Bay oil spill in 1970 to studies on spruce budworm pesticides and the effects of salmon aquaculture chemicals.

It is important in this overview to recognize the outstanding efforts and productivity of the professional support staff. The environmental sciences program has had a small core staff over the past 50 years: Victor Carson, Bill Carson, Dave Sergeant, Vicki Jerome, Brenda Waiwood, Ray Peterson, Hugh Akagi, Art Wilson, Marcel Babineau, Ken MacKeigan, Monica Lyons, and Dawn Sephton, among others; they have been assisted by many contract staff and students. According to Burridge, "Nearly all were actively involved in designing as well as conducting the scientific studies, and have contributed greatly to the success of the program to date."[108]

The St Andrews Biological Station has developed and supported national and international communities of experts in environmental toxicology and environmental chemistry. Its staff has been involved in the Canadian Aquatic Toxicity Workshop (ATW) series since they began and hosted the 1995 ATW in St Andrews. They contributed to five others regionally (Moncton 1986, Charlottetown 2004, and Halifax in 1976, 1983, and 2007). Burridge was made the continuity chairman of this workshop series in 2005. As well, Zitko was an early member of the Society for Environmental Toxicology and Chemistry (SETAC), participating in many of its workshops.

Internationally, Zitko contributed at meetings of the London Dumping Convention (an international maritime organization), and established an international environmental chemistry group, becoming a recognized leader in environmental chemistry. One example of this is the extensive work he conducted on organohalogen compounds. When the station "joined an international study on organochlorine residues in wildlife, organized by the OECD," his was "one of the first papers on PCBs in Canadian fish," and his contributions continued with "the detection of PCTs (polychlorinated terphenyls) in the environment, and the observation that dioxins and furans (chlorinated dibenzodioxins and dibenzofurans) are not present in fish at concentrations exceeding 10–40 nanograms per gram."[109] This was original, state-of-the-art research at the time.

Dave Wildish was a founding member of the Working Group on Environmental Impacts of Mariculture of the International Council for the Exploration of the Seas (ICES), chaired by Professor Harald Rosenthal. Wildish co-convened, with Maurice Heral, an ICES international meeting in St Andrews on the environmental effects of mariculture, and edited the published proceedings.[110] Haya was the Canadian representative to the International Oceanographic Commission's Intergovernmental Panel for Harmful Algal Blooms. He was a member of the ICES Working Group on Environmental Interactions of Mariculture, and the Canada/US National Oceanic and Atmospheric Administration (NOAA) Collaborative Research Working Group.[111] Burridge is also involved with World Wildlife Fund for Nature working groups.

St Andrews scientists entered into numerous collaborative projects with Department of Fisheries and Oceans scientists from other institutions, and with scientists in other Canadian and foreign institutions. Scientists who collaborated closely with St Andrews scientists include O. Hutzinger, S. Safe, A. Defreitas, R.J. Nordstrom, B. Hargrave, R. Ackman, P. Yeats, J. Hellou, G. Fletcher, and C.L. Chou – many working at the National Research Council or Bedford Institute of Oceanography in Halifax, Nova Scotia. A recent product of such collaboration was the monograph *Environmental Effects of Marine Finfish Aquaculture*.[112] St Andrews staff have contributed to groups such as the Marine Environmental Quality Working Group of ICES, the Gulf of Maine Council on the Marine Environment, the Regional Association for Research on the Gulf of Maine, the Bay of Fundy Ecosystem Partnership, aquaculture groups, and numerous scientific societies. Because of the importance of understanding environmental change, SABS scientists also have been involved in numerous biodiversity initiatives, such as the Census of Marine Life, the Ocean Tracking Network, and the Centre for Marine Biodiversity.

Collaboration and outreach, then, have been part of the SABS environmental sciences program for many years. It has also included numerous connections and collaborations with university-based scientists, including: N.J. Poole of Aberdeen University, Scotland; D.D. Kristmanson of the University of New Brunswick; M.P. Miyares of the Universidad de Oviedo, Spain; J.C. Roff of Guelph University; and C. Crawford from the University of Tasmania, Australia.

From 1970, due to the proximity of the Huntsman Marine Science Centre – formerly the Huntsman Marine Laboratory, founded to facilitate university participation in marine science – many graduate students became available for collaborative projects. This collaboration has benefited both groups. Examples include my own work when I was a graduate student at Guelph in 1970–4, but working out of the Huntsman Marine Laboratory. I conducted studies on the sublethal effects of crude oil on lobster larvae and post-larvae,[113] and worked with Dave Scarratt on the *Arrow* oil spill of 4 February 1970, and the Eastport refinery issue in the early 1970s. In addition, this facility enabled Professor Michael Burt of the University of New Brunswick (Fredericton campus) to collaborate with St Andrews scientists on studies on mercury, fish, and parasites. Dave Gaskin and his many graduate students at the University of Guelph worked with St Andrews biologists on studies of marine mammals and pesticides through the 1970s and 1980s. Dave Wildish partnered with Gerhard Pohle of the Huntsman Marine Laboratory to investigate the effects of aquaculture operations on the benthos. Other SABS-Huntsman collaborations included investigations with Bill Hogans on sea lice infestations of salmon and the effectiveness and toxicity of pesticides; with Susan Curry of Mount Allison University to investigate heat shock proteins; and with Tom Moon, former director of the Huntsman Marine Laboratory on fish bioenergetics and fish biochemistry. No doubt there were others, and all of them illustrate the important research linkages between SABS and the university sector.

Scientists at the biological station over the past hundred years, then, have made many kinds of contributions to aquatic environmental science and ecotoxicology. Not only have they helped to create the field and its wider community, they have identified the presence of many industrial chemicals in environmental compartments such as water, sediments, and tissues. They developed new techniques to test the toxicity of chemicals and chemical mixtures, and assisted in understanding the many factors modifying the toxicity of chemical substances. They have identified the presence of chemically induced effects under natural conditions, and determined how to estimate low-level chemical contamination risks in freshwater, estuarine, and coastal waters close to industrial development and activity. Their studies have contributed greatly to developing what is still the nascent field of aquatic science.

In the new century, St Andrews and its region and scientists have faced many challenges. These include the need to protect Passamaquoddy Bay, the greater Bay of Fundy, the Gulf of Maine, and the northwest Atlantic; the need to protect and conserve fisheries and the quality of fish, as mandated in the Fisheries Act; understanding and remediating aquaculture's effects in coastal waters; anticipating the effects of tidal power and liquid natural gas (LNG) transportation developments; and actively implementing the Oceans Act to protect marine environmental quality and the health of Canadian seas. It is vital to continue the interdisciplinary ecosystem approach in the various environmental studies, an approach developed and successfully implemented at St Andrews. Future environmental science and ecotoxicology programs at the St Andrews Biological Station should build on, and benefit from, the knowledge and lessons of its first one hundred years of aquatic science.

## NOTES

Many thanks are due to Dr Robert Stephenson for the invitation to prepare this paper and for support attending the conference in St Andrews, NB, in October 2008. The reviews by Hart, Sprague, Trippel, Wildish, and Zitko were invaluable as key sources to initiate this brief history. I am very grateful to Les Burridge, Bob Cook, Kats Haya, and Dave Wildish for information and their perspectives. Charlotte McAdam and Joanne Cleghorn of the Library, SABS, were most generous with their time and knowledge tracking down key papers, reports, and archival pictures, and showed enthusiastic support of the SABS history project as a whole.

1 See, among others, M. Waldichuk, "The Nature and Extent of Marine Contamination Caused by Land-based Sources in Canada" and J.F. Uthe and V. Zitko, "An Overview of Marine Environmental Quality Issues on the Atlantic Coast of Canada," both in *Canadian Conference on Marine Environmental Quality: Proceedings. Halifax, NS, 29 February to 3 March, 1988*, ed. P.G. Wells and J. Gratwick (Halifax, NS: International Institute for Transportation and Ocean Policy Studies, Dalhousie University, 1988), 75–135 and 199–205; P.G. Wells and S.J. Rolston, eds., *Health of Our Oceans: A Status Report on Canadian Marine Environmental Quality. Conservation and Protection*, 2nd ed. (Ottawa: Environment Canada, March 1991), 1–187; and P.G. Wells, P.D. Keizer, J.L. Martin, P.A. Yeats, K.M. Ellis, and D.W Johnston, "The Chemical Environment of the Bay of Fundy," in *Bay of Fundy Issues: A Scientific Overview. Workshop Proceedings, Wolfville, N.S., Jan. 29–Feb. 1, 1996*, ed. J.A. Percy, P.G. Wells, and A. Evans (Dartmouth, NS, and Sackville, NB: Environment Canada Atlantic Region Occasional Report no. 8, 1997), 37–61.

2 Hubbard, *A Science on the Scales*, 176.

3 E.A. Trippel, "The First Marine Biological Station in Canada: 100 Years of Scientific Research at St Andrews," *Canadian Journal of Fisheries and Aquatic Science* 56 (1999): 2495–507.

4 Also see Uthe and Zitko, "Overview of Marine Environmental Quality Issues"; Wells, Keizer, Martin, Yeats, Ellis, and Johnston, "The Chemical Environment of the Bay of Fundy"; V. Zitko and D.J. Wildish, "The First Marine Biological Station in Canada: Highlights of Environmental Research at St Andrews," *Water Quality Research Journal of Canada* 35 (2000): 809–17. Short reviews of SABS fisheries and environmental science have been written previously, most within the general history of the Fisheries Research Board of Canada and the later Department of Fisheries and Oceans; for example: J.L. Hart, "Fifty Years of Research in Aquatic Biology"; Information Canada, *Review 1969–1970: The Fisheries Research Board of Canada* (Ottawa: Information Canada, 1971), 1–217; Environment Canada, *Review 1971–1972: The Fisheries Research Board of Canada* (Ottawa: Information Canada, 1973), 1–230; Kenneth Johnstone's description of the entire history of the FRBC, up to the 1970s, when the board was disbanded, in *Aquatic Explorers*, 1–342; V. Zitko, "Fifteen Years of Environmental Research," *Marine Pollution Bulletin* 10 (1979): 100–3; A.W.H. Needler, "Seventy-Fifth Anniversary of Two Canadian Biological Stations"; Trippel, "The First Marine Biological Station in Canada: 100 Years of Scientific Research at St Andrews"; J.M. Hubbard, "An Independent Progress"; and Hubbard, *A Science on the Scales*, 1–351. Other recent histories related to biological oceanography and fisheries science in Atlantic Canada include J.E. Stewart and A. Safer, "A Retrospective: Three Quarters of a Century at the Halifax Fisheries Research Laboratory," *Proceedings of the Nova Scotian Institute of Science* 43 (2005): 19–44; R.G. Halliday and L.P. Fanning, "A History of Marine Fisheries Science in Atlantic Canada and Its Role in the Management of Fisheries," *Proceedings of the Nova Scotian Institute of Science* 43 (2006): 159–83; and W.G. Harrison, "Biological Oceanography in Canada (with Special Reference to Federal Government Science)," *Proceedings of the Nova Scotian Institute of Science* 43 (2006): 129–58. Additional information and insights came from the conference at St Andrews Biological Station, October 2008.

5 This chapter focuses on some highlights of the environmental science and ecotoxicology spanning 1908–2008, while complementing the reviews of Trippel (1999) and Zitko and Wildish (2000). Trippel's "The First Marine Biological Station in Canada" surveyed SABS's scientific history in an excellent and readable overview. All of this literature must be read to appreciate

the rich scientific history of St Andrews and of Canadian marine science in the twentieth century, and to understand the context of contemporary environmental studies conducted in the Maritime provinces. General sources concerning the environmental research and ecology/toxicology programs and their outputs at SABS are numerous; remaining archives and report holdings left from the unfortunately recently dismantled library at SABS are invaluable: for example, A.G. Huntsman, "Progress in Fisheries Research in Canada," in *Fifty Years Retrospect, Canada, 1882–1932* (Toronto: Royal Society of Canada / Ryerson Press, 1932); J.L. Hart, "Fifty Years of Research in Aquatic Biology," 159–6; and J.B. Sprague's brief histories about early Canadian aquatic toxicology, "Perspective on a Career: Changing Approaches to Water Pollution Evaluation," *Marine Pollution Bulletin* 25 (1992): 6–13, and "An Informal Look at the Parents of Canadian Aquatic Toxicology," in *Proceedings of the 22nd Annual Aquatic Toxicity Workshop: October 2–4, 1995, St Andrews, NB*, ed. K. Haya and A.J. Niemi, *Canadian Technical Reports of Fisheries and Aquatic Science* 2093 (1996): 2–14.

6 V. Zitko, "Fifteen Years of Environmental Research," *Marine Pollution Bulletin* 10 (1979): 100–3.

7 V. Zitko, and D.J. Wildish, "The First Marine Biological Station in Canada: Highlights of Environmental Research at St Andrews," *Water Quality Research Journal of Canada* 35 (2000): 809–17.

8 L.M. Dickie, "The Quest for Excellence: Funding Marine Science Research and Development," in *The Sea Has Many Voices: Ocean Policy for a Complex World*, ed. C. Lamson (Montreal: McGill-Queen's University Press, 1994), 289–313; also see M. Harris, *Lament for an Ocean: The Collapse of the Atlantic Cod Fishery. A True Crime Story* (Toronto: McClelland & Stewart, 1998); and Hayes, *The Chaining of Prometheus*.

9 Hart, "Fifty Years of Research in Aquatic Biology," 1132–3. Hart's history provides no details on the pollution studies.

10 Sprague, "Parents of Canadian Aquatic Toxicology," 11.

11 A.P. Knight, "The Effects of Polluted Water on Fish Life," *Contributions to Canadian Biology* (1901): 9–18; and A.P. Knight, "The Effects of Dynamite Explosions on Fish Life: A Preliminary Report," *Contributions to Canadian Biology* (1902–5): 21–30.

12 Hart, "Fifty Years of Research in Aquatic Biology," 1133.

13 Huntsman, "Progress in Fisheries Research in Canada," 159–61.

14 Johnstone, *Aquatic Explorers*, 105.

15 Ibid., 122.

16 M.W. Smith, "Physical and Biological Conditions in Heavily Fertilized Waters," *Journal of the Biological Board of Canada* 1 (1934): 67–93.

17 Hart, "Fifty Years of Research in Aquatic Biology," 1127–61.
18 International Passamaquoddy Fisheries Commission, "Report on the Scientific Investigations of the International Passamaquoddy Fisheries Commission," Manuscript, 30 June 1933, 30–3. Huntsman Collection, University of Toronto Archives, B1979–0010, box 9.
19 Hubbard, *A Science on the Scales*, 172–82.
20 Johnstone, *Aquatic Explorers*, 144.
21 D.G. Wilder, "The Relative Toxicity of Certain Metals to Lobsters," *Journal of the Fisheries Research Board of Canada* 8 (1952): 486–7.
22 Fisheries Research Board of Canada, *Annual Report of the Fisheries Research Board of Canada, 1955* (Ottawa: Government of Canada, 1955), 19. The detailed reports mentioned mostly appeared in the annual reports.
23 Johnstone, *Aquatic Explorers*, 225.
24 C.J. Kerswill and P.F. Elson, "Preliminary Observations on Effects of 1954 DDT Spraying on Miramichi Salmon Stocks," *Fisheries Research Board of Canada Progress Reports (Atlantic)* 62 (July 1955): 17–24; and C.J. Kerswill, P.F. Elson, M.H.A. Keenleyside, and J.B. Sprague, "Effects on Young Salmon of Forest Spraying with DDT," *Transactions of the Second Seminar on Biological Problems in Water Pollution, 1959. Technical Report no. W60-3* (Cincinnati: Robert A. Taft Sanitary Engineering Center, 1960), 71.
25 Elson, "Effects on Wild Young Salmon of Spraying DDT"; C.D. Grant, "Effects on Aquatic Insects of Forest Spraying with Phosphamidon in New Brunswick," *Journal of the Fisheries Research Board of Canada* 24 (1967): 823–32; Ide, "Effects of Forest Spraying with DDT on Aquatic Insects"; Keenleyside, "Effects of Forest Spraying with DDT"; C.J. Kerswill, "Studies on Effects of Forest Spraying with Insecticides, 1952–63, on Fish and Aquatic Invertebrates in New Brunswick Streams: Introduction and Summary," *Journal of the Fisheries Research Board of Canada* 24 (1967): 701–8; C.J. Kerswill and H.E. Edwards, "Fish Losses after Forest Spraying with Insecticides in New Brunswick, 1952–62, as Shown by Caged Specimens and Other Observations," *Journal of the Fisheries Research Board of Canada* 24 (1967): 709–29.
26 Hart, "Fifty Years of Research in Aquatic Biology," 1148.
27 D.F. Alderdice and M.E. Worthington, "Toxicity of a DDT Forest Spray to Young Salmon," *Canadian Fish Culturist* 24 (1959): 41–8; J.B. Sprague, "Parents of Canadian Aquatic Toxicology."
28 F. Premdas and J.M. Anderson, "The Uptake and Detoxification of C14-labelled DDT in Atlantic Salmon, *Salmo salar*," *Journal of the Fisheries Research Board of Canada* 20 (1963): 827–37.

29 F.P. Ide, "Effect of Forest Spraying with DDT on Aquatic Insects of Salmon Streams," *Transactions of the American Fisheries Society* 86 (1957): 208–19; F.P. Ide, "Effect of Forest Spraying with DDT on Aquatic Insects, Food of Salmon and Trout, in New Brunswick," *Proceedings of the Entomological Society of Ontario* 91 (1961): 39–40; and Ide, "Effects of Forest Spraying with DDT" (1967).
30 Sprague, "Parents of Canadian Aquatic Toxicology," 11.
31 Johnstone, *Aquatic Explorers*, 245.
32 Ibid., 247, 263, 265.
33 Ibid., 269; J.B. Sprague, "Promising Anti-Pollutant: Chelating Agent NTA Protects Fish from Copper and Zinc," *Nature* 220 (1968): 1345–6.
34 J.R. Weir, quoted in Johnstone, *Aquatic Explorers*, 299.
35 Ibid.
36 Johnstone, *Aquatic Explorers*, 301.
37 F.E.J. Fry, "Effects of the Environment on Animal Activity," *University of Toronto Studies, Biology Series no. 55; Publications of the Ontario Fisheries Research Laboratory* 68 (1947): 1–62; Sprague, "Parents of Canadian Aquatic Toxicology."
38 J.B. Sprague, "Perspective on a Career: Changing Approaches to Water Pollution Evaluation," *Marine Pollution Bulletin* 25 (1992): 6–13.
39 P.G. Wells, E. Pessah, and W.R. Parker, "The Toxicity of Raw and Treated Drainage from Heath Steele Mines, N.B., during Period of September–October, 1974," Environmental Protection Surveillance Report EPS-5-AR-74–14 (1979): 1–27.
40 Zitko and Wildish, "Highlights of Environmental Research at St Andrews," 809–17. The scientific work they were describing here was published in J.B. Sprague, "Effects of Sublethal Concentrations of Zinc and Copper on Migration of Atlantic Salmon," in *Biological Problems in Water Pollution, Third Seminar* (Cincinnati: Robert. A. Taft Sanitary Engineering Center: United States Public Health Service, 1962), Publication no. 999-WP-25, 332–3.
41 Zitko and Wildish, "Environmental Research at St Andrews," 810.
42 Sprague, "Changing Approaches to Water Pollution Evaluation."
43 P.F. Elson, L.M. Lauzier, and V. Zitko, "A Preliminary Study of Salmon Movements in a Polluted Estuary," in *Marine Pollution and Sea Life*, in *FAO Technical Conference on Marine Pollution and Its Effects on Living Resources and Fishing, Rome, Italy*, ed. M. Ruivo (Oxford: Fishing News Books, 1972), 325–30; and P.F. Elson, A.L. Meister, J.W. Saunders, R.L. Saunders, J.B. Sprague, and V. Zitko, "Impact of Chemical Pollution on Atlantic

Salmon in North America," *International Atlantic Salmon Symposium 1972*, International Atlantic Salmon Foundation special publication 4 (1973): 83–110.

44 Zitko and Wildish, "Environmental Research at St Andrews," 811.

45 J.B. Sprague, "Toxicity of Pollution to Aquatic Life: A Summary of Research in Canada," Fisheries Research Board of Canada Manuscript Report 771 (1964): 1–18.

46 J.B. Sprague, "Resistance of Four Freshwater Crustaceans to Lethal High Temperature and Low Oxygen," *Journal of the Fisheries Research Board of Canada* 20 (1963): 387–415; J.B. Sprague, "Factors That Modify Toxicity," in *Fundamentals of Aquatic Toxicology*, ed. G.M. Rand and S.R. Petrocelli (Washington, DC: Hemisphere Publishing Corp., 1985; 2nd ed. Rand, 1995), 124–63, 1012–51.

47 J.B. Sprague, "Lethal Concentrations of Copper and Zinc for Young Atlantic Salmon," *Journal of the Fisheries Research Board of Canada* 21 (1964): 17–26; J.B. Sprague, "Highly Alkaline Water Caused by Asbestos-Cement Pipeline," *Progressive Fish-Culturist* 26 (1964): 111–14; Sprague, "Promising Anti-Pollutant," 1345–6.

48 J.B. Sprague, "Negative Test of Apparent DDT Resistance in an Aquatic Insect after Seven Years' Exposure to Aerial Spraying," Fisheries Research Board of Canada Manuscript Report 908 (1967): 1–28; J.B. Sprague, "Apparent DDT Tolerance in an Aquatic Insect Disproved by Test," *Canadian Entomologist* 100 (1968): 279–84.

49 J.B. Sprague and W.G. Carson, "Toxicity Tests with Oil Dispersants in Connection with Oil Spill at Chedabucto Bay, Nova Scotia," *Fisheries Research Board of Canada Technical Report* 201 (1970): 1–30.

50 J.B. Sprague, "Avoidance of Copper-Zinc Solutions by Young Salmon in the Laboratory," *Journal of the Water Pollution Control Federation* 36 (1964): 990–1004; J.B. Sprague, "Apparatus Used for Studying Avoidance of Pollutants by Young Atlantic Salmon," in *Biological Problems in Water Pollution: Third Seminar, 1962. Robert A. Taft Sanitary Engineering Center*, ed. C.M. Tarzwell (Cincinnati, OH: US Department of Health, Education and Welfare, Public Health Service, Division of Water Supply and Pollution Control, 1965); Sprague, "Effects of Sublethal Concentrations of Zinc and Copper on Migration of Atlantic Salmon," in *Biological Problems in Water Pollution: Third Seminar, 1962. Robert A. Taft Sanitary Engineering Center*, ed. C.M. Tarzwell (Cincinnati, OH: US Department of Health, Education and Welfare, Public Health Service, Division of Water Supply and Pollution Control, 1965), 332–3; J.B. Sprague "Avoidance Reactions of Rainbow Trout to Zinc Sulphate Solutions," *Water Research* 2 (1968): 367–72.

51 J.B. Sprague, "Dissolved Oxygen in Restigouche River Estuary, NB, October 1966," Fisheries Research Board of Canada Manuscript Report 903 (1967): 1–17.
52 Sprague, "Studying Avoidance of Pollutants by Young Atlantic Salmon."
53 J.B. Sprague, "Measurement of Pollutant Toxicity to Fish. I. Bioassay Methods for Acute Toxicity," *Water Research* 3 (1969): 793–821; J.B. Sprague, "Measurement of Pollutant Toxicity to Fish. II. Utilizing and Applying Bioassay Results," *Water Research* 4 (1970): 3–32; J.B. Sprague, "Measurement of Pollutant Toxicity to Fish. III. Sublethal Effects and 'Safe' Concentrations," *Water Research* 5 (1971): 245–66.
54 J.B. Sprague, "The ABC's of Pollutant Bioassay Using Fish," in *Biological Methods for the Assessment of Water Quality*, ed. J. Cairns and K.L. Dickson, ASTM Special Technical Publication 528 (Philadelphia: ASTM, 1973), 6–30; American Public Health Association, *Standard Methods for the Examination of Water and Waste Water*, 12th ed. (Washington, DC: APHA, 1969).
55 Sprague, "Measurement of Pollutant Toxicity to Fish."
56 GESAMP, "The Revised GESAMP Hazard Evaluation Procedure for Chemical Substances Carried by Ships," Reports and Studies of GESAMP, 64 (2004): 1–121.
57 Sprague, "Parents of Canadian Aquatic Toxicology," 2.
58 K. Haya, personal communication; and Zitko and Wildish, "Highlights of Environmental Research at St Andrews."
59 V. Zitko, B.J. Finlayson, D.J. Wildish, J.M. Anderson, and A.C. Kohler, "Methylmercury in Freshwater and Marine Fishes in New Brunswick, in the Bay of Fundy, and on the Nova Scotia Banks," *Journal of the Fisheries Research Board of Canada* 28 (1971): 1285–91.
60 Zitko and Wildish, "Environmental Research at St Andrews."
61 V. Zitko, "Determination of Residual Fuel Oil Contamination of Aquatic Animals," *Bulletin of Environmental Contamination and Toxicology* 5 (1971): 559–64.
62 Sprague and Carson, "Toxicity Tests with Oil Dispersants."
63 V. Zitko, D.E. Aiken, S.N. Tibbo, K.W.T. Besch, and J.M. Anderson, "Toxicity of Yellow Phosphorus to Herring (*Clupea harengus*), Atlantic Salmon (*Salmo salar*), Lobster (*Homarus americanus*), and Beach Flea (*Gammarus oceanicus*)," *Journal of the Fisheries Research Board of Canada* 27 (1970): 21–9.
64 Haya, personal communication.
65 Tom Kosastsky, "Improving Epidemic Control: Lessons from the 1987 Toxic Mussels Affair," *Canadian Medical Association Journal* 147:12 (1992): 1769–72; J.L.C. Wright et al., "Identification of Domoic Acid, a Neuroexcitatory

Amino Acid, in Toxic Mussels from Eastern Prince Edward Island," *Canadian Journal of Chemistry* 67 (1989): 481–90.

66 Wells, Keizer, Martin, Yeats, Ellis and Johnston, "The Chemical Environment of the Bay of Fundy," 37–61.

67 V. Zitko and E. Arsenault, "Chlorinated Paraffins: Properties, Uses, and Pollution Potential," *Fisheries and Marine Service Research and Development Technical Report* 491 (1974): 1–38; V. Zitko and E. Arsenault, "Fate of High-Molecular Weight Chlorinated Paraffins in the Aquatic Environment," *American Chemical Society, Division of Environmental Chemistry* 15 (1975): 174–6; V. Zitko and E. Arsenault, "Fate of High-Molecular Weight Chlorinated Paraffins in the Aquatic Environment,' in *Advances in Environmental Science and Technology*, vol. 8.2, ed. I.H. Suffet (New York: Wiley-Interscience, 1977), 409–18; and V. Zitko, "Uptake and Excretion of Chlorinated and Brominated Hydrocarbons by Fish," *Fisheries and Marine Service Technical Report* 737 (1977): 1–14.

68 Zitko and Wildish, "Highlights of Environmental Research at St Andrews," 811.

69 V. Zitko and W.V. Carson, "Accumulation of Thallium in Clams and Mussels," *Bulletin of Environmental Contamination and Toxicology* 14 (1975): 530–3; V. Zitko, W.V. Carson, and W.G. Carson, "Thallium: Occurrence in the Environment and Toxicity to Fish," *Bulletin of Environmental Contamination and Toxicology* 13 (1975): 23–30.

70 Zitko and Wildish, "Highlights of Environmental Research at St Andrews," 812.

71 T. Yamagishi, T. Miyazaki, S. Horii, and K. Akiyama, "Synthetic Musk Residues in Biota and Water from Tama River and Tokyo Bay (Japan)," *Archives of Environmental Contamination and Toxicology* 12:1 (1983): 83–9; and T. Yamagishi, T. Miyazaki, S. Horii, and S. Kaneko, "Identification of Musk Xylene and Musk Ketone in Freshwater Fish Collected from the Tama River," *Tokyo Bulletin of Environmental Contamination and Toxicology* 6:5 (1981): 656–62.

72 D.W. McLeese, "Behaviour of Lobsters Exposed to Bleached Kraft Mill Effluent," *Journal of the Fisheries Research Board of Canada* 27 (1970): 731–6; "Response of Lobsters, *Homarus americanus*, to Odor Solution in the Presence of Bleached Kraft Mill Effluent," *Journal of the Fisheries Research Board of Canada* 30 (1973): 279–82; "Olfactory Response and Fenitrothion Toxicity in American Lobsters (*Homarus americanus*)," *Journal of the Fisheries Research Board of Canada* 31 (1974): 1127–31; and "Chemosensory Response of American Lobsters (*Homarus americanus*) in the Presence of Copper and Phosphamidon," *Journal of the Fisheries Research Board of Canada* 32 (1975): 2055–60.

73 B.L. Olla, J. Atema, R. Forward, J. Kittredge, R.J. Livingston, D.W. McLeese, D.C. Miller, W.B. Vernberg, P.G. Wells, and K. Wilson, "The Role of Behavior in Marine Pollution Monitoring. Behavior Panel Report," *Rapports et procès-verbaux des réunions du Conseil International pour l'Exploration de la Mer* 179 (1980): 174–81.
74 D.W. McLeese, "Toxicity of Phosphamidon to American Lobsters (*Homarus americanus*) held at 4 and 12° C," *Journal of the Fisheries Research Board of Canada* 31 (1974): 1556–8; and D.W. McLeese and C.D. Metcalfe, "Toxicity of Mixtures of Phosphamidon and Methidathion to Lobsters (*Homarus americanus*)," *Chemosphere* 8 (1979): 59–62.
75 D.W. McLeese, "Fenitrothion Toxicity to the Freshwater Crayfish, *Orconectes limosus*," *Bulletin of Environmental Contamination and Toxicology* 16 (1976): 411–16.
76 D.W. McLeese, C.D. Metcalfe, and V. Zitko, "Lethality of Permethrin, Cypermethrin and Fenvalerate to Salmon, Lobster and Shrimp," *Bulletin of Environmental Contamination and Toxicology* 25 (1980): 950–5.
77 D.W. McLeese, D.B. Sergeant, C.D. Metcalf, V. Zitko, and L.E. Burridge, "Uptake and Excretion of Aminocarb, Nonylphenol, and Pesticide Diluent 585 by Mussels (*Mytilus edulis*)," *Bulletin of Environmental Contamination and Toxicology* 24 (1980): 575–81.
78 D.A. Lord, R.A.F. Matheson, L. Stuart, J.J. Swiss, and P.G. Wells, "Environmental Monitoring of the 1976 Spruce Budworm Spray Program in New Brunswick, Canada," Fisheries and Environment Canada, Environmental Protection Service, Surveillance Report EPS-5-AR-78–3 (1978): 1–161; P.G. Wells, W.R. Parker, and D. Vaughan, "The Acute Toxicity of Four Insecticides and Their Formulations to Fingerling Rainbow Trout, *Salmo gairdneri*," in "Environmental Monitoring of the 1976 Spruce Budworm Spray Program in New Brunswick, Canada," Fisheries and Environment Canada, EPS, Surveillance Report EPS-5-AR-78–3 (1978): 84–110; and P.G. Wells, R.A. Matheson, D.A. Lord, and K.G. Doe, "Environmental Monitoring of the 1978 Spruce Budworm Spray Program in New Brunswick, Canada: Field Sampling and Aquatic Toxicity Studies with Fish," Environment Canada, Environmental Protection Service Surveillance Report EPS-5-AR-79–1 (1979): 1–14.
79 Zitko and Wildish, "Highlights of Environmental Research at St Andrews."
80 D.W. McLeese, "Uptake and Excretion of Cadmium by Marine Organisms from Sea Water with Cadmium at Low Concentrations: A Review," in "Cadmium Pollution of Belledune Harbour, New Brunswick, Canada," ed. J.F. Uthe and V. Zitko, *Canadian Technical Report of Fisheries and Aquatic Science* 963 (1980): 55–63; D.W. McLeese, S. Ray, and L.E. Burridge, "Lack of Excretion of Cadmium from Lobsters," *Chemosphere* 10 (1981): 775–8.

81 R.H. Peterson, J.L. Metcalfe, and S. Ray, "The Effects of Cadmium on Yolk Utilization, Growth, and Survival of Atlantic Salmon Alevins and Newly Feeding Fry," *Archives of Environmental Contamination and Toxicology* 12 (1983): 37–44.
82 S. Ray, "Bioaccumulation of Cadmium in Marine Organisms," *Experientia* 40 (1984): 14–23.
83 McLeese, Ray, and Burridge, "Lack of Excretion of Cadmium from Lobsters"; Peterson, Metcalfe, and Ray, "The Effects of Cadmium on Yolk Utilization, Growth, and Survival"; S. Ray and W. White, "Selected Aquatic Plants as Indicator Species for Heavy Metal Pollution," *Journal of Environmental Science and Health* A11 (1976): 717–25; S. Ray and W.J. White, "Some Observations on Heavy Metal Concentration in Northeastern New Brunswick Estuarine Surficial Sediments," *Fisheries and Marine Service Technical Report* 696 (1977); S. Ray and W. White, "*Equisetum arvense* – An Aquatic Vascular Plant as a Biological Monitor for Heavy Metal Pollution," *Chemosphere* 8 (1979): 125–8; S. Ray and J. Coffin, "Ecological Effects of Cadmium Pollution in the Aquatic Environment: A Review," *Fisheries and Marine Service Technical Report* 734 (1977); S. Ray, D.W. McLeese, and C.D. Metcalfe, "Heavy Metals in Sediments and in Invertebrates from Three Coastal Areas in New Brunswick, Canada: A Natural Bioassay," *ICES CIM 1979/E:29* (1979): 1–6; and S. Ray, D.W. McLeese, and D. Pezzack, "Chelation and Interelemental Effects on the Bioaccumulation of Heavy Metals by Marine Invertebrates," *Proceedings of the International Conference for Management and Control of Heavy Metals in the Environment, London, U.K. (1979)* (London: Pergamon Press, 1983), 35–8.
84 D.J. Wildish and N.A. Lister, "Biological Effects of Fenitrothion in the Diet of Brook Trout," *Bulletin of Environmental Contamination and Toxicology* 10 (1973): 333–9.
85 Zitko and Wildish, "Environmental Research at St Andrews."
86 Ibid., 810.
87 Ibid. A review paper on L'Etang Inlet research by P.M. Strain, D.J. Wildish, and P.A. Yeats appeared in 1995 ("The Application of Simple Models of Nutrient Loading and Oxygen Demand to the Management of a Marine Tidal Inlet," *Marine Pollution Bulletin* 30 [1995]: 253–61). This paper summarized earlier research, including the following by scientists at the biological station: D.D. Kristmanson, D.J. Wildish, and N.J. Poole, "Mixing of Pulp Mill Effluents in the Upper L'Etang," Fisheries Research Board of Canada Manuscript Report 1416 (1976): 1–36; N.J. Poole, D.J. Wildish, and N.A. Lister, "The Use of Micro-ecosystem Models to Investigate Pollution by Pulp Mill Effluent," Fisheries Research Board of Canada

Manuscript Report 1403 (1976); N.J. Poole, D.J. Wildish, and N.A. Lister, "Effects of a Neutral-Sulfite Pulp Effluent on Some Chemical and Biological Parameters in the L'Etang Inlet, New Brunswick. L'Etang Inlet Survey III," Fisheries Research Board of Canada Manuscript Report 1404 (1976): 1–27; N.J. Poole, R.J. Parkes, and D.J. Wildish, "The Reactions of the Estuarine Ecosystem to Effluent from the Pulp and Paper Industry," *Helgolander Wissenschaftliche Meeresuntersuchungen* (now *Helgoland Marine Research*) 30 (1977): 622–32; N.J. Poole, D.J. Wildish, and D.D. Kristmanson, "The Effects of the Pulp and Paper Industry on the Aquatic Environment," *CRC Critical Reviews in Environmental Control* 8 (1978): 153–95; D.J. Wildish, H. Akagi, and N.J. Poole, "Avoidance by Herring of Sulphite Pulp Mill Effluents," International Council for the Exploration of the Sea, Fisheries Improvement Committee, *C.M. 1976/E:26*: 1–8; D.J. Wildish, N.J. Poole, and D.D. Kristmanson, "The Effect of Anaerobiosis on Measurement of Sulfite Pulp Mill Effluent Concentration in Estuarine Water by U.V. Spectrophotometry," *Bulletin of Environmental Contamination and Toxicology* 16 (1976): 208–13; D.J. Wildish, H. Akagi, and N.J. Poole, "Avoidance by Herring of Dissolved Components in Pulp Mill Effluents," *Bulletin of Environmental Contamination and Toxicology* 18 (1977): 521–5; and D.J. Wildish, N.J. Poole, and D.D. Kristmanson, "Temporal Changes of Sublittoral Macrofauna in L'Etang Inlet Caused by Sulfite Pulp Mill Pollution," Fisheries and Marines Service Technical Report 718 (1977): 1–13. D.J. Wildish and G.W. Pohle published a further overview in 2005: "Benthic Macrofaunal Changes Resulting from Finfish Mariculture," in *The Handbook of Environmental Chemistry*, vol. 5, part M, *Water Pollution: Environmental Effects of Marine Finfish Aquaculture*, ed. B.T. Hargrave (Berlin: Springer-Verlag, 2005), 275–304.

88 Poole, Wildish, and Kristmanson, "The Effects of the Pulp and Paper Industry."

89 Robert H. Cook, personal communication.

90 D.J. Wildish and M.L.H. Thomas, "Effects of Dredging and Dumping on Benthos of Saint John Harbour," *Marine Environmental Research* 15 (1985): 45–57.

91 D.J. Wildish, D.L. Peer, and D.A. Greenberg, "Benthic Macrofauna Production in the Bay of Fundy and Possible Effects of a Tidal Power Barrage at Economy Point–Cape Tenny," *Canadian Journal of Fisheries and Aquatic Science* 43 (1986): 2410–17.

92 Zitko and Wildish, "Highlights of Environmental Research at St Andrews."

93 D.J. Wildish, P.D. Keizer, A.J. Wilson, and J.L. Martin, "Seasonal Changes of Dissolved Oxygen and Plant Nutrients in Seawater near Salmonid

Net Pens in the Macrotidal Bay of Fundy," *Canadian Journal of Fisheries and Aquatic Science* 50 (1993): 303–11; B.T. Hargrave, D.E. Duplisea, E. Pfeiffer, and D.J. Wildish, "Seasonal Changes in Benthic Fluxes of Dissolved Oxygen and Ammonia Associated with Marine Cultured Atlantic Salmon," *Marine Ecology Progress Series* 96 (1993): 249–57; and B.T. Hargrave, G.A. Phillips, L.I. Doucette, M.J. White, T.G. Milligan, D.J. Wildish, and R.E. Cranston, "Assessing Benthic Impacts of Organic Enrichment from Marine Aquaculture," *Water, Air and Soil Pollution* 99 (1997): 641–50.

94 Hargrave et al., "Assessing Benthic Impacts of Organic Enrichment"; and D.J. Wildish, H.M. Akagi, N. Hamilton, and B.T. Hargrave, "A Recommended Method for Monitoring Sediments to Detect Organic Enrichment from Mariculture in the Bay of Fundy," *Canadian Technical Report of Fisheries and Aquatic Science* 2286 (1999): 1–31.

95 Zitko and Wildish, "Environmental Research at St Andrews," 814.

96 R.H. Peterson, "Temperature Selection of Juvenile Atlantic Salmon (*Salmo salar L.*) Exposed to Some Pesticides," Fisheries Research Board of Canada Manuscript Report 1251 (1973): 1–9; R.H. Peterson, "Temperature Selection of Atlantic Salmon (*Salmo salar*) and Brook Trout (*Salvelinus fontinalis*) as Influenced by Various Chlorinated Hydrocarbons," *Journal of the Fisheries Research Board of Canada* 30 (1973): 1091–7; R.H. Peterson, "Influence of Fenitrothion on Swimming Velocities of Brook Trout (*Salvelinus fontinalis*)," *Journal of the Fisheries Research Board of Canada* 31 (1974): 1757–62; R.H. Peterson, "Lethal Responses of Brook Trout (*Salvelinus fontinalis* [Mitch.]) to Dissolved Copper as Affected by Prior Sublethal Exposure to the Metal," Fisheries Research Board of Canada Manuscript Report 1309 (1974): 1–9; R.H. Peterson, "Temperature Selection of Juvenile Atlantic Salmon (*Salmo salar*) as Influenced by Various Toxic Substances," *Journal of the Fisheries Research Board of Canada* 33 (1976): 1722–30; R.H. Peterson, "Variations in Aquatic Insect Densities Associated with Copper-Zinc Concentrations," Fisheries Marine Service Manuscript Report 1470 (1978): 1–3; R.H. Peterson, "Effects on Fishes," in *Workshop on Long-Range Transport for Air Pollution and Its Impacts on the Atlantic Region* (Environment Canada, Atmospheric Environmental Services, Atlantic Region, 1979), 83–5; R.H. Peterson and V. Zitko, "Variations in Insect Drift Associated with Operational and Experimental Contamination with Fenitrothion in New Brunswick," Fisheries Research Board of Canada Technical Report 439 (1974): 1–23; R.H. Peterson, P.G. Daye, and J.L. Metcalfe, "Inhibition of Atlantic Salmon (*Salmo salar*) Hatching at Low pH," *Canadian Journal of Fisheries and Aquatic Science* 37 (1980):

770–4; R.H. Peterson, J.L. Metcalfe, and S. Ray, "The Effects of Cadmium on Yolk Utilization, Growth, and Survival of Atlantic Salmon Alevins and Newly Feeding Fry," *Archives of Environmental Contamination and Toxicology* 12 (1983): 37–44; R.H. Peterson, D.J. Martin-Robichaud, and J. Power, "Toxicity of Potash Brines to Early Developmental Stages of Atlantic Salmon (*Salmo salar*)," *Bulletin of Environmental Contamination and Toxicology* 41 (1988): 391–7.

97 G.L. Lacroix, "Survival of Eggs and Alevins of Atlantic Salmon (*Salmo salar*) in Relation to the Chemistry of Interstitial Water in Redds in Some Acidic Streams of Atlantic Canada," *Canadian Journal of Fisheries and Aquatic Science* 42 (1985): 292–9; and G.L. Lacroix and D. Knox, "Acidification Status of Rivers in Several Regions of Nova Scotia and Potential Impacts on Atlantic Salmon," *Canadian Technical Report of Fisheries and Aquatic Science* 2573 (2005): 1–71.

98 See, for example, D.E. Aiken and E.H. Byard, "Histological Changes in Lobsters (*Homarus americanus*) Exposed to Yellow Phosphorus," *Science* 176 (1972): 1434–5; D.E. Aiken and V. Zitko, "Effect of Iranian Crude Oil on Lobsters (*Homarus americanus*) Held in Floating Crates," *ICES C.M. 1977/E:45* (1977): 1–13; D.J. Scarratt, "Lobster Larvae off Pictou, Nova Scotia, Not Affected by Bleached Kraft Mill Effluent," *Journal of the Fisheries Research Board of Canada* 26 (1969): 1931–4; D.J. Scarratt, "Bleached Kraft Mill Effluent near Pictou, N.S., and Its Effect on the Marine Flora and Fauna with a Note on the Pictou Co. Lobster Landings," Fisheries Research Board of Canada Manuscript Report 1037 (1969): 1–24; D.J. Scarratt, "Impact of Spills and Clean-up Technology on Living Natural Resources and Resource-Based Industry," in "Summary of Physical, Biological, Socio-Economic and Other Factors Relevant to Potential Oil Spills in the Passamaquoddy Region of the Bay of Fundy," Fisheries Research Board of Canada Technical Report 428 (1974): 141–58; D.J. Scarratt and G.E. Raine, "Avoidance of Low Salinity by Newly Hatched Lobster Larvae," *Journal of the Fisheries Research Board of Canada* 24 (1967): 1403–6; D.J. Scarratt and A.J. Wilson, "Experiments with Rotenone in Northumberland Strait and Stomach Analysis of Fish Collected," Fisheries Research Board of Canada Manuscript Report 1107 (1970): 1–8; D.J. Scarratt and V. Zitko, "Bunker C Oil in Sediments and Benthic Animals from Shallow Depths in Chedabucto Bay, NS," *Journal of the Fisheries Research Board of Canada* 29 (1972): 1347–50; D.J. Scarratt and V. Zitko, "Sublittoral Sediment and Benthos Sampling and Littoral Observations in Chedabucto Bay, April 1973," in *Proceedings of Conference "Oil and the Canadian Environment," May 16, 1973*, ed. D. MacKay and W. Harrison

(Toronto: Institute of Environmental Science and Engineering, University of Toronto, 1975), 78–9; D.J. Scarratt, J.B. Sprague, D.J. Wilder, V. Zitko, and J.M. Anderson, "Some Biological and Chemical Investigations of a Major Winter Oil Spill on the Canadian East Coast," *International Council for the Exploration of the Sea C.M. 1970/E:14* (1970): 1–7; S.L. Waddy, L.E. Burridge, M.N. Hamilton, S.M. Mercer, D.E. Aiken, and K. Haya, "Emamectin Benzoate Induces Molting in American Lobster, *Homarus americanus*," *Canadian Journal of Fisheries and Aquatic Science* 59 (2002): 1096–9; S.L. Waddy, M.N. Hamilton, L.E. Burridge, S.M. Mercer, D.E. Aiken, and K. Haya, "Molting Response of Female Lobsters (*Homarus americanus*) to Emamectin Benzoate Varies with Reproductive Stage," in *Aquaculture Canada 2002, Proceedings of the Contributed Papers of the19th Annual Meeting of the Aquaculture Association of Canada, Charlottetown, PEI*, ed. C.I. Hendry, *Aquaculture Association of Canada Special Publication* 6 (2003): 75–7; S.L. Waddy, S.M. Mercer, M.N. Hamilton-Gibson, D.E. Aiken, and L.E. Burridge, "Feeding Response of Female American Lobsters, *Homarus americanus*, to SLICE®-Medicated Salmon Feed," *Aquaculture* 269 (2007): 123–9; S.L. Waddy, V.A. Merritt, M.N. Hamilton-Gibson, D.E. Aiken, and L.E. Burridge, "Relationship between Dose of Emamectin Benzoate and Molting Response of Ovigerous American Lobsters (*Homarus americanus*)," *Ecotoxicology and Environmental Safety* 67 (2007): 95–9.

99 Zitko and Wildish, "Environmental Research at St Andrews," 812.

100 K. Haya and B.A. Waiwood, "Acid pH and Chorionase Activity of Atlantic Salmon (*Salmo salar*) Eggs," *Bulletin of Environmental Contamination and Toxicology* 27 (1981): 7–12.

101 K. Haya, B.A. Waiwood, and L.Van Eeckhaute, "Disruption of Energy Metabolism and Smoltification during Exposure of Juvenile Atlantic Salmon (*Salmo salar*) to Low pH," *Comparative Biochemistry and Physiology* 82C (1985): 323–9.

102 L.E. Burridge and K. Haya, "The Lethality of Ivermectin, a Potential Agent for Treatment of Salmonids against Sea Lice, to the Shrimp *Crangon Septemspinosa*," *Aquaculture* 117 (1993): 9–14; L.E. Burridge and K. Haya, "The Lethality of Pyrethrins to Larvae and Post-Larvae of the American Lobster (*Homarus americanus*)," *Ecotoxicology and Environmental Safety* 38 (1997): 150–4; and L.E. Burridge, K. Haya, S.L. Waddy, and J. Wade, "The Lethality of Anti-Sea Lice Formulations Salmosan (Azamethiphos) and Excis (Cypermethrin) to Stage IV and Adult Lobsters (*Homarus americanus*) during Repeated Short-Term Exposures," *Aquaculture* 182 (2000): 27–35.

103 K. Haya, L.E. Burridge, and B.D. Chang, "Environmental Impact of Chemical Wastes Produced by the Salmon Aquaculture Industry," *ICES Journal of Marine Science* 58 (2001): 492; and K. Haya, L.E. Burridge, I.M. Davies, and A. Ervik, "A Review and Assessment of Environmental Risk of Chemicals Used for the Treatment of Sea Lice Infestations of Cultured Salmon," in *The Handbook of Environmental Chemistry*, vol. 5, part M, *Water Pollution: Environmental Effects of Marine Finfish Aquaculture*, ed. B.T. Hargrave (Berlin: Springer-Verlag, 2005), 305–41.
104 Haya, personal communication.
105 Cook, personal communication.
106 Ibid.
107 Burridge, personal communication.
108 Ibid.
109 These were D. Wildish, "The Toxicity of Polychlorinated Biphenyls (PCB) to *Gammarus oceanicus*," *Bulletin of Environmental Contamination and Toxicology* 5 (1970): 202–4, and V. Zitko, "Polychlorinated Biphenyls and Organochlorine Pesticides in Some Freshwater and Marine Fishes," *Bulletin of Environmental Contamination and Toxicology* 6 (1971): 464–70. See Zitko and Wildish, "Highlights of Environmental Research at St Andrews," 810.
110 D.J. Wildish and M. Heral, eds., *Environmental Effects of Mariculture*, International Council for the Exploration of the Sea, *Journal of Marine Science* 58 (2001): 363–530.
111 Haya, personal communication.
112 B.T. Hargrave, *Environmental Effects of Marine Finfish Aquaculture* (Berlin: Springer-Verlag, 2005).
113 P.G. Wells and J.B. Sprague, "Effects of Crude Oil on American Lobster (*Homarus americanus*) Larvae in the Laboratory," *Journal of the Fisheries Research Board of Canada* 33 (1976): 1604–14; and P.G. Wells, "Effects of Venezuelan Crude Oil on Young Stages of the American Lobster, *Homerus americanus*," Ph.D. thesis, University of Guelph, 1976, 1–274.

# 12 Aquaculture Research and Development at the St Andrews Biological Station, 1908–2008

ROBERT H. COOK

American oyster culture and the use of hatcheries for salmonid enhancement have been in place since the late 1800s. However, aquaculture development in Atlantic Canada, indeed Canada as a whole, progressed slowly, in part because of the need for a better scientific understanding of the conditions required for the successful culture of fish. Researchers at the St Andrews Biological Station, provided with ready access to sea water and fresh water, have been able to study the physiology, behaviour, and culture of live fish and shellfish under experimental conditions. As a consequence, gains in the understanding of fish biology and ecology from investigations at the biological station over the past century have been substantial. This research created the scientific underpinnings for current aquaculture programs in Atlantic Canada, including the culture of oyster, lobster, and scallop, Atlantic salmon, halibut, haddock, and other freshwater and marine fish. Until 1978, however, aquaculture was generally considered to be of academic interest only, with little commercial relevance. In that year, the rather slow pace of aquaculture development changed dramatically when an experimental trial initiated by the St Andrews Biological Station successfully demonstrated that Atlantic salmon could survive over winter in sea cages in the Bay of Fundy, and that such production was economically viable. After further trials, now with active private-sector involvement, commercial salmon farming was soon to become a profitable industry. St Andrews scientists developed a comprehensive scientific basis for the biology and culture of Atlantic salmon. Later, environmental studies related to site locations and impacts allowed station scientists to provide advice vital to a rapidly expanding industry. This advice related to the culture of Atlantic salmon has been used not only in Atlantic

Canada, but also in British Columbia, Norway, and Chile. Furthermore, the success of salmon aquaculture brought about a number of related developments: the first National Conference on Aquaculture was held in St Andrews in 1983 at which time the Aquaculture Association of Canada was created; the Salmonid Demonstration and Development Farm was established at Lime Kiln Bay, New Brunswick, in 1986; the first annual "Atlantic Aquaculture Fair" was held in St Andrews in 1988; and the importance of aquaculture was politically recognized in a report prepared by the Parliamentary Standing Committee on Fisheries and Oceans entitled "Aquaculture in Canada," presented to the House of Commons in 1988.[1] St Andrews Biological Station scientists working on aquaculture developed close working relationships with the Atlantic Salmon Federation and the neighbouring Huntsman Marine Science Centre, and collaborated with research centres, universities, governmental agencies, and the private sector. More recently they have undertaken research partnerships on the genome of cultured halibut and cod, and developed new techniques to apply integrated multi-trophic aquaculture (IMTA). Station scientists, heavily involved in national and international aquaculture organizations, have established the biological station as a leader in aquaculture research. The St Andrews Biological Station was designated as the national Centre for Integrated Aquaculture Science (CIAS). This review highlights noteworthy contributions to aquaculture by SABS on the occasion of its 100th anniversary.

## History of Different Aquaculture Programs and the St Andrews Biological Station, 1908–1960s

The availability of a reliable source of sea water for the biological station's wet laboratory facilities has been an important factor in the station's success in pursuing research on topics of importance to aquaculture. Shortly after the permanent land-based laboratory facilities were established in 1908, biological observations were under way on the experimental holding of finfish and shellfish. The experience gained from maintaining fish and shellfish under captivity created the beginnings of the knowledge base required for the development of commercial aquaculture. An early contribution to Canadian aquaculture was research on the cultivation of Prince Edward Island oysters, which was well under way in the 1920s. A field station was established at Ellerslie, Prince Edward Island, to provide ongoing support of these studies.

During this period, there were also investigations on the biology of Atlantic salmon, both on the performance of wild stocks and to improve the fish culture production at federal hatcheries, used for salmon enhancement programs, such as those on the Miramichi River in New Brunswick and the Margaree River in Nova Scotia. It was not until the 1960s that research on the physiology of fish and shellfish became more focused, with the objective of providing the science required for aquaculture. Initial research was on the culture of oysters, trout, Atlantic salmon, and the American lobster; a few decades later the studies expanded to the culture of scallops and various species of marine fish.

In its earliest years, the Atlantic Biological Station (as SABS was originally called) was at the forefront of research on the biology of the American oyster (*Crassostrea viriginica*) with the object of finding methods for its culture and improved production. These studies began in 1903 in Prince Edward Island, when the seasonal mobile station was located there, and expanded over the years. In 1917, scientists first observed a "blight" that affected oysters in Malpeque Bay, Prince Edward Island; this led to intensified research on the biology and culture of oysters. In 1928, the federal government assumed jurisdiction over the oyster-growing areas of Prince Edward Island, a step required to more effectively manage and develop oyster culture on the island. Soon thereafter the Biological Board established the Oyster Culture Station at Ellerslie, a permanent field substation of the biological station situated on Malpeque Bay. Its purpose was to allow scientists to provide advice on oyster culture; the substation became operational in the early 1930s. Malpeque Bay is a prime natural habitat with a favourable temperature regime for oyster growth and spat production. Dr Alfred W.H. Needler, who started his research career on oyster culture, was the first scientist in charge of the Ellerslie operation, where he was assisted by his wife Alfreda. He later became the director of the Atlantic Biological Station (1941–54) and after holding other senior appointments within the Fisheries Research Board, served as the deputy minister of fisheries. Dr J.C. Medcof joined the oyster research team at the biological station in the late 1930s, and worked out of Ellerslie. Carl Medcof's research on the larval life of the oyster and his 1961 Fisheries Research Board Bulletin entitled "Oyster Farming in the Maritimes" were among his many contributions,[2] some of which are described in greater depth by Jennifer Martin elsewhere in this volume. Roy Drinnan, a noted oyster biologist, was the last resident manager of the Ellerslie substation and worked on the problem of rehabilitating disease-affected oyster stocks;[3]

his ongoing contributions to aquaculture after leaving Ellerslie in the late 1960s will be described later.

The culture of oysters has a long history, particularly in Prince Edward Island, where leases were established over a century ago to enable fishers to hold oysters in sub-tidal areas until they were marketed. The Elleslie station was not created until later because the dominion government ran into provincial opposition over ownership of the oyster bottoms, a matter not resolved for a couple of decades.[4] Oyster production was traditionally a fishery in which oysters were harvested from the wild and sold. The trend in the history of oyster production, however, was to introduce culture techniques into the production cycle. Harvesting oysters in the wild and holding them within leased growing areas until marketing has become common practice. Often, wild oysters collected from less productive growing areas or mildly contaminated waters are transferred to clean leased sites and marketed at a later date. Oyster production is being steadily improved through the application of aquaculture technologies on collecting, propagating, and holding spat and applying various grow-out strategies until marketing. The use of wild oyster spat still predominates in the oyster culture industry; increasingly, however, aquaculture technologies are being used to enhance spat production, collection, and holding. The use of controlled grow-out conditions is expanding.

Early oyster studies investigated oyster diseases and their impact on oyster stocks; the highly infectious "Malpeque Disease" posed a severe threat to oyster production in the Maritimes since first being observed in 1917 in Prince Edward Island's waters. Dr Alfred Needler was unsuccessful in identifying the cause of the disease;[5] in the 1970s and 1980s, research on the causative agent of Malpeque Disease was continued by Dr Ming Lee of the Disease and Nutrition Section at the Halifax Laboratory (a component of the Fisheries and Environmental Sciences Division headed by the director of the St Andrews Biological Station).[6] To this day the causative agent of this disease is unknown, but infected oysters introduced into oyster beds in unaffected areas can spread the disease, with a mortality rate in the previously unexposed oysters of over 90 per cent. Despite this continuing mystery, molluscan shellfish researchers working out of St Andrews and Ellerslie, such as Dr J.C. Medcof, achieved marked success in researching other problems confronting the shellfish industry.[7] Dr Medcof and other early investigators studied the incidence of toxic algal blooms or Paralytic Shellfish Poison (PSP), caused by the dinoflagellate *Gonyaulax tamarensis*, in shellfish stocks in the region, work described in greater detail by Jennifer Martin in her chapter in this volume.

PSP investigations were a high priority in the 1940s. A collaborative arrangement was developed between the biological station and the public health laboratories where the toxin analyses were performed. St Andrews researchers monitored shellfish growing areas, collected specimens, and prepared extractions for analysis by the Department of Health and Welfare laboratories in Ottawa for PSP. Not only was PSP a threat to the harvesting of soft-shell clams and mussels, but its presence also posed a threat to oyster culture operations. To speed up PSP analysis, a mouse bioassay unit was later established at the Fish Inspection Laboratory at Black's Harbour, New Brunswick, to test samples collected in the Maritimes. The development and rapid expansion of the mussel aquaculture industry in Nova Scotia and Prince Edward Island in the late 1960s required monitoring for PSP and, later in the 1980s to ensure health safety, for domoic acid and other toxins in shellfish growing areas. From the beginning of this research in the late 1930s to the present day, toxic algal blooms have been monitored and investigated at the St Andrews Biological Station as part of the Applied Ecology program.

Like the oyster fishery, the fishery for American lobster (*Homarus americanus*) has been of great importance in the Maritimes since the 1800s. Research on the biology and culture of the lobster began soon after the Atlantic Biological Station was established. In 1915, Dr A.P. Knight of Queen's University, assisted by other university researchers, studied the early life biology of lobster at several Maritime locations; their main efforts were in Prince Edward Island. They investigated live-holding culture techniques to improve the viability of the canner fishery. Lobster hatcheries had been established in the late 1890s, and, by the early 1900s, there were fifteen such operations in the Maritime provinces, releasing large numbers of early juvenile lobsters to growing areas as a means of resource enhancement. Dr Knight studied the culture of lobster and the biological requirements of the early life stages. He used hatchery boxes to maintain berried females, collected the larvae, tested various holding conditions, and observed their growth under culture conditions. These culture experiments showed that the use of hatcheries to augment wild lobster populations was not effective. When Dr Knight became the chairman of the Biological Board of Canada in 1917, he closed the lobster hatcheries, a decision based largely on his own research experience.[8]

The biological station became increasingly involved in research on lobster biology, ecology, and stock abundance, as fishery managers requested scientific advice on regulations to manage the lobster fishery. The station also contributed important advice on the holding requirements of lobster in response to questions of how to improve on the best methods for holding, transferring, and shipping live captured lobsters. In the lobster fishery, captured lobsters are often not marketed immediately. They can be stored for the short term in wooden crates alongside fishermen's wharves. Alternatively, they can be maintained in larger facilities, such as lobster pounds, consisting of onshore tanks or a tidal embayment. Technically, while lobsters are being stored under such conditions, they are, in fact, being cultured. To respond to questions of how to improve the live holding of lobsters, scientists needed to consider such factors as water temperature, salinity, oxygen concentration, water exchange rate, loading density, nutrition, and disease control under experimental conditions. This research provided a useful background for future lobster aquaculture initiatives. The 1964 *Bulletin of the Fisheries Research Board* by Drs Don McLeese and Dick Wilder entitled "Lobster Storage and Shipment" provided an excellent background on this topic.[9]

Salmon and trout were the other commercially important species that were the targets of attempts to enhance the natural populations in the early twentieth century. Hatcheries to augment the wild stocks of trout and salmon were being tested in Ontario several years before Confederation in 1867. The first government Atlantic salmon hatchery in the Maritimes was established on the Miramichi River at South Esk, New Brunswick, in 1873. In the decades that followed, salmon hatcheries were constructed and became operational in all the Maritime provinces. Thus, when the Atlantic Biological Station was established in 1908, there was practical information available on the culture of Atlantic salmon gained from these hatchery operations. The researchers working out of the biological station conducted research on the biology of Atlantic salmon; often this work was in response to problems observed at hatcheries in the field. For example, Dr A.G. Huntsman reported on a massive salmon mortality event on the Restigouche River in the 1930s; he later studied salmon migration on the Margaree River in Cape Breton, Nova Scotia.[10] Dr Paul Elson's studies on Atlantic salmon, initiated in the 1950s, provided comprehensive observations on salmon biology, particularly those factors affecting the growth, survival, and migration of salmon in New Brunswick rivers.[11] Other notable St Andrews researchers on Atlantic salmon over the years have included Drs William Hoar, Miles Keenlyside, C.J. Kerswill, Aivars Stasko, and Phil Symons. Research on trout was carried out by Mordon Smith and Wilfred Saunders.

## Aquaculture Research at the St Andrews Biological Station, 1960s–present

Roy Drinnan, a noted oyster biologist, was the last resident manager of the Ellerslie station; after leaving in the late 1960s, he relocated to Cape Breton, Nova Scotia, where he directed the resource development activities of the Cape Breton Development Corporation (DEVCO), a federal Crown corporation. The principal focus of his program was to promote commercial aquaculture in the Bras d'Or Lakes, a prime oyster-growing region.[12] Under Drinnan, DEVCO funding supported aquaculture studies in other areas. For example, at St Andrews he supported a study on the rearing of salmonids in sea pens by providing a research grant to Dr Saunders and his team to carry out this work. Drinnan subsequently was appointed the first aquaculture coordinator for the Maritimes Region, in 1977. The aquaculture coordinator, and the Disease and Nutrition Section, although located at the Halifax Laboratory, reported to the director of the St Andrews Biological Station (who also served as chief of the Fisheries and Environmental Sciences Division, Resource Branch, Maritime Region).

In 1974, St Andrews scientist Dr. David Aiken led the development of the first and only lobster grow-out facility to operate as an integrated culture system. This facility uniquely combined all phases of lobster culture into a single integrated process, from broodstock to the marketable product, all in a continuous production system.[13] The system's elements involved maintaining lobster broodstock that would be induced to produce seed on a monthly basis. This provided a predictable supply of newly hatched larvae, in turn allowing a continuous source of post-larval lobsters to supply the critical juvenile grow-out phase.[14] The science resulting from this culture facility was substantial and provided important insight into rearing lobster, their growth, and the physiology of the moulting process. Scientists studied the effect of photoperiod, temperature, water management (quality and flow), health and nutrition, population density, and holding conditions.

The core element needed for aquaculture technology was to *understand* how the biological systems worked, such as the physiology of lobster moulting, in order to successfully manipulate the biology of a candidate species for aquaculture production. Several key results from the lobster culture program led directly to understanding central biological processes, which yielded information that in turn could be used in circumstances related not only to lobster cultivation and husbandry, but also to the management of the lobster fishery. For example, the elucidation of the role of water temperature in lobster reproduction allowed scientists to predict the likely impacts of climate change on the reproductive potential of nearshore populations.[15] The improved understanding of lobster reproduction has also allowed us to argue effectively against pressure from special-interest groups to open offshore and deep-water areas to increased exploitation of vulnerable stocks. Furthermore, aquaculture technology for the year-round production of both eggs and larvae from pre-market lobsters allowed enhancement efforts to proceed without further impact on the natural production of eggs and larvae, as egg-bearing females from wild populations no longer needed to be captured for this purpose.[16]

Research on the bacteriological lobster disease "Gaffkemia," and studies on the nutrition of lobster when held under culture conditions, were carried out in the 1970s and 1980s under the direction of Drs Jim Stewart and John Castell, members of the Department of Fisheries and Oceans' Halifax Fisheries Research Laboratory. They headed the Fisheries and Environmental Division's aquaculture research component of this laboratory's Disease and Nutrition Section.[17] As a result of their ongoing lobster culture research, interest grew in developing commercialized lobster culture. Marine Lobster Farms, a private lobster culture project under the direction of Mr Peter Patton, received federal financial support (NRC Pilot Industrial Laboratory Program [PILP]) in 1978. This land-based lobster culture facility was located in Victoria, Prince Edward Island. Dr Robert Cook was the scientific authority for the administration of the contract and Drs Aiken and Castell served as scientific advisers to this pilot-scale operation. They made considerable progress on holding systems and water management regimes for rearing lobsters through the seed, larval, and post-larval stages. Unfortunately, local water quality problems and the impact of the lobster disease "Gaffkemia" took a heavy toll and the project ended in 1983. Unpublished monthly reports provided the only recorded technical information on this project.[18] During the same period, attempts to rear lobsters individually in trays and maintain them in a lagoon-based grow-out in the Magdalen Islands of Quebec also failed because of problems with "Gaffkemia" and poor growth.

The Lobster Culture Laboratory at the St Andrews Biological Station was fully operational until 1983. Shortly thereafter, it was discontinued – primarily as a result of regional organizational actions and reduced financial resources for the project. Research on the culture and biology of lobster continues at St Andrews, but is no longer carried out within a fully integrated culture facility.

Scallops, like lobsters, have also been the target of aquaculture programs. The St Andrews Biological Station had acquired a reasonable understanding of the biology of the sea scallop (*Placopecten magellanicus*) over the years since its foundation. Researchers primarily focused on questions concerning its habitat and ecology, and investigations to determine the cause of mass mortalities among scallops. In the 1950s and 1960s, Drs Carl Medcof, Neil Bourne, and Lloyd Dickie were involved in these investigations; later, Dr John Caddy undertook research on the biology and factors affecting the dynamics of scallop populations, which he discusses elsewhere in this volume. Attention to culturing scallops did not begin in earnest until the mid-1980s, when the biological station assigned Dr Michael Dadswell to undertake scallop research. Dadswell saw the potential for scallop culture while studying wild populations on Georges Bank and in Passamaquoddy Bay.[19] The relative success of scallop aquaculture in Newfoundland, spearheaded by research carried out at Memorial University's Marine Sciences Research Laboratory, clearly supported the view that scallop culture had commercial potential.

Scientists had ready access to scallops and abundant scallop spat collected from the Tongue Shoal near St Andrews; this provided a prime opportunity to investigate a wide variety of spat collection and grow-out technologies. In the 1980s, a group of station scientists, including Dr Dadswell (who later joined the staff at Acadia University, Wolfville, Nova Scotia) and Drs David Wildish and Shawn Robinson, along with highly motivated graduate students Cyr Couturier and Jay Parsons, began to make significant progress. They made advances in spat collection techniques, understanding the intermediate grow-out of spat to juvenile stages and assessing the final grow-out to market size using techniques such as floating lantern devices, various tray designs, or the "ear hanging" of scallops from floats.[20] These studies provided a basis to evaluate the economic potential for scallop aquaculture.

It is important to note that other shellfish species also have been cultured at the biological station in support of enhancement or aquaculture objectives. In the 1990s, Dr Shawn Robinson carried out research on culturing sea urchins and investigated conditions that would enhance roe quality and production. Dr John Castell developed the sea urchin diets and biologist David Robichaud conducted field assessments.[21] The key objective was to enhance roe quality to meet the requirements of Japanese buyers. In addition, Dr Robinson has cultured soft-shell clams and studied the use of cultured juvenile clams for both enhancement and planting development strategies.[22]

Obviously, various diseases posed a major challenge to the culture of different species. In 1985, the biological station hosted and co-sponsored the Third International Conference on Toxic Dinoflagellates in St Andrews. Dr Alan White, who led the biological station's research on PSP during the 1970s and 1980s, was a key organizer of this conference.[23] Research on toxic algal blooms continues today by the St Andrews Biological Station biologist, Jennifer Martin. This project in the first decade of the 2000s developed techniques to provide salmon farmers with an early warning of phytoplankton blooms, and trained sea-farm staff on how to identify key phytoplankton species, which Martin highlights in her chapter in this volume.

## The Science behind Farming Atlantic Salmon since 1960

Research on the biology of salmon and trout was conducted over many years at the biological station and mainly focused on enhancing wild populations and resolving problems at hatcheries involved in enhancement programs. In the 1960s, research began on the biology of salmonids in support of aquaculture development and became an important area of research at the St Andrews Biological Station. In the 1960s and 1970s, scientists at the station also contributed substantially to understanding the impact of forestry harvesting practices and hydroelectric developments on salmon and salmon habitat. In addition Dr Dick Peterson and others investigated the effects of spruce budworm control operations on salmon and trout; this and the research by Drs John Sprague and Vlado Zitko on the effects of pollution from the mining and the pulp and paper industries on salmon and salmon habitat is discussed elsewhere by Peter Wells in this volume.

Research on the physiology and culture of the Atlantic salmon began in earnest at the biological station in the 1960s under the scientific leadership of Dr Richard Saunders.[24] His research on Atlantic salmon smolt, with emphasis on the environmental control of the smolting process, has provided a critical understanding of this important factor for both salmon enhancement and aquaculture.[25] An experimental salmon hatchery was developed at the station that allowed research on a broad spectrum of factors affecting the growth and physiology of salmon. Dr Saunders was joined in this research on salmonid physiology and culture by Drs Richard Peterson and Arnie Sutterlin, and received capable technical assistance from Gene Henderson and Paul Harmon.[26] Their advice and support to the salmon farming industry was significant and appreciated.

Consideration was also given in the 1960s to experimental sea-ranching trials, with a land site being acquired by the biological station at nearby Birch Cove, New Brunswick, where cultured salmon smolts were to be released into the sea and recaptured as adult salmon upon their return from the sea. Unfortunately, this project never got under way.

The sea-ranching concept was eventually developed and implemented in the 1970s when the International Atlantic Salmon Foundation, later to be renamed the Atlantic Salmon Federation, relocated its headquarters from Gaspé, Quebec, to Chamcook, near St Andrews, and constructed a research salmon hatchery. In 1973, the Atlantic Salmon Federation entered into a funding agreement with the Government of Canada, represented by the St Andrews Biological Station, to conduct a research program on the genetics of Atlantic salmon with two key aims: to improve the broodstock of salmon and to select for traits that would enhance stocks of Atlantic salmon. The Atlantic Salmon Federation provided the capital for research facilities and equipment; federal funding covered operational costs. The Huntsman Marine Science Centre collaborated in managing the program. The research hatchery proceeded to produce salmon smolts selectively bred to demonstrate characteristics with a high rate of return to their site of release to marine waters. Dr R.L. Saunders, on an executive exchange from the biological station, served as the research director of this program from 1975–9, a program that later became the Salmon Genetics Research Program (SGRP). Later in the 1980s, the SGRP played a major role in the production and improvement of seed-stock to support the Bay of Fundy salmon aquaculture industry.

In 1978, the Atlantic Salmon Federation initiated the Aquaculture Technicians Training Program to provide practical hatchery training to students interested in aquaculture. This was administered by the Huntsman Marine Science Centre and offered on-site training and practical work experience at various hatcheries. Graduates of this 48-week course were soon to be employed in the newly developing aquaculture industry. Another important feature of the Aquaculture Technicians Training Program was the lectures and workshops it convened, which featured leading researchers in aquaculture. Biological station researchers benefited from these opportunities to meet visiting experts. During the 1980s, Dr Saunders organized international salmon smolt workshops every second year, mainly attended by Canadian and Norwegian researchers.

As a result of the St Andrews Biological Station investigations on salmon physiology, and proactive collaboration with salmon researchers in Norway, salmon aquaculture in the Maritimes began to look more attractive. However, two key developments in the early 1970s dimmed these prospects. First, an ambitious and well-funded land-based salmon culture venture, Sea Pool Fisheries Limited, which started in Nova Scotia in 1969, went bankrupt by 1972, representing the loss of an investment of almost five million dollars. The collapse of the Sea Pool project, owing to managerial and technical shortfalls, had a strong negative impact on the further development of salmon aquaculture in the Maritimes. The second, and perhaps more critical, factor was that the winter temperature of Maritime coastal waters was lethal to salmon.[27] Several experiments to test the overwintering of Atlantic salmon in sea cages were carried out by the Resource Development Branch of the federal fisheries department in the 1970s. Salmon were maintained in sea pens at Arichat, Cape Breton, and later at Polly Cove, near Halifax, to test their survival under winter conditions. In all cases, the salmon did not survive the cold water temperatures. Similar results occurred with sea trout held in cages in the Bras d'Or Lakes in Cape Breton, and Brandy Cove, the site of the St Andrews Biological Station. The clear message was that salmon aquaculture and the sea cages successfully deployed in the coastal waters of Norway could not be applied in the Maritimes.

These setbacks, and others in the development of Canadian aquaculture, presented serious impediments to funding support for aquaculture research programs. Indeed, in Canada the term "aquaculture" was linked with hatchery practices used to enhance trout and salmon for the recreational and commercial fisheries and with notions of an "academic" scientific pursuit, with the exception of oyster culture in Prince Edward Island. Aquaculture was considered to have limited economic potential. This situation was reflected in a 1974 Fisheries Research Board Bulletin entitled "Aquaculture in Canada: The Practice and the Promise."[28] The report concluded that freshwater aquaculture and the marine culture of shellfish and fish had potential, and that Canada possessed considerable scientific knowledge about aquaculture, but lacked actual demonstrations and trials to confirm its economic viability. With limited commercial aquaculture and no industry representation, aquaculture research and development was not considered to be a departmental priority. The situation became worse in the late 1970s, with the declaration of extended jurisdiction for Canadian offshore fisheries. Resources were diverted from programs such as culture research, and new resources were allocated to expand the research effort on offshore marine fish populations and stock assessments.

The image of Atlantic Canadian aquaculture changed dramatically as a result of the actions of the salmon research team at St Andrews. In the early 1970s, salmon researchers at the biological station were introduced to the methods of salmon culture used in Norway from experienced visiting Norwegian scientists. Subsequently, Dr Arnold (Arne) Sutterlin went to Tromsø University in Norway, on sabbatical, to observe the marine cage culture of Atlantic salmon in waters north of the Arctic Circle.[29] Upon his return to the station, he immediately prepared plans to establish a pilot-scale salmon sea-cage culture in the Fundy Isles region of the Bay of Fundy. His project team involved biological station salmon researchers R.L. (Dick) Saunders, Gene B. Henderson, and Paul Harmon, a private-sector site owner and biologist, Art McKay, and a contracted biologist, Susan Merrill, who had experience with the Norwegian industry. Their efforts eventually proved that overwintering Atlantic salmon in sea cages was technically possible in the Bay of Fundy. These trials clearly demonstrated that culturing Atlantic salmon in sea cages had a strong economic potential. From those historic trials at Lord's Cove on Deer Island, New Brunswick, the technical basis for the salmon aquaculture industry was established. The trial was reported in a 1981 technical report by Sutterlin et al.[30] This marked the beginning of a series of developments involving the biological station, including building partnerships with Bay of Fundy fishermen (mainly herring weir fisherman) and others interested in salmon aquaculture. The biological station director and Dr Saunders's team worked closely with Bill Groom, the local provincial fisheries representative, and senior staff of the fisheries department of the Province of New Brunswick for technical and financial support; and collaborated with scientists in the Disease and Nutrition Section at the Halifax Laboratory, namely, Dr Robin McKelvie, John Cornick, and Betty Zwicker on matters of fish health, and Dr Santosh Lall on salmon diet formulations and nutrition.

The first National Conference on Aquaculture was held in St Andrews in 1983 to celebrate the seventy-fifth anniversary of the founding of the St Andrews Biological Station. This occasion also served as the starting point to create the Aquaculture Association of Canada (AAC). Former biological station director Dr John Anderson served as the first AAC president in 1984.[31]

In the 1980s, as achievements mounted in developing salmon aquaculture in the Bay of Fundy, participants decided that the selection of strains of Atlantic salmon with improved performance in sea cages should become the prime objective of selective breeding within the Salmon Genetics Research Program. This action proved timely, as the availability, quality, and performance of salmon smolt used in sea cages was a factor of increasing economic importance. From the Atlantic Salmon Federation perspective at that time, the improved availability of local farmed Atlantic salmon in food markets reduced the market for those who fished and sold wild salmon illegally (i.e., those poaching the resource). It also had a positive influence in meeting the market demand once provided by the commercial harvest of salmon on rivers; the numbers of salmon returning from their marine offshore migrations were in serious decline. The scientific contributions of the Salmon Genetics Research Program to the development of salmon aquaculture were significant. The research of Atlantic Salmon Federation geneticists Drs Gerry Friars and John Bailey allowed them to provide solid advice to the salmon industry. In 1993 Dr Friars prepared a practical primer for salmon farmers entitled *Breeding Atlantic Salmon*.[32]

In response to the rapid development of salmon aquaculture in the Bay of Fundy, the federal and provincial fisheries departments, in the early 1980s, established the Southern New Brunswick Aquaculture Development Committee. It was co-chaired by Mr David McMinn of the New Brunswick Department of Fisheries and Dr James Stewart of the Bedford Institute of Oceanography. Membership on this committee included the director of the biological station, the federal area manager for Southern New Brunswick, the director of aquaculture for the New Brunswick Department of Fisheries, and a fisheries representative from St Stephen, NB. This committee was created to advise government on issues related to salmon aquaculture development. Mr Eugene (Gene) Henderson was assigned as an adviser to the committee so that it could benefit from his practical knowledge on salmon aquaculture. His participation had long-lasting effects on the industry. For example, he chaired a working group to develop guidelines for the physical separation of salmon farms, which the committee provided to the salmon growers.[33] Although the industry took strong exception to the recommended guidelines, subsequent sea lice infestations and serious outbreaks of infectious diseases eventually led to more stringent measures for site separation, site fallowing, and bay management.

In 1986, the St Andrews Biological Station initiated the Salmonid Demonstration and Development Farm, a concept modelled after the long-established program of the Canadian Department of Agriculture and its network of experimental farms across Canada. Each experimental farm was established with a specific crop or animal production focus based on the agriculture priority of a given region. The Salmonid Demonstration and Development Farm was created to support Atlantic salmon aquaculture by establishing an experimental cage site with shore-based facilities in the south-western Bay of Fundy area, in general proximity to the newly established private salmon farms. Initial funding was provided by a grant from the Canada–New Brunswick Cooperation Agreement on Fisheries Development. Dr Robert Cook, the biological station director, was responsible for the project. Gene Henderson provided technical assistance and Walter Ross from the station provided administrative support. They selected a coastal land and marine site in Lime Kiln Bay, New Brunswick. The land and aquaculture site belonged to Ian Hamilton and Chris Saulnier, who became the contract "recipient" for funding under the terms of the Cooperation Agreement. A building, with meeting room, laboratories, and offices, was constructed, along with an array of sea cages to be anchored nearby.

The original design of the farm consisted of twelve sea cages. Four of the salmon cages were wooden, using an octagonal frame design developed by one of the first salmon farmers in the Bay of Fundy, John Malloch of Campobello Island, New Brunswick. The Malloch cage design was promoted, with funding support from the Province of New Brunswick, as a model to be used by new participants entering the industry and was in common use in the early years. The second set of four cages was square, with a plastic tube frame; the third set of four cages was metal, imported from Norway and in common use by Norwegian salmon farmers (the Jamek cage). Test trials using salmon smolts placed in these sea cages were used to answer early questions raised by the new salmon growers: what type of cage should be used (wood, plastic, or metal); what type of diet should be fed the fish – a "moist" diet prepared from ground up fresh fish or a "dry" diet using commercially prepared fish pellets; and what age of smolt should be used to stock the cages: one- or two-year-olds? The research findings on these and other topics of salmon culture were able to be effectively communicated to the salmon growers. With Gene Henderson as manager, the Salmonid Demonstration and Development Farm became the principal source of advice and technical support to existing and new salmon growers in the industry; it was a focus for disseminating practical information on the sea-cage culture of salmonids. The meeting room there became the hub for growers seeking and sharing technical advice. The Salmonid Demonstration and Development Farm also provided a site for scientists to conduct their own research on salmon culture. In particular, in collaboration with the Salmon Genetics Research Program, it provided an important test site where salmon smolts produced for aquaculture, through the Atlantic Salmon Federation hatchery's selective breeding and broodstock development programs, could be observed for their performance in the sea cages.

The Salmonid Demonstration and Development Farm proved to be a cornerstone in advancing the development of salmon aquaculture. In 1989, the New Brunswick Salmon Growers' Association was formed and Gene Henderson, upon his retirement from the biological station, became the first general manager. At this time, the New Brunswick Salmon Growers' Association assumed responsibility for the Salmonid Demonstration and Development Farm operation. Its building became the association's headquarters and, over the years, the site was expanded to also become a shore-based service centre in support of the industry. Currently, the New Brunswick Salmon Growers' Association has its main office in the nearby town of St George, New Brunswick.

As the number of sea-cage sites expanded, so too did the demand for salmon smolts. The limited number produced at the biological station was soon inadequate to assist the pioneer group of salmon farmers. A temporary arrangement was made with the federal Fish Culture Station at Mactaquac, near Fredericton, New Brunswick, on the Saint John River, to increase its production of salmon smolt above the level normally dedicated for salmon enhancement, and to make these extra smolt available to Bay of Fundy aquaculture. A salmon seed-stock program was introduced to distribute these salmon from the Mactaquac facility to meet aquaculture requests. At the same time, private salmon hatcheries were under construction and, by the late 1980s, salmon growers had access to private sources of salmon smolts. Owing to this success, larger corporations soon entered the salmon aquaculture industry. Several of these operations had not only their own hatcheries and grow-out sites, but processing plants as well. Aquaculture, at the end of the 1980s, was recognized as a major economic activity and no longer simply a promising one.

This recognition was assisted by astute promotion of the idea of aquaculture as a valid regional economic opportunity. In 1987, Dr Robert Cook was appointed by the deputy minister to serve as the department's scientific adviser to the Parliamentary Standing Committee on Fisheries and Oceans to assist in the preparation of their report entitled "Aquaculture in Canada"[34] (released in 1988). This submission to the House of Commons enhanced the status and credibility of aquaculture in the formulation of future government policy. In addition, in 1988, interested parties planned to develop an event to celebrate aquaculture similar to the long-standing tradition of agriculture fairs held to celebrate the harvest in farming communities across Canada. This idea took hold with strong industry interest. The station director, Dr Robert Cook, supported by former station director Dr John Anderson and local merchant Sheila Simpson, made it happen. It was named the Atlantic Aquaculture Fair and Dr Cook became its first president. The fair venue provided a technical forum for fish farmers, with technical lectures, invited presentations from international experts, public food demonstrations to promote the aquaculture product, and an international trade show featuring aquaculture equipment and services. At the closing banquet, an annual award was inaugurated to recognize the "Aquaculturist of the Year." The First Atlantic Aquaculture Fair was opened by the premier of New Brunswick, Frank McKenna. Dr Arne Sutterlin was the first to be named "Aquaculturist of the Year." Other Department of Fisheries and Oceans winners of the award have been Roy Drinnen (1990), Betty Zwicker (1991), Dr Richard Saunders and Gene Henderson (1993), and Dr Brian Glebe (2001).

## Marine Fish Culture and the St Andrews Biological Station since the 1980s

The commercial success in salmon aquaculture raised the question of what other finfish species could be developed. Hatchery production of striped bass had been used for many years to stock freshwater lakes. This species appeared to be a natural candidate for aquaculture. In the late 1980s, Dr Dick Peterson – author of the chapter on salmon fieldwork research in this volume – tested the culture requirements of striped bass at an experimental hatchery he set up at St Andrews, as the technique to produce seed stock was reasonably well understood.[35] Research to determine the optimum environmental conditions for the grow-out of striped bass – for example, in ponds, sea cages, or shore-based tanks – was now required. In addition, studies on market potential of the cultured striped bass product were needed.

The Atlantic halibut was another prime candidate for aquaculture research and development given its high market price and relative low availability from the traditional fishery. Research on halibut culture began in Norway and Scotland in the early 1980s. In the late 1980s at St Andrews, Dr Ken Waiwood led a team, composed of Maria Buzeta, Ken Howes, John Reid, and several graduate students, to culture the species through its several developmental stages to adulthood.[36] They discovered that the critical period for larval halibut survival is during their early pre-metamorphosis stage, and before reaching the juvenile stage. Once the halibut had become juveniles, various saltwater grow-out regimes were tested, including tanks and cages within the modified "Tide Pool" facility at the biological station. Industry partners who also participated in these tests included herring weir fisherman, salmon farmer John Malloch on Campobello Island, and Ralph Cline of Maritime Mariculture Inc. Wild juvenile halibut, collected off Sable Island, Nova Scotia, and reared in a modified herring weir, eventually served as a source of broodstock to supply eggs and larvae for research and to assist private hatcheries when they lacked broodstock to produce eggs for their own operations. With help from Norwegian colleagues, the biological station produced the first cultured Atlantic halibut, as well as haddock, in North America. These halibut juveniles also became the first cultured Atlantic halibut broodstock and are still producing gametes for research at the biological station.

In 1995, Debbie Martin-Robichaud assumed responsibility for the marine finfish broodstock program. A collaborative research project with Dr Tillman Benfey, at the University of New Brunswick, and Dr Mike Reith, at the Institute of Marine Biotechnology of the National Research Council, developed the first halibut broodstock capable of producing all-female stocks for improved performance.[37] The industry partner, Scotia Halibut Ltd., has been able to sell some of these all-female juvenile halibut internationally at a premium price. The team of molecular geneticists at the Institute of Marine Biotechnology has also developed the first genetic map for Atlantic halibut and has identified the key molecular markers now used in breeding programs to select broodstock with improved genetic traits for growth. These projects are identifying key factors that must be addressed if the culture of Atlantic halibut in sea cages is to become economically viable.

The experience and ability to culture marine fish has provided St Andrews Biological Station with an opportunity to participate in Genome Canada. Dr Ed Tripple of St Andrews and Dr Sharen Bowmen of Dalhousie University, Halifax, initiated a research program in 2005 entitled "Atlantic Cod Genomics Brood Stock Development." This research is designed to identify genes from cod tissues and, through genetic modelling, to relate specific genes to factors such as growth rate, size, disease resistance, and resistance to thermal or handling stress. The program requires the production of 100,000 juveniles from 200 cod families, at a commercial sea-cage site in Back Bay, New Brunswick. These fish will be selected for performance in the enhancement of wild stocks as well as in aquaculture. Cod will be collected in the wild, and from tissue analyses a determination will be made as to what stocks are best to use as broodstock. A parallel study is being carried out at the Ocean Sciences Centre of Memorial University of Newfoundland and Labrador, in partnership with researchers at Guelph University in Ontario. Initial results have been positive, with high return rates of stocking as a result of using cultured larval cod for stocking. Progress is also being made on the development of genomic tools to select for traits in cod culture. This research, along with the ongoing marine finfish broodstock development activities at the biological station, is advancing our understanding of how genetic factors govern the growth of fish under culture conditions. The resulting technology has assisted in the development of culture techniques and broodstock development for Atlantic cod. Future plans are to develop techniques to produce monosex populations of Atlantic cod for improved grow-out potential, and to mitigate the environmental impact of eggs released from cage-reared cod. St Andrews researchers have also worked to develop haddock culture technology with industry partner Heritage Salmon Ltd.

In terms of connections with the larger world community in aquaculture research, the biological station has provided three presidents of the World Aquaculture Society (WAS): Dr David Aiken in 1987, Dr John Castell in 1989, and Susan Waddy in 1993. In addition, Dr Aiken initiated the launch of the WAS publication *World Aquaculture*, and served as editor from 1987 to 1997; he was assisted by Susan Waddy as science editor during this period. Many scientists at the biological station have participated in the science programs of the International Council for the Exploration of the Sea (ICES), serving on working groups and committees. Dr John Anderson, during his term as station director (1962–7), served as a Canadian delegate. Dr Robert Cook, station director (1977–92), participated in ICES beginning in 1979. He was also the Canadian representative on the Mariculture Committee beginning in 1987 and served as its chairman from 1993–6. Dr Tom Sephton, station director (1997–2005), also served as the Canadian representative on the ICES Mariculture Committee.

## Research on Environmental Interactions of Aquaculture

With the rapid expansion of salmon aquaculture also came disease and environmental problems and interaction problems with the traditional fisheries. St Andrews Biological Station scientists were called upon to provide scientific advice to address these environmental effects and interaction problems.

The commercial development of salmon aquaculture in the mid-1980s resulted in noticeable impacts on the marine environment. Too many salmon grow-out sites were being allocated in a relatively limited area, that is, the Fundy Isles area of the Bay of Fundy, and the numbers of cages being placed on each site, as well as cage size (holding capacity), were increasing. Some growers were stocking their sea cages with smolt densities well in excess of levels adopted by the international salmon farming community based on scientific advice. This combination of more sites, more salmon per site, and higher densities of salmon per cage, led to a number of adverse effects on the environment and salmon production. The federal-provincial guidelines developed by Gene Henderson on the physical separation of salmonid aquaculture farms, which had been intended to reduce these risks, were not being heeded. In the late 1980s, measures to control the spread of the sea lice parasite were required to deal with sea lice infestations both within individual cages and between sea-cage sites. The overcrowding of sea cages and the degradation of environmental conditions also increased the incidence of infectious diseases. Major outbreaks soon appeared.

The St Andrews Biological Station, in collaboration with Dr Ron Trites at the Bedford Institute of Oceanography, Dartmouth, NS, carried out circulation and current studies in the L'Etang Inlet, New Brunswick. Dr David Wildish of the biological station first carried out benthic environmental monitoring in 1985. These benthic studies continued through to the early 2000s in collaboration with Dr Barry Hargrave of the Bedford Institute. They also assessed sedimentary changes that resulted from the fallowing of a salmon site.[38] They went on to develop a benthic monitoring technique based on sulphide levels in the sediments under salmon cages,[39] which was adopted by the Province of New Brunswick for its environmental surveillance of the industry. Dr Fred Page's further research in 2005 on the spatial and temporal variability of sediments under salmon cages has led to changes in the New Brunswick monitoring program.[40]

In the years 1995–8, oceanographer Dr Page and his team conducted research on the dispersion of the chemicals used to control sea lice. The team used seabed drifters to predict the movement of bath treatment chemicals following their release into the marine environment. Dye studies were also carried out in collaboration with Environment Canada. Oceanographic studies predicted the potential spread of Infectious Salmon Anaemia (ISA) disease among salmon farms in the Fundy Isles area using a tidal circulation model with confirmation from field measurements.[41] Dr Page's research helped New Brunswick and the salmon aquaculture industry to develop a Bay Management Area system for stocking and fallowing farms in light of the ISA crisis, a critical step in developing new policies governing salmon aquaculture in the Bay of Fundy that were implemented in 2006. A three-year crop rotation with fallowing is now required between successive crops. Oceanographic research is under way on the potential for offshore aquaculture development within a coastal zone management framework, and is assessing factors such as currents and waves that stress cage structures operating at exposed sites.

Many St Andrews Biological Station researchers have also studied the interaction of aquaculture with traditional fisheries. Dr Rob Stephenson investigated the interaction of salmon aquaculture and the Bay of Fundy herring weir fishery in the late 1980s. In the early 1990s, Dr Peter Lawton studied how salmon aquaculture affected nearby lobster populations. In 2001–2, his research considered lobster abundance in relation to salmon cage sites on Grand Manan, New Brunswick, and in 2002–3, he studied the impact of salmon aquaculture on juvenile lobsters in Maces Bay, New Brunswick.

A program entitled Integrated Multi-Trophic Aquaculture (IMTA) was started in 2001 as a means of developing sustainable, multi-species systems, with a view of manipulating the ecosystem to optimize production. This involved research on the trophic relationships of the species to be cultured and on their energy requirements and contributions. The approach is to combine the culture species that are artificially fed

– for instance, caged finfish – with species that either extract inorganic products, such as seaweed, or consume organic material, such as shellfish. To test the Integrated Multi-Trophic Aquaculture concept, research on factors affecting the production of blue mussel at salmon cage sites is ongoing. Dr Shawn Robinson led this research program at the biological station, in collaboration with Dr Thierry Chopin at the University of New Brunswick, Saint John. Environment Canada, the Canadian Food Inspection Agency, and Cooke Aquaculture Ltd. were also key partners in this program. Expressions of interest internationally are now under consideration by the industry membership of the IMTA partnership.

In 2007, the assistant minister of science of the Department of Fisheries and Oceans – and former station director – Dr Wendy Watson-Wright (1992–7), established the Centre for Integrated Aquaculture Science (CIAS), to be located at the St Andrews Biological Station. The Centre for Integrated Aquaculture Science is, in effect, a virtual centre of expertise for aquaculture science involving research staff from Department of Fisheries and Oceans facilities across Canada and a focus for knowledge development for aquaculture within the department. The intent of the CIAS is to coordinate research within an integrated, ecosystem-based, management framework, enabling all facets of departmental knowledge including oceans, fish habitat, and ecosystems, so that aquaculture production science can provide comprehensive advice on matters pertaining to aquaculture development. The CIAS secretariat at St Andrews is headed by Dr Fred Page.

## Conclusion

Throughout its history, the Biological Station at St Andrews and its scientists have been noted for their leadership in the aquatic sciences. In the field of aquaculture, the research of pioneering scientists at St Andrews on such species as the American oyster, Atlantic salmon, lobster, scallop and halibut over the past century has contributed significantly to advancing our understanding of how to culture these species. During the period between 1977 and the late 1980s, the station director was also a program division chief with regional responsibilities for aquaculture programs at the Halifax Laboratory, including disease and nutrition research, administration of the fish health protection regulations, and the Aquaculture Coordination office. The critical mass of aquaculture science expertise at the St Andrews Biological Station was further enhanced by work at the Huntsman Marine Science Centre on arctic char and haddock and the testing of the spherical "Kiel cage" in Brandy Cove. The research activities of the Atlantic Salmon Federation and its co-sponsorship of the Salmon Genetics Research Program also added to this "critical mass." Complementary research activities at the biological station on environmental and fishery interactions, oceanographic studies, benthic effects, and toxic algal blooms completed the aquaculture picture. As a result of these research programs and interactions, the St Andrews Biological Station assumed a leadership role, nationally and internationally, in aquaculture. Partnerships with other research centres, the private sector, and governments also contributed to the development of the aquaculture industry.

The research role of St Andrews has recently shifted towards studies on the ecological interactions of aquaculture, the assessment of integrated multi-trophic culture systems, marine fish genome definition, and brood-stock development, and to serving as focus for the coordination and implementation of the Department of Fisheries and Oceans aquaculture research activities. It is clear that research on the environmental interactions involved in aquaculture will continue to be essential as new species become economically viable for commercial development and require space in the coastal zone. The biological station is well positioned with its complement of fish biologists, aquaculture specialists, oceanographers, and researchers on the marine environment, to provide sound scientific advice to support the development of an environmentally sustainable aquaculture industry.

## NOTES

In September 2008, the station director, Dr Rob Stephenson, invited me to present a paper on the history of aquaculture research at the St Andrews Biological Station to the "Evolution of Marine Science in Canada Workshop" commemorating the biological station's 100th anniversary. This paper is based partially on the reflections of a group of station aquaculture researchers, past and present, who were consulted in September 2008 to consider the history of aquaculture at St Andrews: Dr John Anderson, Dr Dick Saunders, Dr David Aiken, Dr Ken Waiwood, Dr Brian Glebe, Dr Shawn Robinson, Dr Ed Tripple, Dr Fred Page, Blythe Chang, and Debbie Martin-Robichaud. I wish to thank all those consulted, including former regional Resource Branch director Dr James Stewart, for their contributions in the preparation of this chapter, and also am grateful for the editorial review provided by Dr David N. Nettleship and Jennifer Hubbard.

1 House of Commons Standing Committee on Fisheries and Oceans, "Aquaculture in Canada," Report to the House pursuant to Standing Order 96(2), 2nd Session, Thirty-third Parliament (Ottawa: Communication Canada, 1988).
2 J.C. Medcof, "Oyster Farming in the Maritimes," *Bulletin of the Fisheries Research Board of Canada* 131 (1961): 1–158.
3 See, for example, R.E. Drinnan and J.C. Medcof, "Progress in Rehabilitating Disease-Affected Oyster Stocks," *Fisheries Research Board of Canada, General Series Circular* no. 34 (1961): 1–3.
4 For further information on these problems and earlier attempts to culture oysters, see Jennifer Hubbard, "The Commission of Conservation and the Canadian Atlantic Fisheries," *Scientia Canadensis* 12:1 (Spring-Summer 1988), 22–52.
5 A.W.H. Needler and R.R. Logie, "Serious Mortalities in Prince Edward Island Oysters Caused by a Contagious Disease," *Transactions of the Royal Society of Canada*, ser. 3, vol. 41:5 (1947): 73–89.
6 M.F. Li, G.S. Traxler, S. Clyburne, and J.E. Stewart, "Malpeque Disease: Isolation and Morphology of a Labyrinthomyxa-like Organism from Diseased Oysters," *International Council for Exploration of the Sea C.M.1980/F 15* (1980): 1–9.
7 A few of Medcof's many articles include J.C. Medcof, "Dark-meat and the Shell Disease of Scallops," Fisheries Research Board of Canada, Atlantic Biological Station Progress Report 45 (1949); A. Prakash and J.C. Medcof, "Hydrographic and Meteorological Factors Affecting Shellfish Toxicity at Head Harbour, New Brunswick," *Journal of the Fisheries Research Board of Canada* 19 (1962): 11–112; L.M. Dickie and J. C. Medcof, "Causes of Mass Mortalities of Scallops (*Placopecten magellanicus*) in the Southwestern Gulf of St Lawrence," *Journal of the Fisheries Research Board of Canada* 20 (1963): 451–82.
8 A.P. Knight's work on the Maritime lobster industry is recounted in more detail in Jennifer Hubbard, *A Science on the Scales: The Rise of Canadian Atlantic Fisheries Biology, 1898–1939* (Toronto: University of Toronto Press, 2006), 98–100, 111–19.
9 D.W. McLeese and D.G. Wilder, "Lobster Storage and Shipment," *Bulletin of the Fisheries Research Board of Canada* 147 (1964): 1–69.
10 See, for example, A.G. Huntsman, "The Cause of Periodic Scarcity in Atlantic Salmon," *Transactions of the Royal Society of Canada*, series 3, vol. 31 (1937), section 5: 17–27; A.G. Huntsman, "Report of the Margaree Salmon and Trout Investigations," Biological Board of Canada Annual Report (1935): 47–8; A.G. Huntsman, "Report of the Margaree Salmon and Trout

Investigations for 1937," Biological Board of Canada Annual Report (1937): 64–6.

11 P.F. Elson and C.J. Kerswill, "Studies on Canadian Atlantic Salmon," *Transactions of the 20th North American Wildlife Conference* (1955): 415–526; P.F. Elson, "Predator–prey Relationships between Fish-eating Birds and Atlantic Salmon, with a Supplement on Fundamentals of Merganser Control," *Bulletin of the Fisheries Research Board of Canada* 133 (1962): 1–87.

12 R.E. Drinnan, "Oysters – Disease, Predation, Parasites and Competitors," in *The Proceedings of the Bras D'Or Lakes Aquaculture Conference, held in Sydney, Cape Breton, June 1975*, ed. Gregory McKay (Sydney, NS: College of Cape Breton Press, 1976), 125–9.

13 D.E. Aiken and W.W. Young-Lai, "Dactylotomy, Chelotomy and Dactylostasis: Methods for Enhancing Survival and Growth of Small Lobsters (*Homarus americanus*) in Communal Conditions," *Aquaculture* 22 (1981): 45–52.

14 D.E. Aiken and S.L. Waddy, "Production of Seed Stock for Lobster Culture," *Aquaculture* 44:2 (1985): 103–14; S.L. Waddy and D.E. Aiken, "Broodstock Management for Year-round Production of Larvae for Culture of the American Lobster," *Canadian Technical Report of Fisheries and Aquatic Science* 1272 (1984): 1–14.

15 S.L. Waddy, D.E. Aiken, and D.P.V. De Kleijn, "Control of Growth and Reproduction," in *Biology of the Lobster* Homarus americanus, ed. J.R. Factor (Toronto: Academic Press, 1995), 217–66; and S.L. Waddy and D.L. Aiken, "Temperature Regulation of Reproduction in Female American Lobsters (*Homarus americanus*)," *ICES Marine Science Symposia* 199 (1995): 54–60.

16 S.L. Waddy and D.E. Aiken, "Lobster (*Homarus americanus*) Culture and Resource Enhancement: The Canadian Experience," *Canadian Industry Report of Fisheries and Aquatic Science* 244 (1998): 9–18.

17 J.E. Stewart and J.D. Castell, "Various Aspects of Culturing the American Lobster (*Homarus americanus*)," in Proceedings of the FAO Conference on Aquaculture, Kyoto, Japan 1976. FIR: AQ/Cont/76/E: 11, and in *Advances in Aquaculture*, ed. T.V.R. Pillay and W.A. Dill (Farnham, Surrey, Eng.: Fishing News Books, 1979), 314–19 .

18 D.E. Aiken and S.L. Waddy, "Culture of the American Lobster, *Homerus americanus*," in *Cold Water Aquaculture in Atlantic Canada*, 2nd ed., ed. A.D. Boghan (Sackville, NB: Canadian Institute for Research on Regional Development, 1995), 156–7.

19 G.J. Parsons and M.J. Dadswell, "Effect of Stocking Density on Growth, Production, and Survival of the Giant Scallop, *Placopecten magellanicus*,

Held in Intermediate Suspension Culture in Passamaquoddy Bay, New Brunswick," *Aquaculture* 103:3–4 (1992): 291–309.

20 C. Couturier, P. Dabinett, and M. Lanteigne, "Scallop Culture in Atlantic Canada," in *Cold-Water Aquaculture in Atlantic Canada, 2nd ed.*, ed. A.D. Boghen (Moncton, NB: University of Moncton, 1996), 297–340; C.Y. Coutuier, "Scallop Aquaculture in Canada: Fact or Fantasy," *World Aquaculture* 21:2 (1990): 54–62; G.J. Parsons and M. J. Dadswell, "Evaluation of Intermediate Culture Techniques, Growth, and Survival of the Giant Scallop, *Placopecten magellanicus,* in Passamaquoddy Bay, New Brunswick," Canadian Technical Report of Fisheries and Aquatic Sciences 2012 (1994): 1–29; Parsons and Dadswell, "Effect of Stocking Density"; S.M.C. Robinson, "Shellfish Culture in the Bay of Fundy," *Aquaculture Association of Canada Special Publication* 2 (1997): 85–93; S.M.C. Robinson, "The Shellfish Industry in the Gulf of Maine: Status and Possible Future Directions," *Gulf of Maine News*, Summer 1996: 7–9; S.M.C. Robinson, "A Review of the Biological Information Associated with Enhancing Scallop Production," *World Aquaculture* 24 (1993): 61–7; D.J. Wildish, A.J. Wilson, W.W. Young-Lai, A.M. DeCoste, D.E. Aiken, and J.M. Martin, "Biological and Economic Feasibility of Four Grow-out Methods for the Culture of Giant Scallops in the Bay of Fundy," *Canadian Technical Report of Fisheries and Aquatic Science* 1658 (1988): 1–21; D.J. Wildish, D.D. Kristmanson, R.L. Hoar, A.M. DeCoste, S.D. McCormick, and A.W. White, "Giant Scallop Feeding and Growth Responses to Flow," *Journal of Experimental Marine Biology and Ecology* 113 (1987): 207–20; and W.W. Young-Lai and D.E. Aiken, "Biology and Culture of the Giant Scallop, *Placopecten magellanicus*: A Review," *Canadian Technical Report of Fisheries and Aquatic Science* 1478 (1986): 1–21.

21 S.M.C. Robinson and A. MacIntyre, "Biological Fishery Information for the Rational Development of the Green Sea Urchin Industry. Final Report for the New Brunswick Department of Fisheries and Aquaculture and the Canada–New Brunswick Co-operation Agreement on Economic Diversification," St Andrews, NB, Biological Station, 1995: 1–90; S.M.C. Robinson, J.D. Castell, and E.J. Kennedy, "Developing Suitable Colour in the Gonads of Cultured Green Sea Urchins (*Strongylocentrotus droebachiensis*)," *Aquaculture* 206 (2002): 289–303.

22 S.M.C. Robinson, "Clam Enhancement Trials in the Bay of Fundy," *Department of Fisheries and Oceans Science Review, 1994–5* (1996): 1–16; and Robinson, "The Shellfish Industry in the Gulf of Maine," 7–9.

23 Donald M. Anderson, Alan W. White, and Daniel G. Baden, eds., *Proceedings of the Third International Conference on Toxic Dinoflagellates* (New York: Elsevier Science Publishing, 1985).

24 For example, R.L. Saunders, "Adjustment of Buoyancy in Young Atlantic Salmon and Brook Trout by Changes in Swimbladder Volume," *Journal of the Fisheries Research Board of Canada* 22 (1965): 335–52; R.L Saunders, "Heated Effluent for the Rearing of Fry – For Farming and For Release," in *Harvesting Polluted Waters*, ed. O. Devik (New York: Plenum Press,1976), 213–36.
25 R.L. Saunders and K.R. Allen, "Effects of Tagging and Fin-clipping on the Survival and Growth of Atlantic Salmon between Smolt and Adult Stages," *Journal of the Fisheries Research Board of Canada* 24 (1967): 2595–611; R.L. Saunders and E.B. Henderson, "Growth of Atlantic Salmon Smolts and Post-Smolts in Relation to Salinity, Temperature, and Diet," *Fisheries Research Board of Canada Technical Report* 149 (1969): 1–20; R.L. Saunders and E.B. Henderson, "Influence of Photoperiod on Smolt Development and Growth of Atlantic Salmon (*Salmo solar*)," *Journal of the Fisheries Research Board of Canada* 27 (1970): 1295–311; R.L. Saunders and E.B. Henderson, "Changes in Gill ATPase Activity and Smolt Status of Atlantic Salmon (*Salmo solar*)," *Journal of the Fisheries Research Board of Canada* 35 (1978): 1542–6.
26 For example, R.L. Saunders and E.B. Henderson, "Atlantic Herring as a Dietary Component for Culture of Atlantic Salmon," *Aquaculture* 3:4 (1974): 369–85; A.M. Sutterlin, R.L. Saunders, E.B. Henderson, and P.R. Harmon, "The Homing of Atlantic Salmon (*Salmo salar*) to a Marine Site," *Canadian Technical Reports of Fisheries and Aquatic Science* 1058 (1982): 1–6.
27 R.L. Saunders, B.C. Muise, and E.B. Henderson, "Mortality of Salmonids Cultured at Low Temperature in Sea Water," *Aquaculture* 5:3 (1975): 243–52; R.L. Saunders, "Winterkill! The Reality of Lethal Winter Sea Temperature in East Coast Salmon Farming," *Bulletin of the Aquaculture Association of Canada* 1 (1987): 36–40; R.L. Saunders, "The Thermal Biology of Atlantic Salmon: Influence of Temperature on Salmon Culture with Particular Reference to Constraints Imposed by Low Temperature," *Institute of Freshwater Research, Drottningholm* 63 (1986): 77–90.
28 H.R. MacCrimmon, J.E. Stewart, and J.R. Brett, "Aquaculture in Canada: The Practice and the Promise," *Bulletin of the Fisheries Research Board of Canada* 188 (1974): 1–84; also published as *Fisheries and Marine Service Bulletin* 188 (Ottawa: Department of the Environment, 1974): 1–84.
29 A.M. Sutterlin and S.P. Merrill, "Norwegian Salmonid Farming," *Fisheries and Marine Service Technical Report* 779 (1978): 1–47.
30 A.M. Sutterlin, E.B. Henderson, S.P. Merrill, R.L. Saunders, and A.A. MacKay, "Salmonid Rearing Trials at Deer Island, New Brunswick, with

Some Projections on Economic Viability," *Canadian Technical Report of Fisheries and Aquatic Sciences* 1011 (1981): 1–32.

31 Other AAC presidents with current or past affiliations with the biological station were Dr David Aiken (1985), Dr Neil Bourne (1986), Susan Waddy (1991), former students Cyr Couturier (1995, 2001) and Jay Parsons (1997), and Dr Shawn Robinson (2002).

32 G.W. Friars, *Breeding Atlantic Salmon: A Primer* (St Andrews, NB: Atlantic Salmon Federation, 1993), 1–13.

33 J.M. Anderson, *The Salmon Connection: The Development of Salmon Aquaculture in Canada* (Tantallon, NS: Glen Margaret Publishing, 2007), 40–1.

34 MacCrimmon, Stewart, and Brett, "Aquaculture in Canada."

35 R.H. Peterson, D.J. Martin-Robichaud, P. Harmon, and A. Berge, "Notes on Striped Bass Culture, with Reference to the Maritime Provinces," Dept. of Fisheries and Oceans, Communications Branch, P.O. Box 550, Halifax, N.S. (1996): ii + 35 pp.

36 K.G. Waiwood and M.I. Buzeta, "Reproductive Biology of Southwest Scotian Shelf Haddock (*Melanogrammus aeglefinus*)," *Canadian Journal of Fisheries and Aquatic Science* 46 (1989): 153–70; K.G. Waiwood, "Haddock," *Bulletin of the Aquaculture Association of Canada* 94:1 (1994): 16–21; K.G. Waiwood, K.G. Howes, and J. Reid, "Halibut Aquaculture Research at the St Andrews Biological Station," in *Science Review 1992 & '93 of the Bedford Institute of Oceanography, Halifax Fisheries Research Laboratory, and the St Andrews Biological Station*, ed. A. Fiander (Dartmouth, NS: Department of Fisheries and Oceans, Scotia-Fundy Region, 1994), 43–6.

37 H.B. Tvedt, Tillmann J. Benfey, D.J. Martin-Robichaud, C. McGowan, and M. Reith, "Gynogenesis and Sex Determination in Atlantic Halibut (*Hippoglossus hippoglossus*)," *Aquaculture* 252:2–4 (2006): 573–83.

38 D.J. Wildish, H.M. Akagi, and N. Hamilton, "Sedimentary Changes at a Bay of Fundy Salmon Farm Associated with Site Fallowing," *Bulletin of the Aquaculture Association of Canada* 101 (2001): 49–54; D.J. Wildish, H.M. Akagi, N. Hamilton, and B.T. Hargrave, "A Recommended Method for Monitoring Sediments to Detect Organic Enrichment from Mariculture in the Bay of Fundy," *Canadian Technical Report of Fisheries and Aquatic Science* 2286 (1999): 1–31; D.J. Wildish, V. Zitko, H.M. Akagi, and A.J. Wilson, "Sedimentary Anoxia Caused by Salmonid Mariculture Wastes in the Bay of Fundy and Its Effects on Dissolved Oxygen in Seawater," *Proceedings of Canada-Norway Finfish Aquaculture Workshop, Sept. 11–14, 1989*, ed. R.L. Saunders, *Canadian Technical Report of Fisheries and Aquatic Science* 1761 (1990): 11–18.

39 D.J. Wildish, B.T. Hargrave, and G. Pohle, "Cost Effective Monitoring of Organic Enrichment Resulting from Salmon Mariculture," *International Council for the Exploration of the Sea Journal of Marine Science* 58 (2001): 469–76.

40 E. Black, T. Chopin, J. Grant, F. Page, N. Ridler, and J. Smith, "Aquaculture and Integrated Resource Management: A Canadian Perspective," in *The Role of Aquaculture in Integrated Coastal and Ocean Management: An Ecosystem Approach*, ed. J. McVey, C.-S. Lee, D. Jang, and P. O'Bryen (Baton Rouge: World Aquaculture Society, 2006); B.D. Chang, F.H. Page, and B. Hill, "Preliminary Analysis of Coastal Marine Resource Use and the Development of Open Ocean Aquaculture in the Bay of Fundy," *Canadian Technical Reports of Fisheries and Aquatic Science* 2585 (2005); F.H. Page, R. Losier, P. McCurdy, D. Greenberg, J. Chaffey, and B. Chang, "Dissolved Oxygen and Salmon Cage Culture in the Southwestern New Brunswick Portion of the Bay of Fundy," in *Handbook of Environmental Chemistry*, ed. B.T. Hargrave (Berlin: Springer-Verlag, 2005), 1–28; F.H. Page, R.L. Stephenson, and B.D. Chang, "A Framework for Addressing Aquaculture–Environment–Fisheries Interactions in the Southwestern New Brunswick Portion of the Bay of Fundy," *Aquaculture Association of Canada Special Publication* 8 (2004): 69–72; F. Page, J. Piercey, B. Chang, B. MacDonald, and D. Greenberg, "An Introduction to the Oceanographic Aspects of Integrated Multi-Trophic Aquaculture," *Bulletin of the Aquaculture Association of Canada* 104–3 (2004): 35–43; and F.H. Page, B.D. Chang, R.J. Losier, D.A. Greenberg, and P. McCurdy, "Water Circulation and Management of ISA in the Southwest New Brunswick Salmon Culture Industry," *Aquaculture Association of Canada Special Publication* 8 (2004): 64–8.

41 F.H. Page, B.D. Chang, R.J. Losier, D.A. Greenberg, J.D. Chaffey, and P. McCurdy, "Water Circulation and Management of Infectious Salmon Anemia in the Salmon Aquaculture Industry of Southern Grand Manan Island, Bay of Fundy," *Canadian Technical Report of Fisheries and Aquatic Science* 2595 (2005); and F.H. Page, B.D. Chang, R.J. Losier, D.A. Greenberg, and P. McCurdy, "Water Circulation and Management of ISA in the Southwest New Brunswick Salmon Culture Industry," *Aquaculture Association of Canada Special Publication* 8 (2004): 64–8.

# Bibliography

**Archival Sources**

Bedford Institute of Oceanography Archives, Accession 2006-41.
Huntsman Collection, University of Toronto Archives.
National Archives of Canada, RG 23, vol. 1200, file 726-1–11 [1].
St Andrews Biological Station Archives: Transcripts of the directors' panel and related archival materials

**Unpublished Sources**

"Annual Report of the Atlantic Biological Station." St Andrews, New Brunswick. Unpublished reports, 1913–49.
"Annual Report of the St Andrews Biological Station." St Andrews, New Brunswick. Unpublished reports, 1950–77.
Atlantic Oceanographic Group. "Canadian Drift Bottle Data, Atlantic Coast." Fisheries Research Board of Canada Manuscript Report (Oceanography and Limnology) 15 (1958): 1–156.
Bailey, W.B. "Oceanographic Observations in the Bay of Fundy since 1952." Fisheries Research Board of Canada Manuscript Report of the Biological Stations 633 (1957): 1–72.
Bailey, W.B. "Some Features of the Oceanography of the Passamaquoddy Region." Fisheries Research Board of Canada Manuscript Report (Oceanography and Limnology) 2 (1957): 1–60.
Bumpus, D.F., J.R. Chevrier, F.D. Forgeron, W.D. Forrester, D.G. MacGregor, and R.W.Trites. "International Passamaquoddy Fisheries Board Report to the International Joint Commission (1959)." Appendix 1, Oceanography: ii–iii,1–223. Available at www.ijc.org/files/publications/AA2.pdf.

Caddy, J.F. "Some Recommendations for Conservation of Georges Bank Scallop Stocks." International Commission for the Northwest Atlantic Fisheries Research Document 72/6 (1972): 1–4.

Caddy, J.F., and R.A. Chandler. "Georges Bank Scallop Survey, August 1966: A Preliminary Study of the Relationship between Research Vessel Catch, Depth, and Commercial Effort." Fisheries and Marine Services Manuscript Report 1054 (1969): 1–13.

Campbell, N.J., and L.M. Lauzier. "Ice Studies of the Atlantic Oceanographic Group." Fisheries Research Board of Canada Manuscript Report (Oceanography and Limnology) 60. St Andrews Biological Station, 1960.

Carrothers, P.J.G. "Fishing Gear Engineering Research at Sea." Film, 1966. Remastered camcording. Available at www.mar.dfo-mpo.gc.ca/sabs/.

Carrothers, P.J.G., and T.J. Foulkes. "Measured Towing Characteristics of Canadian East Coast Otter Trawls." International Commission for the Northwest Atlantic Fisheries Research Document 71/39 (1971): 1–21. Available at http://icnaf.nafo.int/docs/1971/res-39.pdf.

Clemens, W.A. "Education and Fish: An Autobiography." Fisheries Research Board of Canada Manuscript Report 974 (1968): 1–102.

Department of Fisheries and Oceans. "Assessment of Georges Bank Scallops (*Placopecten magellanicus*)." Department of Fisheries and Oceans Canada, Scientific Advisory Section, Scientific Advisory Report 32 (2006): 1–11. Available at http://www.dfo-mpo.gc.ca/csas/Csas/status/2006/SAR-AS2006_032_REVISE_E.pdf.

Dickie, L.M. "Fluctuations of the Giant Scallop, *Placopecten magellanicus* (Gmelin), in the Digby Area of the Bay of Fundy." Ph.D. thesis, University of Toronto,1953.

Dickie, L.M., and C.D. MacInnes. "Gulf of St Lawrence Scallop Explorations – 1957." Fisheries Research Board of Canada Manuscript Report Series 650 (1958): 1–62.

Elson, P.F. "Adult Salmon Required to Maintain Stocks." Fisheries Research Board of Canada, Biological Station, St Andrews, NB, Investigators' Summaries, no. 66 (1955–6): 173–6.

Elson, P.F. "Eels as a Limiting Factor in Salmon Production." Fisheries Research Board of Canada Manuscript Report of the Biological Station 213 (1941).

Elson, P.F. "The Fundamentals of Merganser Control for Increasing Production of Salmon Smolts from Streams." Fisheries Research Board of Canada Manuscript Report 694 (1960): 1–10. Available at http://www.dfo-mpo.gc.ca/Library/38221.pdf.

Elson, P.F. "Increasing Salmon Stocks by Control of Mergansers and Kingfishers." Fisheries Research Board of Canada, Progress Reports of the Atlantic Coast Stations 51 (1950): 12–15.

Elson, P.F. "Magnitude of Smolt Runs Measured by Sampling." Fisheries Research Board of Canada, Report of the Atlantic Biological Station for 1951: 95–7.

Environment Canada. "Review for 1971–1972 of the Fisheries Research Board of Canada." Ottawa: Information Canada, 1973: 1–230.

Finley, Carmel. "The Tragedy of Enclosure: Fish, Fisheries Science, and Foreign Policy, 1920–1958." Doctoral dissertation, University of California, San Diego, 2007.

Fisheries Research Board of Canada. "Annual Report of the Fisheries Research Board of Canada 1971." Ottawa: Information Canada, 1971. 1–22.

Fisheries Research Board of Canada. "Biological Station, St Andrews, NB, Annual Report and Investigators' Summaries 1957–58" (1958).

Foulkes, T.J., S. Polar, M. Babineau, and P.J.G. Carrothers. "Instruction Manual (for) modified E G & G Model 200A-210 Automatic Camera System." Fisheries Research Board of Canada Manuscript Report 1304 (1974): 1–35 + illustrations and appendices. Available at http://www.dfo-mpo.gc.ca/Library/23086.pdf.

Gavaris, E.D., and R.L. Stephenson. "Report of the Commemoration of the 100th Anniversary of Permanent Buildings of the St Andrews Biological Station – 2008." Manuscript Report of Fisheries and Aquatic Science 2943 (2010): iv + 27 pp.

GESAMP. "The Revised GESAMP Hazard Evaluation Procedure for Chemical Substances Carried by Ships." Reports and Studies of GESAMP 64 (2002). Available at http://www.gesamp.org/publications/publicationdisplaypages/rs64.

Hachey, H.B. "History of the Fisheries Research Board of Canada." Fisheries Research Board of Canada Manuscript Report Series (Biological) 843 (1965): 1–499.

Hachey, H.B., and W.B. Bailey. "The General Hydrography of the Waters of the Bay of Fundy." Fisheries Research Board of Canada Manuscript Report 455 (1952).

Hart, J.L. "Report of the Atlantic Biological Station for 1954." Fisheries Research Board of Canada, 1954.

Hart, J.L. "Report of the Atlantic Biological Station for 1955." Fisheries Research Board of Canada, 1955.

House of Commons Standing Committee on Fisheries and Oceans. "Aquaculture in Canada." Report to the House Pursuant to Standing Order 96(2), 2nd Session, 33rd Parliament. Ottawa, 1988.

Hubbard, J.M. "An Independent Progress: The Development of Marine Biology on the Atlantic Coast of Canada, 1898–1939." Doctoral dissertation, Institute for the History and Philosophy of Science and Technology, University of Toronto, 1992.

Huntsman, A.G. "Preserving Scallops: Fisheries Experimental Station (Atlantic) Note No. 1." Canadian Biological Board, Progress Reports of the Atlantic Biological Station and Fisheries Experimental Station (Atlantic) no. 1 (1931): 1–15.

Huntsman, A.G. "Report of the Atlantic Biological Station for 1931." Biological Board of Canada, 1931.

Huntsman, A.G. "Report of the Atlantic Biological Station for 1932." Biological Board of Canada, 1932.

Huntsman, A.G. "Report of the Margaree Salmon and Trout Investigations." Biological Board of Canada Annual Report 1935: 47–8.

Huntsman, A.G. "Report of the Margaree Salmon and Trout Investigations for 1937." Biological Board of Canada Annual Report 1937: 64–6.

Information Canada. "Review for 1969–1970 of the Fisheries Research Board of Canada." Ottawa: Information Canada, 1971. 1–217.

International Joint Commission, United States and Canada. "Report of the International Joint Commission on the International Passamaquoddy Tidal Power Project." Washington, DC, and Ottawa, Canada: International Joint Commission, 1961. 1–21. Available at http://www.ijc.org/files/publications/AA13.pdf.

International Passamaquoddy Fisheries Commission. "Report of the International Passamaquoddy Fisheries Commission." Washington, DC: US House of Representatives, 73rd Congress, 2nd Session, Document no. 200 (1934): 1–24.

Jamieson, G.S., M. Etter, and R.A. Chandler. "The Effect of Scallop Fishing on Lobsters in the Western Northumberland Strait." Canadian Atlantic Fisheries Scientific Advisory Committee Resource Document 81/71 (1981): 1–19.

Jamieson, G.S., N.B. Witherspoon, and R.A. Chandler. "Bay of Fundy Scallop Stock Assessment – 1980." Canadian Atlantic Fisheries Scientific Advisory Committee Research Document 81/27 (1981): 1–25.

Joint Group of Experts on the Scientific Aspects of Marine Environmental Protection. Session Reports 1969–74. Available at http://www.gesamp.org/publications/gesamp-session-reports-1969---1974; Reports and Studies 1-90. Available at http://www.gesamp.org/publications/gesamp-reports-and-studies.

Kenchington, E., D.L. Roddick, and M.J. Lundy. "Bay of Fundy Scallop Analytical Stock Assessment and Data Review 1981–1994: Digby Grounds." Department of Fisheries and Oceans Atlantic Fisheries Research Document 95/10 (1995): 1–70.

Kerswill, C.J. "Some Environmental Factors Limiting Growth and Distribution of the Quahaug, *Venus mercenaria L.*" Fisheries Research Board of Canada Manuscript Report 187 (1941): 1–104.

Klugh, A.B. "A Hydrographic Reconnaissance of the Lower St John and the Kennebecasis, New Brunswick." Biological Board of Canada Manuscript Report 48 (1910): 1–4.

Kristmanson, D.D., D.J. Wildish, and N.J. Poole. "Mixing of Pulp Mill Effluents in the Upper L'Etang." Fisheries Research Board of Canada Manuscript Report 1416 (1976): 1–36.

Lauzier, L.M., J.G. Clark, and A.W. Brown 1964. "Canadian Drift Bottle Program, 1960–1963, Atlantic Coast." Fisheries Research Board of Canada Manuscript Report (Oceanography and Limnology) 178 (1964): 1–35.

Lauzier, L.M., and J.H. Hull. "Coastal Station Data Temperatures along the Canadian Atlantic Coast, 1921–1969." Fisheries Research Board of Canada Technical Report 150 (1969): 1–25.

Lord, D.A., R.A.F. Matheson, L. Stuart, J.J. Swiss, and P.G. Wells. "Environmental Monitoring of the 1976 Spruce Budworm Spray Program in New Brunswick, Canada." Fisheries and Environment Canada, EPS, Surveillance Rep. EPS-5-AR-78–3 (1978): 1–161.

MacGregor, D.G. "Water Conditions in the Strait of Canso." Fisheries Research Board of Canada Manuscript Report of the Biological Stations 552 (1954).

MacPhail, J.S. "The Inshore Scallop Fishery of the Maritime Provinces." Fisheries Research Board of Canada, Atlantic Biological Station, St Andrews, NB, General series 22 (1984): 1–4.

Martin, J.L., D.J. Wildish, M.M. LeGresley, and M.M. Ringuette. "Phytoplankton Monitoring in the Southwestern Bay of Fundy during 1990–1992." Canadian Manuscript Report of Fisheries and Aquatic Science 2277 (1995): 1–154. Available at http://www.dfo-mpo.gc.ca/Library/181492.pdf.

McGonigle, R.H., and A.W.H. Needler. "Report of the Atlantic Biological Station for 1940." Fisheries Research Board of Canada.

Medcof, J.C. "Dark-meat and the Shell Disease of Scallops." Fisheries Research Board of Canada, Atlantic Biological Station Progress Report 45 (1949): 3–6.

Medcof, J.C., and J.F. Caddy. "Underwater Observations on Performance of Clam Dredges of Three Types." Fisheries Research Board of Canada Manuscript Report 1313 (1974): 1–9. Available at http://www.dfo-mpo.gc.ca/Library/23094.pdf.

Mohn, R.K., and G. Robert. "Comparison of Two Harvesting Strategies for the Georges Bank Scallop Stock." Canadian Atlantic Fisheries Scientific Advisory Committee Research Document 84:10 (1984).

Mohn, R.K., G. Robert, and G.A.P. Black. "Georges Bank Scallop Stock Assessment – 1988." Canadian Atlantic Fisheries Scientific Advisory Committee Research Document 89:21 (1989).

Mohn, R.K., G. Robert, and D.L. Roddick. "Georges Bank Scallop Stock Assessment – 1986." Canadian Atlantic Fisheries Scientific Advisory Committee Research Document 87:9 (1987).

Naidu, K.S. "An Analysis of the Meat Count Regulation." Canadian Atlantic Fisheries Scientific Advisory Committee Research Document 84:73 (1984).

Naidu, K.S., and Cahill, F.M. "Culturing Giant Scallops in Newfoundland Waters." Canadian Manuscript Report of Fisheries and Aquatic Science 1867 (1986): i–iv + 1–23.

Needler, A.W.H. "Report of the Atlantic Biological Station for 1941." Fisheries Research Board of Canada (1941).

Needler, A.W.H. "Report of the Atlantic Biological Station for 1942." Fisheries Research Board of Canada (1942).

Needler, A.W.H. "Report for 1946 of the Atlantic Biological Station, St Andrews, N.B." Fisheries Research Board of Canada (1946).

Needler, A.W.H. "Report for 1947 of the Atlantic Biological Station, St Andrews, N.B." Fisheries Research Board of Canada (1947).

Needler, A.W.H. "Report of the Atlantic Biological Station for 1949." Fisheries Research Board of Canada (1949).

Needler, A.W.H. "Report for 1950 of the Atlantic Biological Station, St Andrews, N.B." Unpublished report to the Fisheries Research Board of Canada (1950).

Needler, A.W.H. "Report for 1951 of the Atlantic Biological Station, St Andrews, N.B." Unpublished report to the Fisheries Research Board of Canada (1951).

Needler, A.W.H. "Report of the Atlantic Biological Station for 1952." Unpublished report to the Fisheries Research Board of Canada (1952).

Peterson, R.H. "Lethal Responses of Brook Trout (*Salvelinus fontinalis* (*Mitch.*)) to Dissolved Copper as Affected by Prior Sublethal Exposure to the Metal." Fisheries Research Board of Canada Manuscript Report 1309 (1974): 1–9. Available at http://www.dfo-mpo.gc.ca/Library/23090.pdf.

Peterson, R.H. "Temperature Selection of Juvenile Atlantic Salmon (*Salmo salar* L.) Exposed to Some Pesticides." Fisheries Research Board of Canada Manuscript Report 1251 (1973): 1–9.

Peterson, R.H. "Variations in Aquatic Insect Densities Associated with Copper-Zinc Concentrations." Fisheries and Marine Services Manuscript Report 1470 (1978): 1–3.

Peterson, R.H., D.J. Martin-Robichaud, P. Harmon, and A. Berge. "Notes on Striped Bass Culture, with Reference to the Maritime Provinces." Fisheries and Oceans Canada, Canadian Stock Assessment Secretariat Research Document 99/004 (1994): ii + 35 pp. Available at http://www.dfo-mpo.gc.ca/csas/csas/docrec/1999/pdf/99_004e.pdf.

Picard-Aitken, M., D. Campbell, and G. Côté. "Bibliometric Study in Support of Fisheries and Oceans Canada's International Science Strategy." Report to Fisheries and Oceans Canada by Science-Metrix (2009): 1–66.

Poole, N.J., D.J. Wildish, and N.A. Lister. " Effects of a Neutral-sulfite Pulp Effluent on Some Chemical and Biological Parameters in the L'Etang Inlet,

New Brunswick. L'Etang Inlet Survey III." Fisheries Research Board of Canada Manuscript Report 1404 (1976): 1–27. Available at http://www.dfo-mpo.gc.ca/Library/73876.pdf.

Poole, N.J., D.J. Wildish, and N.A. Lister. "The Use of Microecosystem Models to Investigate Pollution by Pulp Mill Effluent." Fisheries Research Board of Canada Manuscript Report 1403 (1976): 1–21. Available at http://www.dfo-mpo.gc.ca/Library/73875.pdf.

Rathbun, M.J. "Canadian Atlantic Fauna 10: Arthropoda. 10 m. Decapoda." Atlantic Biological Station, St Andrews, NB, Biological Board of Canada, 1929: 1–38.

Rigby, M.S., and A.G. Huntsman. "Materials Relating to the History of the Fisheries Research Board of Canada (Formerly the Biological Board of Canada) for the Period 1898–1924." Fisheries Research Board of Canada Manuscript Report 660 (1958): 1–272.

Robert, G., G.A.P. Black, M.A.E. Butler, and S.J. Smith. "Georges Bank Scallop Stock Assessment – 1999." Department of Fisheries and Oceans Canadian Stock Assessment Secretariat Response Document 2000/016 (2000).

Robert, G., G.S. Jamieson, and M.J. Lundy. "Profile of the Canadian Offshore Scallop Fishery on Georges Bank, 1978–1981." Canadian Atlantic Fisheries Science Advisory Committee Research Document 82/15 (1982).

Robert, G., M.J. Lundy, and M.A.E. Butler-Connolly. "Scallop Fishing Grounds on the Scotian Shelf." Canadian Atlantic Fisheries Scientific Advisory Committee Research Document 85/28 (1958).

Robinson, S.M.C., and A. MacIntyre. "Biological Fishery Information for the Rational Development of the Green Sea Urchin Industry. Final Report for the New Brunswick Department of Fisheries and Aquaculture and the Canada–New Brunswick Co-operation Agreement on Economic Diversification." St Andrews, NB, Biological Station (1995): 1–90.

Scarratt, D.J. "Bleached Kraft Mill Effluent near Pictou, N.S., and Its Effect on the Marine Flora and Fauna with a Note on the Pictou Co. Lobster Landings." Fisheries Research Board of Canada Manuscript Report 1037 (1969): 1–24.

Scarratt, D.J., and A.J. Wilson. "Experiments with Rotenone in Northumberland Strait and Stomach Analysis of Fish Collected." Fisheries Research Board of Canada Manuscript Report 1107 (1970): 1–8.

Smith, S.J. "Atlantic Salmon Sport Catch and Effort Data, Maritimes Region, 1951–79." Canadian Data Report of Fisheries and Aquatic Science 258 (1981): 1–267.

Sprague, J.B. "Dissolved Oxygen in Restigouche River Estuary, NB., October 1966." Fisheries Research Board of Canada Manuscript Report 903 (1967): 1–17. Available at http://www.dfo-mpo.gc.ca/Library/31898.pdf.

Sprague, J.B. "Negative Test of Apparent DDT Resistance in an Aquatic Insect after Seven Years' Exposure to Aerial Spraying." Fisheries Research Board of Canada Manuscript Report 908 (1967): 1–28.

Sprague, J.B. "Toxicity of Pollution to Aquatic Life: A Summary of Research in Canada." Fisheries Research Board of Canada Manuscript Report 771 (1964): 1–18. Available at http://www.dfo-mpo.gc.ca/Library/59356.pdf.

Stevenson, J.A. "Bivalve Larvae Observations." St Andrews Library reprint.

Stevenson, J.A. "Growth of the Giant Scallop (*Placopecten grandis*)." Manuscript Report of the Biological Stations 420 (1932): 1–33.

Stevenson, J.A. "Growth in the Giant Scallop (*Placopecten grandis*)." Biological Board of Canada Manuscript Report 823 (1933).

Wells, P.G., R.A. Matheson, D.A. Lord, and K.G. Doe. "Environmental Monitoring of the 1978 Spruce Budworm Spray Program in New Brunswick, Canada. Field Sampling and Aquatic Toxicity Studies with Fish." Environment Canada, Environmental Protection Service Surveillance Report EPS-5-AR-79–1 (1979): 1–14.

Wells, P.G., W.R. Parker, and D. Vaughan. "The Acute Toxicity of Four Insecticides and Their Formulations to Fingerling Rainbow Trout, *Salmo gairdneri*." In "Environmental Monitoring of the 1976 Spruce Budworm Spray Program in New Brunswick, Canada." Fisheries and Environment Canada, Environmental Protection Service Surveillance Report EPS-5-AR-78–3 (1978): 84–110.

Wells, P.G., E. Pessah, and W.R. Parker. "The Toxicity of Taw and Treated Drainage from Heath Steele Mines, N.B., during Period of September–October, 1974." Canada, Environmental Protection Service Report Series EPS-5-AR-74–14 (1974): 1–27.

White, A.W., and H.M. Akagi. "A Compilation of Total Releases and Recoveries of Drift Bottles and Sea-bed Drifters in Continental Shelf Waters of the Canadian Atlantic Coast from 1960 through 1973." Fisheries Research Board of Canada Manuscript Report 1281 (1974): 1–49. Available at http://www.dfo-mpo.gc.ca/Library/26443.pdf.

Wildish, D.J., H. Akagi, and N.J. Poole. "Avoidance by Herring of Sulphite Pulp Mill Effluents." International Council for the Exploration of the Sea, Fisheries Improvement Committee, C.M. 1976/E:26 (1976): 1–8.

Wildish, D.J., W.G. Carson, W.V. Carson, and J.H. Hull. "Effects of a Neutral-sulphite, Pulp Effluent on Some Chemical and Biological Parameters in the L'Etang Inlet, New Brunswick. L'Etang Inlet Survey I." Fisheries Research Board of Canada Manuscript Report 1177 (1972): 1–18.

Wildish, D.J., W.V. Carson, A.J. Wilson, and J.H. Hull. "Effects of a Neutral-sulphite, Pulp Effluent on Some Chemical and Biological Parameters in the

L'Etang Inlet, New Brunswick. L'Etang Inlet Survey II." Fisheries Research Board of Canada Manuscript Report 1295 (1974): 1–56. Available at http://www.dfo-mpo.gc.ca/Library/23078.pdf.

**Published Sources**

Aiken, D.E., and E.H. Byard. "Histological Changes in Lobsters (*Homarus americanus*) Exposed to Yellow Phosphorus." *Science* 176 (1972): 1434–5.

Aiken, D.E., and S.L. Waddy. "Production of Seed Stock for Lobster Culture." *Aquaculture* 44: (1985): 103–14.

Aiken, D.E., and W.W. Young-Lai, "Dactylotomy, Chelotomy and Dactylostasis: Methods for Enhancing Survival and Growth of Small Lobsters (*Homarus americanus*) in Communal Conditions." *Aquaculture* 22 (1981): 45–52.

Aiken, D.E., and V. Zitko. "Effect of Iranian Crude Oil on Lobsters (*Homarus americanus*) Held in Floating Crates." *International Council for the Exploration of the Sea C.M./E:45* (1977): 1–13.

Ainley, M. "A Family of Women Scientists." *Concordia University Newsletter* 7 (1986): 4–11.

Ainley, M.G. "Gendered Careers: Women Science Educators at Anglo-Canadian Universities, 1920–1980." In *Historical Identities: The Professoriate in Canada*, ed. P. Stortz and E.L. Panayotidis. Toronto: University of Toronto Press, 2005. 248–70.

Ainley, M.G. "Last in the Field? Canadian Women Natural Scientists, 1815–1965." In *Despite the Odds: Essays on Canadian Women and Science*, ed. M.G. Ainley. Montreal: Véhicule Press, 1990. 25–62.

Ainley, M.G. "Marriage and Scientific Work in Twentieth-Century Canada: The Berkeleys in Marine Biology and the Hoggs in Astronomy." In *Creative Couples in the Sciences*, ed. H.M. Pycior, N.G. Slaack, and P.G. Abir-am. New Brunswick, NJ: Rutgers University Press, 1996. 143–55.

Ainley, M.G., ed. *Despite the Odds: Essays on Canadian Women and Science.* Montreal: Véhicule Press, 1990.

Ainley, M.G., and A.C. Millar. "A Select Few: Women and the National Research Council of Canada, 1916–1991." *Scientia Canadensis* 15:2 (1991): 105–16.

Alderdice, D.F., and M.E. Worthington. "Toxicity of a DDT Forest Spray to Young Salmon." *Canadian Fish Culture* 24 (1959): 41–8.

Allen, K.R., and J.K. Lindsey. "Commercial Catches of Atlantic Salmon in the Maritimes Area, 1949–1965." *Fisheries Research Board of Canada Technical Report* 29 (1967): 1–160.

Alston, Richard M. *The Individual vs. the Public Interest: Political Ideology and National Forest Policy*. Boulder, CO: Westview Press,1983.

Anderson, D.M., A.W. White, and D.G. Baden, eds. *Toxic Dinoflagellates*. New York: Elsevier, 1985.

Anderson, F. "The Demise of the Fisheries Research Board of Canada: A Case Study of Canadian Research Policy." *Journal of the History of Canadian Science, Technology, and Medicine* 8 (1984): 151–6.

Anderson, J.M. *The Salmon Connection: The Development of Atlantic Salmon Aquaculture in Canada*. Tantallon, NS: Glen Margaret Publishing, 2007.

Angelini, Ronaldo, and Coleen L. Malony. "Fisheries Ecology and Modelling: An Historical Perspective." *Pan-American Journal of Aquatic Sciences* 2:2 (2007): 75–85.

Anisef, P., and J. Lennards. "University." In *The Canadian Encyclopedia*, vol. 3. Edmonton: Hurtig Publications, 1985.

Anonymous. "Dickie, L.M." 2009. Available at http://www.thecanadianencyclopedia.ca/en/article/lloyd-m-dickie/.

Anonymous. "Edward Ernest Prince." *Proceedings of the Transactions of the Royal Society of Canada*, 3rd series, 31 (1937): xx–xxiii.

Anonymous. "Eminent Living Geologists: Joseph Frederick Whiteaves, LL.D., F.G.S., F.R.S. (Canada)." *Geology Magazine*, new series, *Decade V* (1906) 3: 433–42.

Anonymous. *From Leadline to Laser*. Centennial conference of the Canadian Hydrographic Service. Ottawa: Canadian Hydrographic Service / Canadian Hydrographic Association. 1983.

Anonymous. "Prince, Edward Ernest." In *The Canadian Men and Women of the Times: A Handbook of Canadian Biography of Living Characters*, ed. H.J. Morgan. Toronto: William Briggs, 1912. 918–19.

Aoyama, S. "The Mutsu Bay Scallop Fisheries: Scallop Culture, Stock Enhancement, and Resource Management." In *Marine Invertebrate Fisheries: Their Assessment and Management*, ed. J.F. Caddy. Toronto: John Wiley and Sons Inc., 1989. 525–39.

Arai, M.N. "Charles McLean Fraser (1872–1946): His Contributions to Hydroid Research and to the Development of Fisheries Biology and Academia in British Columbia." *Hydrobiologia* 530/531 (2004): 3–11.

Arai, M.N. "Publications of Edith and/or Cyril Berkeley." *Journal of the Fisheries Research Board of Canada* 28 (1971): 1365–72.

Arai, M.N. "Research on Coelenterate Biology in Canada through the Early Twentieth Century. *Archives of Natural History* 19:1 (1992): 55–68.

Arnold, A.F. *The Sea-beach at Ebb-tide: A Guide to the Study of the Seaweeds and the Lower Animal Life Found between Tide-marks.* New York: The Century Co. / Dover, 1901, 1968.

Bailey, W.B., D.G. MacGregor, and H.B. Hachey. "Annual Variations of Temperature and Salinity in the Bay of Fundy." *Journal of the Fisheries Research Board of Canada* 11 (1954): 32–47.

Baird, F.T. "Migration of the Deep Sea Scallop (*Pecten magellanicus*)." *Maine Department of the Sea and Shore Fisheries Circular*, 1953: 1–8.

Baird, F.T. "Observations on the Early Life History of the Giant Sea Scallop (*Pecten magellanicus*)." *Maine Department of the Sea and Shore Fisheries Research Bulletin* 14 (1953):1–7.

Balech, E. "The Genus *Alexandrium* or *Gonyaulax* of the *Tamarensis* Group." In *Toxic Dinoflagellates*, ed. D.M. Anderson, A.W. White, and D.G. Baden. New York: Elsevier, 1985. 33–8.

Balech, E. "The Genus *Alexandrium Halim*. (*Dinoflagellata*). Sherkin Island Marine Station, Sherkin Island, Co. Cork, Ireland. 1995. In *WoRMS World Register of Marine Species* http://www.marinespecies.org/aphia.php?p=taxdetails&id=109711.

Balogh, Brian. "Scientific Forestry and the Roots of the Modern American State: Gifford Pinchot's Path to Progressive Reform." *Environmental History* 7:2 (2002): 198–225.

Beamish, F.W.H. "Swimming Capacity." In *Fish Physiology*, vol. 7, ed. W.S. Hoar and D.J. Randall. New York: Academic Press, 1978. 101–87.

Beaulnes, A. "Government and Science Policy in the 1970s." *Journal of the Fisheries Research Board of Canada* 31 (1974):1278–80.

Benson, K.R. "Why American Marine Stations? The Teaching Argument." *American Zoology* 28:1 (1988): 7–14.

Berger, C. *Honour and the Search for Influence: A History of the Royal Society of Canada*. Toronto: University of Toronto Press, 1996.

Berger, C. *Science, God, and Nature in Victorian Canada*. Toronto: University of Toronto Press, 1988.

Berkeley, A.A. "Sex Reversal in *Pandalus danae*." *American Naturalist* 63 (1929): 1–3.

Bjerkan, P. "Results of the Hydrographical Observations Made by Dr. John Hjort in the Canadian Atlantic Waters during the Year 1915." In *Canadian Fisheries Expedition, 1914–1915, in the Gulf of St Lawrence and Atlantic Waters of Canada*. Ottawa: Canadian Department of the Naval Services, 1919. 348–403. Available at https://archive.org/details/canadianfisherie00cana.

Blackadar, R.G. "Geological Survey of Canada." In *The Canadian Encyclopedia*, 2nd ed., vol. 2. Edmonton: Hurtig Publishers, 1988. 889–90.

Blogoslawski, W.J., M.E. Stewart, J.W. Hurst, and F.G. Kern. "Ozone Detoxification of Paralytic Shellfish Poison in the Softshell Clam (*Mya arenaria*)." *Toxicon* 17 (1979): 650–4.

Boghen, A.D. *Cold-Water Aquaculture in Atlantic Canada*. 2nd ed. Sackville, NB: Canadian Institute for Research on Regional Development / Tribune Press Ltd., 1995.

Bond, R.M., and J.C. Medcof. "Epidemic Shellfish Poisoning in New Brunswick." *Canadian Medical Association Journal* 79 (1957): 19–24.

Bonnevie, K. "Heteropoda Collected during the *Michael Sars* North Atlantic Deep-sea Expedition 1910." In *Reports on the Scientific Results of the Michael Sars Expedition 1910*, vol. 3(2). Bergen: The Trustees of the Bergen Museum, 1920. 1–16, pl. I–IV.

Bonnevie, K. "Pelagic Nudibranchs from the "Michael Sars" North Atlantic Deep-sea Expedition 1910." In *Reports on the Scientific Results of the Michael Sars Expedition 1910*, vol. 5(2). Bergen: The Trustees of the Bergen Museum, 1929. 1–9, pl. I–IV.

Bonnevie, K. "Pteropoda from the *Michael Sars* North Atlantic Deep-sea Expedition 1910." In *Reports on the Scientific Results of the Michael Sars Expedition 1910*, vol. 3(1). Bergen: The Trustees of the Bergen Museum, 1913. 1–69, pl. I–IX.

Bottom, Daniel L. "To Till the Water: A History of Ideas in Fisheries Conservation." In *Pacific Salmon & Their Ecosystems: Status and Future Options*, ed. D.J. Stouder, P.A. Bisson, and R.J. Naiman. Detroit: Chapman and Hill, 1997. 569–97.

Bourgeois, M., J.C. Brêthes, and M. Nadeau. "Substrate Effects on Survival, Growth and Dispersal of Juvenile Sea Scallops, *Placopecten magellanicus* (*Gmelin* 1791)." *Journal of Shellfish Research* 25 (2006): 43–9.

Bourne, N. "Paralytic Shellfish Poison in Sea Scallops (*Placopectin magallanicus Gmelin*)." *Journal of the Fisheries Research Board of Canada* 22 (1965): 1137–49.

Bourne, N. "Relative Efficiency of Catches by 4- and 5-inch Rings on Offshore Scallop Drags." *Journal of the Fisheries Research Board of Canada* 22 (1965): 313–33.

Bourne, N. "Relative Fishing Efficiency of Three Types of Scallop Drags." *International Commission for the Northwest Atlantic Fisheries Bulletin* 3 (1966): 15–25.

Bourne, N. "Scallops and the Offshore Fishery of the Maritimes." *Bulletin of the Fisheries Research Board of Canada* 145 (1964): 1–60.

Bourne, N., and A. McIver. "Gulf of St Lawrence Scallop Explorations – 1961." *Fisheries Research Board of Canada, St Andrews, NB, Biological Station General Series Circular* 35 (March 1962).

Bourne, N.F., and S.M.C. Robinson. "In Memoriam: John Carl Medcof 1911–1997." *Journal of Shellfish Research* 19 (2000): 1–5.

Bowler, P.J. "The Early Development of Scientific Societies in Canada." In *The Pursuit of Knowledge in the Early American Republic*, ed. A. Oleson and S.C. Brown. Baltimore: Johns Hopkins University Press, 1976. 326–39.

Brady, T. "The Algae of Acrimony." *Minnesota* (January–February 2008). Available at http://www.minnesotaalumni.org/s/1118/content.aspx?pgid=1077.

Brander, L., and D.L. Burke. "Rights-based vs. Competitive Fishing of Sea Scallops *Placopecten magellanicus* in Nova Scotia." *Aquatic Living Resources* 8 (1995): 279–88.

Brown, C., ed. *The Illustrated History of Canada*. Toronto: Key Porter Books, 2002.

Brown, P.S. "Early Women Ichthyologists." *Environmental Biology of Fishes* 41 (1994): 9–30.

Buerkle, U. "Detection of Trawling Noise by Atlantic Cod (*Gadus morhua L.*)." *Marine Behavior and Physiology* 4 (1977): 233–42.

Buerkle, U. "Estimation of Fish Length from Acoustic Target Strengths." *Canadian Journal of Fisheries and Aquatic Science* 44 (1987): 1782–5.

Buerkle, U., and A. Sreedharan. "Acoustic Target Strengths of Cod in Relation to Their Aspect in the Sound Beam." In *Meeting on Hydroacoustical Methods for the Estimation of Marine Fish Populations, 25–29 June 1979*, ed. J.B. Suomala. Cambridge, MA: Charles Stark Draper Laboratory, 1981. 229–47.

Bugden, G.L. "Oceanographic Observations from the Bay of Fundy for the Pre-operational Environmental Monitoring Program for the Point Lepreau, NB, Nuclear Generating Station." *Canadian Data Report of Hydrography and Ocean Sciences* 27 (1985): 1–41.

Bumpus, D.F., and L.M. Lauzier. *Serial Atlas of the Marine Environment*. Folio 7: *Surface Circulation on the Continental Shelf off Eastern North America between Newfoundland and Florida*. New York: American Geographical Society of New York, 1965.

Burridge, L.E., and K. Haya. "The Lethality of Ivermectin, a Potential Agent for Treatment of Salmonids against Sea Lice, to the Shrimp *Crangon septemspinosa*." *Aquaculture* 117 (1993): 9–14.

Burridge, L.E., and K. Haya. "The Lethality of Pyrethrins to Larvae and Postlarvae of the American Lobster (*Homarus americanus*)." *Ecotoxicology and Environmental Safety* 38 (1997): 150–4.

Burridge, L.E., K. Haya, S.L. Waddy, and J. Wade. "The Lethality of Anti-sea Lice Formulations Salmosan (*azamethiphos*) and Excis (*cypermethrin*) to Stage IV and Adult Lobsters (*Homarus americanus*) during Repeated Short-term Exposures." *Aquaculture* 182 (2000): 27–35.

Bush, K.J. "Catalogue of Mollusca and Echinodermata Dredged on the Coast of Labrador by the Expedition under the Direction of Mr. W.A. Stearns in 1882." *Proceedings of the U.S. National Museum* 6 (1884): 236–47, pl. 9.

Byrne, J.M., F.W.H. Beamish, and R.L. Saunders. "Influence of Salinity, Temperature and Exercise on Plasma Osmolality and Ionic Concentration in Atlantic Salmon." *Journal of the Fisheries Research Board of Canada* 29 (1972): 1217–20.

Caddy, J.F. "Efficiency and Selectivity of the Canadian Offshore Scallop Dredge." *International Council for the Exploration of the Sea C.M. 1971/K:25* (1971): 1–8.

Caddy, J.F. "Long-term Trends and Evidence for Production Cycles in the Bay of Fundy Scallop Fishery. *Rapports et Procès-verbaux des Réunions du Conseil International pour l'Exploration de la Mer* 175 (1979): 97–108.

Caddy, J.F. *Marine Habitat and Cover: Oceanographic Methodology Series*. Paris, France: UNESCO Publishing, 2007.

Caddy, J.F. "A Method of Surveying Scallop Populations from a Submersible." *Journal of the Fisheries Research Board of Canada* 27 (1970): 535–49.

Caddy, J.F. 1989. "A Perspective on the Population Dynamics and Assessment of Scallop Fisheries, with Special Reference to the Sea Scallop, *Placopecten magellanicus Gmelin*." In *Marine Invertebrate Fisheries: Their Assessment and Management*, ed. J. Caddy. Toronto: John Wiley, Interscience, 1989. 559–90.

Caddy, J.F. "Practical Considerations for Quantitative Estimations of Benthos from a Submersible." In *Underwater Research*, ed. E.A. Drew, J.N. Lythgoe, and J.D. Woods. New York: Academic Press, 1976. 285–98.

Caddy, J.F. "Progressive Loss of Byssal Attachment with Size in the Sea Scallop, *Placopecten magellanicus* (*Gmelin*)." *Journal of Experimental Marine Biology and Ecology* 9 (1972): 179–90.

Caddy, J.F. "Recent Scallop Recruitment and Apparent Reduction in Cull Size by the Canadian Fleet on Georges Bank." *International Commission for the Northwest Atlantic Fisheries Redbook, Part III* (1971): 147–55.

Caddy, J.F. "Size Selectivity of the Georges Bank Offshore Dredge and Mortality Estimate for Scallops from the Northern Edge of Georges in the Period June 1970 to 1971." *International Commission for the Northwest Atlantic Fisheries Redbook* (1972): 79–85.

Caddy, J.F. "Spatial Model for an Exploited Population, and Its Application to the Georges Bank Scallop Fishery." *Journal of the Fisheries Research Board of Canada* 32 (1975): 1305–28.

Caddy, J.F. "Underwater Observations on Scallop (*Placopecten magellanicus*) Behaviour and Drag Efficiency." *Journal of the Fisheries Research Board of Canada* 25 (1968): 2123–41.

Caddy, J.F. "Underwater Observations on Tracks of Dredges and Trawls and Some Effects of Dredging on a Scallop Ground." *Journal of the Fisheries Research Board of Canada* 30 (1973): 173–80.

Caddy, J.F. "Use of Manned Underwater Vehicles for Fisheries Research." Oceanology International Conference, 1972. University of California, BPS Exhibitions Ltd., 1972. 278–81.

Caddy, J.F., ed. *Marine Invertebrate Fisheries: Their Assessment and Management*. Toronto: John Wiley, Interscience, 1989.

Caddy, J.F., and D.J. Agnew. "An Overview of Global Experience to Date with Recovery Plans for Depleted Marine Resources and Suggested Guidelines for Recovery Planning." *Fish and Fisheries* 14 (2004): 43–112.

Caddy, J.F., and F. Carocci. "The Spatial Allocation of Fishing Intensity by Port-based Inshore Fleets: A GIS Application." *International Council for the Exploration of the Sea Journal of Marine Science* 56 (1999): 388–403.

Caddy, J.F., and J.A. Carter. "Macro-epifauna of the Lower Bay of Fundy: Observations from a Submersible and Analysis of Faunal Adjacencies." *Canadian Technical Report of Fisheries and Aquatic Science* 1254 (1984): 1–35.

Caddy, J.F., and R.A. Chandler. "Accumulation of Paralytic Shellfish Poison by the Rough Whelk (*Buccinum undatum L.*)." *Proceedings of the National Shellfisheries Association* 58 (1968): 46–50.

Caddy, J.F., R.A. Chandler, and D.G. Wilder. "Biology and Commercial Potential of Several Unexploited Molluscs and Crustacean on the Atlantic Coast of Canada." In *Proceedings of Government-Industry Meeting on the Utilization of Atlantic Marine Resources*, Queen Elizabeth Hotel, Montreal, QC, 5–7 February 1974. Ottawa: Environment Canada, 1974. 57–106.

Caddy, J.F., and O. Defeo. "Enhancing or Restoring the Productivity of Natural Populations of Shellfish and Other Marine Invertebrate Resources." *FAO Fisheries Technical Papers 448*. Rome: FAO, 2003: 1–159.

Caddy, J.F., and J.A. Gulland. 1983. "Historical Patterns of Fish Stocks." *Marine Policy* 7:4 (1983): 267–78.

Caddy, J.F., and J.-C. Seijo. "Application of a Spatial Model to Explore Rotating Harvest Strategies for Sedentary Species." *Canadian Special Publication of Fisheries and Aquatic Science* 125 (1999): 359–65.

Caddy, J.F., and A. Sreedharan. "The Effect of Recent Recruitment to the Georges Bank Scallop Fishery on Meat Sizes Landed by the Offshore Fleet in the Summer of 1970." *Fisheries Research Board Technical Report* 256 (1971).

Caddy, J.F., and J. Watson. "Submersibles for Fisheries Research." *Hydrospace* 2 (1969): 12–16.

Cairns, J., and K.L. Dickson, eds. *Biological Methods for the Assessment of Water Quality*. Philadelphia: American Society for Testing and Materials Special Technical Publication 528.

Cameron, A.T., and I. Mounce. "Some Physical and Chemical Factors Influencing the Distribution of Marine Flora and Fauna in the Strait of Georgia

and Adjacent Waters." *Contributions of the Canadian Biological Stations*, new series, 1:4 (1922): 39–72.
Campbell, N.J. "An Historical Sketch of Physical Oceanography in Canada." *Journal of the Fisheries Research Board of Canada* 32 (1976): 2155–67.
Canada. Department of the Naval Services. *Canadian Fisheries Expedition, 1914–1915, in the Gulf of St Lawrence and Atlantic Waters of Canada*. Ottawa: J. de Labroquerie, 1919.
Canadian Hydrographic Service / Canadian Hydrographers' Association. *From Leadline to Laser: Centennial Conference of the Canadian Hydrographic Service. Canadian Special Publication of Fisheries and Aquatic Science* 67. Ottawa: Fisheries and Oceans, Scientific Information and Publications Branch, 1983.
Carrothers, P.J.G., T.J. Foulkes, M.P. Connors, and A.G. Walker. "Data on Engineering Performance of Canadian East Coast Groundfish Otter Trawls." *Fisheries Research Board of Canada Technical Report* 125 (1969): 1–100.
Carson, R. *Silent Spring*. Boston: Houghton Mifflin Co., 1962.
Cawood, J. "The Magnetic Crusade: Science and Politics in Early Victorian England." *Isis* 70 (1979): 493–518.
Centre for Marine Biodiversity. "TOWCAM Survey System." 2008. Available from http://www.marinebiodiversity.ca/cmb/research/pockmarks/towcam-survey-system/.
Chambers, R., and B. Mossop. "A Report on Cross-fertilization Experiments (*Astrias* X *Solaster*)." *Transactions of the Royal Society of Canada, Section 4* (1918): 145–7.
Chang, B.D., R.J. Losier, F.H. Page, D.A. Greenberg, and J.D. Chaffey. "Use of a Water Circulation Model to Predict the Movements of Phytoplankton Blooms Affecting Salmon Farms in the Grand Manan Island Area, Southwestern New Brunswick." *Canadian Technical Report of Fisheries and Aquatic Science* 2703 (2007): iii + 64 pp.
Chang, B.D., F.H. Page, and B. Hill. "Preliminary Analysis of Coastal Marine Resource Use and the Development of Open Ocean Aquaculture in the Bay of Fundy." *Canadian Technical Report of Fisheries and Aquatic Science* 2585 (2005): iv + 36 pp. Available at http://www.dfo-mpo.gc.ca/Library/327939.pdf.
Chang, B.D., F.H. Page, R.J. Losier, D.A. Greenberg, J.D. Chaffey, and E.P. McCurdy. "Application of a Tidal Circulation Model for Fish Health Management of Infectious Salmon Anemia in the Grand Manan Island Area, Bay of Fundy." *Bulletin of the Aquaculture Association of Canada* 105:1 (2005): 22–33.
Chang, B.D., F.H. Page, J.L. Martin, G. Harrison, E. Horne, L.E. Burridge, M.M. LeGresley, A. Hanke, and P. McCurdy. "Phytoplankton Early

Warning Approaches for Salmon Farms in Southwestern New Brunswick." *Aquaculture Association of Canada Special Publication* 9 (2005): 20–3.

Chang, B.D., and W.M. Watson-Wright. "Aquaculture Research at the St Andrews Biological Station: Past, Present and Future." *Aquaculture Association of Canada Special Publication* 2 (1997): 105–8.

Chartrand, L., R. Duchesne, and Y. Gingras. *Histoire des Sciences au Québec.* Montreal: Éditions du Boréal, 1987.

Canadian Pulp and Paper Association. "Forestry Conditions in Sweden, Norway, Great Britain and France." *Bulletin Number Thirty-four*, article 6, 15 October 1921. Internet archive: https://archive.org/stream/forestryconditio00beckrich#page/4/mode/2up.

Chambers, R., and B. Mossop. "A Report on Cross Fertilization Experiments (*Asterias* X *Solaster*)." *Transactions of the Royal Society of Canada XII, Section 4* (1918): 145–7.

Chase, A. *In a Dark Wood: The Fight over Forests and the Myth of Nature*. New Brunswick, NJ: Transaction Publishers, 1995, 2001.

Chase, F.A. "Mary J. Rathbun (1860–1943)." *Journal of Crustacean Biology* 10 (1990): 165–7.

Chevrier, J.R., and R.W. Trites. "Drift-bottle Experiments in the Quoddy Region, Bay of Fundy." *Journal of the Fisheries Research Board of Canada* 17 (1960): 743–62.

Chopin T. "Integrated Multi-trophic Aquaculture. What It Is, and Why You Should Care … and Don't Confuse It with Polyculture." *Northern Aquaculture* 12:4 (2006): 4.

Clemens, W.A. 1958. "Reminiscences of a Director." *Journal of the Fisheries Research Board of Canada* 15(5): 779–96.

Clemens, W.A., and L.S. Clemens. "A Contribution to the Biology of the Muttonfish (*Zoarces anguillaris*)." *Contributions to Canadian Biology 1918–1920* (1921): 69–83.

Clepper, H. *Professional Forestry in the United States*. Baltimore: Johns Hopkins University Press, 1971.

Clewson, M., and R. Sedjo. "History of Sustained Yield Concept and Its Application to Developing Countries." In *History of Sustained Yield Forestry: A Symposium*, ed. H.K. Steen. Portland, OR: Forestry History Society, 1984. 3–15.

Cole, D., and B. Lochner, eds. *To the Charlottes: George Dawson's Survey 1878 of the Queen Charlotte Islands*. Vancouver, BC: UBC Press,1993.

Cook, R. "The Triumph and Trials of Materialism." In *The Illustrated History of Canada*, ed. C. Brown. Toronto: Key Porter Books, 2002. 377–472.

Copeland, G.G. "The Temperatures and Densities and Allied Subjects of Passamaquoddy Bay and Its Environs: Their Bearing on the Oyster Industry." *Contributions to Canadian Biology 1906–1910* (1912): 281–94.

Couturier, C.Y. "Scallop Aquaculture in Canada: Fact or Fantasy." *World Aquaculture* 21:2 (1990): 54–62.

Couturier, C.Y., P. Dabinett, and M. Lanteigne, "Scallop Culture in Atlantic Canada." In *Cold-Water Aquaculture in Atlantic Canada*, 2nd ed., ed. A.D. Boghen. Moncton, NB: University of Moncton, 1996. 297–340.

Craigie, E.H. *A History of the Department of Zoology of the University of Toronto up to 1962*. Toronto: University of Toronto Press, 1966.

Craigie, E.H. "Hydrographic Investigations in the St Croix River and Passamaquoddy Bay in 1914." *Contributions to Canadian Biology 1914–1915* (1916): 151–61.

Craigie, E.H. "A Hydrographic Section of the Bay of Fundy in 1914." *Contributions to Canadian Biology 1914–1915* (1916): 163–7.

Craigie, E.H., and W.H. Chase. "Further Hydrographic Investigations in the Bay of Fundy." *Contributions to Canadian Biology 1917–1918* (1918): 127–48.

Creese, M.R.S. *Ladies in the Laboratory? American and British Women in Science, 1800–1900*. London: The Scarecrow Press, 1998.

Creese, M.R.S. *Ladies in the Laboratory II: West European Women in Science, 1800–1900: A Survey of Their Contributions to Research*. Lanham, MD: The Scarecrow Press, 2004.

Currie, M.E. "Exuviation and Variation of Planktonic Copepods with Special Reference to *Calanus finmarchicus*." *Transactions of the Royal Society of Canada, Section 4*, 12 (1918): 207–33.

Dadswell, M.J., and M.S. Sinclair. "Aquaculture of Giant Scallop *Placopecten magellanicus* in the Canadian Maritimes: Its Use as an Experimental Tool to Investigate Recruitment and Growth." *International Council for the Exploration of the Sea ENEM/63* (1989): 1–17.

Dame, R.F., ed. *Bivalve Filter Feeders. NATO ASI Series G. Ecological Sciences 33*. Berlin: Springer-Verlag, 1992.

Damkaer, D.M. "Harriet Richardson (1874–1958): First Lady of Isopods." *Journal of Crustacean Biology* 20 (2000): 803–11.

Davidson, V.M. "Fluctuations in the Abundance of Planktonic Diatoms in the Passamaquoddy Region, New Brunswick from 1924 to 1931." *Contributions to Canadian Biology and Fisheries*, new series, 8 (1934): 357–407.

Dawson, W.B. *The Currents at the Entrance to the Bay of Fundy and on the Steamship Routes in Its Approaches off Southern Nova Scotia from Investigations of the Tidal and Current Survey in the Season of 1904*. Ottawa: Canada, Dept. of Marine and Fisheries, 1905.

Dawson, G.M. "Report on the Queen Charlotte Islands, by Mr. G.M. Dawson, with Appendices A. to G." In *Geological Survey of Canada. Report of Progress for 1878–79*. Montreal: Dawson Brothers (1880). 1–239.

Dawson, W.B. *Tables of Hourly Direction and Velocity of the Currents and Time of Slack Water in the Bay of Fundy and Its Approaches as far as Cape Sable: From Investigations of the Tidal and Current Survey in the Seasons of 1904 and 1907.* Ottawa: Canada, Dept. of Marine and Fisheries, 1908.

Dawson, W.B. *Temperatures and Densities of the Waters of Eastern Canada, Including the Atlantic from the Bay of Fundy to Newfoundland, the Gulf of St Lawrence, and the Straits Connecting It with the Ocean: From Investigations by the Tidal and Current Survey in the Seasons of 1894 to 1896, and 1903 to 1911.* Ottawa: Canada, Dept. of the Naval Service, 1922.

De Beer, G.R. "Edwin Stephen Goodrich, 1868–1946." *Obituary Notices of Fellows of the Royal Society* 5:15 (1947): 477–90.

Denny, M.W. *Air and Water: The Biology and Physics of Life's Media.* Princeton, NJ: Princeton University Press, 1993.

Derick, C.M. "Notes on the Development of the Holdfasts of Certain *Florideae.*" *Botanical Gazette* 28 (1899): 246–63.

DeVecchi, V. "The Dawning of a National Scientific Community in Canada, 1878–1896." *Scientia Canadensis* 18 (1984): 32–58.

DeVecchi, V. "The Pilgrim's Progress, the BAAS, and Research in Canada: From Montreal to Toronto." *Transactions of the Royal Society of Canada,* series 4, 20 (1982): 519–32.

Dickie, L.M. "Fluctuations in Abundance of the Giant Scallop, *Placopecten magellanicus* (*Gmelin*) in the Digby Area of the Bay of Fundy." *Journal of the Fisheries Research Board of Canada* 12 (1955): 797–857.

Dickie, L.M. "The Quest for Excellence: Funding Marine Science Research and Development." In *The Sea Has Many Voices: Ocean Policy for a Complex World,* ed. C. Lamson. Kingston, ON: McGill-Queen's University Press, 1994. 289–313.

Dickie, L.M., and L.P. Chiasson. "Offshore and Newfoundland Scallop Explorations." *Fisheries Research Board of Canada General Series Circular* 25 (1955): 1–4.

Dickie, L.M., and J.C. Medcof. "Causes of Mass Mortalities of Scallops (*Placopecten magellanicus*) in the Southwestern Gulf of St Lawrence." *Journal of the Fisheries Research Board of Canada* 20 (1963): 451–82.

Ditt, Karl, and J. Rafferty. "Nature Conservation in England and Germany 1900–70: Forerunner of Environmental Protection?" *Contemporary European History* 5:1 (1996): 1–28.

Doern, B., and J. Kinder. *Strategic Science in the Public Interest: Canada's Government Laboratories and Science-Based Agencies.* Toronto: University of Toronto Press, 2007.

Doucette, G.J., A.D. Cembella, J.L. Martin, J. Michaud, T.V.N. Cole, and R.M. Rolland. "Paralytic Shellfish Poisoning (PSP) Toxins in North Atlantic Right

Whales *Eubalaena glacialis* and Their Zooplankton Prey in the Bay of Fundy, Canada." *Marine Ecology Progress Series* 306 (2006): 303–13.

Dowd, M. "Oceanography and Shellfish Production: A Bio-physical Synthesis Using a Simple Model." *Bulletin of the Aquaculture Association of Canada* 100:2 (2000): 3–9.

Dowd, M. "Seston Dynamics in a Tidal Inlet with Shellfish Aquaculture: A Model Study Using Tracer Equations." *Estuarine, Coastal and Shelf Science* 57 (2003): 523–37.

Drew, E.A., J.N. Lythgoe, and J.D. Woods, eds. *Underwater Research*. New York: Academic Press, 1976.

Drinnan, R.E. "Oysters: Disease, Predation, Parasites and Competitors." In *The Proceedings of the Bras D'Or Lakes Aquaculture Conference, held in Sydney, Cape Breton, June 1975*, ed. G. McKay. Sydney, NS: College of Cape Breton Press, 1976. 125–9.

Drinnan, R.E., and J.C. Medcof. "Progress in Rehabilitating Disease Affected Oyster Stocks." *Fisheries Research Board of Canada General Series Circular* 34 (1961): 1–3.

Druett, J. *Hen Frigates: Wives of Merchant Captains under Sail*. New York: Simon and Schuster, 1988.

Duff, D. "Investigation of the Haddock Fishery, with Special Reference to the Growth and Maturity of the Haddock." *Contributions to Canadian Biology 1914–1915* (1916): 95–102.

Dunbar, M.J. "Eastern Arctic Waters." *Bulletin of the Fisheries Research Board of Canada* 88 (1951): 1–131.

Dymond, J.R. "One Hundred Years of Science in Canada. 10. Zoology." In *The Royal Canadian Institute Centennial Volume 1849–1949*, ed. W.S. Wallace. Toronto: Royal Canadian Institute, 1949. 108–20.

Dymond, M.J. "Zoology in Canada." In *A History of Science in Canada*, ed. H.M. Tory. Toronto: Ryerson Press, 1939. 41–57.

Eggleston, W. *National Research in Canada: The NRC 1916–1966*. Toronto: Clarke, Irwin Co., 1978.

Elner, R.W., and G.S. Jamieson. "Predation of Sea Scallops, *Placopecten magellanicus*, by the Rock Crab, *Cancer irroratus*, and the American Lobster, *Homarus americanus*." *Journal of the Fisheries Research Board of Canada* 36 (1979): 537–43.

Elson, P.F. "Effects on Wild Young Salmon of Spraying DDT over New Brunswick Forests." *Journal of the Fisheries Research Board of Canada* 24 (1967): 731–67.

Elson, P.F. "Impact of Recent Economic Growth and Industrial Development on the Ecology of Northwest Miramichi Atlantic Salmon (*Salmo salar*)." *Journal of the Fisheries Research Board of Canada* 31:5 (1974): 521–44.

Elson, P.F. "Number of Salmon Needed to Maintain Stocks." *Canadian Fish Culturist* 21 (1957): 19–23.

Elson, P.F. "Predator–Prey Relationships between Fish-eating Birds and Atlantic Salmon, with a Supplement on Fundamentals of Merganser Control." *Bulletin of the Fisheries Research Board of Canada* 133 (1962): 1–87.

Elson, P.F., and C.J. Kerswill. "Impact on Salmon of Spraying Insecticide over Forests." *Advances in Water Pollution Research* 1 (1966): 55–74.

Elson, P.F., and C.J. Kerswill. "Studies on Canadian Atlantic Salmon." *Transactions of the 20th North American Wildlife Conference* (1955): 415–526.

Elson, P.F., L.M. Lauzier, and V. Zitko. "A Preliminary Study of Salmon Movements in a Polluted Estuary." In *Marine Pollution and Sea Life,* in *FAO Technical Conference on Marine Pollution and Its Effects on Living Resources and Fishing, Rome, Italy,* ed. M. Ruivo. London: Fishing News Books, 1972. 325–30.

Elson, P.F., A.L. Meister, J.W. Saunders, R.L. Saunders, J.B. Sprague, and V. Zitko. "Impact of Chemical Pollution on Atlantic Salmon in North America." *International Atlantic Salmon Symposium 1972, St Andrews, N.B.* International Atlantic Salmon Foundation special publication 4 (1973): 83–110.

Elson, P.F., and A. Tuomi. *The Foyle Fisheries: New Basis for Rational Management.* Lurgan, Ireland: LM Press, 1975.

Factor, J.R., ed. *Biology of the Lobster* Homarus americanus. Toronto: Academic Press, 1995.

Fernow, B. *A Brief History of Forestry in Europe*. Toronto: University of Toronto Press, 1913.

Ferrier, M., J.L. Martin, and J.N. Rooney-Varga. "Stimulation of *Alexandrium fundyense* Growth by Bacterial Assemblages from the Bay of Fundy." *Journal of Applied Microbiology* 92 (2002): 706–16.

Fiander, A., ed. *Science Review 1992 & '93 of the Bedford Institute of Oceanography, Halifax Fisheries Research Laboratory, and the St Andrews Biological Station.* Dartmouth, NS: Dept. of Fisheries and Oceans, Scotia-Fundy Region, 1994.

Fillmore, S., and R.W. Sandilands. *The Chartmakers: The History of Nautical Surveying in Canada*. Toronto: NC Press, 1983.

Fingard, J. "Gender and Inequality at Dalhousie: Faculty Women before 1950." *Dalhousie Review,* Winter 1984–5: 687–703.

Finley, C. *All the Fish in the Sea: Maximum Sustainable Yield and the Failure of Fisheries Management*. Chicago: University of Chicago Press, 2011.

Finley, C. "The Social Construction of Fishing, 1949." *Ecology and Society* 14:1, article 6 (2009). Available at http://www.ecologyandsociety.org/vol14/iss1/art6/.

Fisheries Research Board of Canada. "Summary of Physical, Biological, Socio-economic and Other Factors Relevant to Potential Oil Spills in the Passamaquoddy Region of the Bay of Fundy." *Fisheries Research Board of Canada Technical Report* 428 (1974): 141–58. Available at http://www.dfo-mpo.gc.ca/Library/23142.pdf.

Forbes, J.R., ed. *Proceedings of the Fourth Canadian Workshop on Harmful Marine Algae, Canadian Technical Reports of Fisheries and Aquatic Science* 201 (1994)

Ford, A.R. *A Path not Strewn with Roses: One Hundred Years of Women at the University of Toronto 1884–1984.* Toronto: University of Toronto Press, 1985.

Forrester, W.D. "Current Measurements in Passamaquoddy Bay and the Bay of Fundy 1957 and 1958." *Journal of the Fisheries Research Board of Canada* 17 (1960): 727–9.

Foster, Janet. *Working for Wildlife: The Beginning of Preservation in Canada.* Toronto: University of Toronto Press, 1978.

Foulkes, T.J. "Design and Development of 'TUACS,' a Towed Underwater Automatic Camera Sled." *Fisheries Research Board of Canada Technical Report* 292 (1972).

Foulkes, T.J. "The Development of a Bottom-Referencing Underwater Towed Instrument Vehicle, BRUTIV, for Fisheries Research 1973–79." In *Underwater Photography: Scientific and Engineering Applications,* ed. P.F. Smith. New York: Van Nostrand Reinhold, 1984: 95–107.

Foulkes, T.J., and J.F. Caddy. "'DUCCS,' a Drifting Underwater Collapsible Camera Sled." *Fisheries Research Board of Canada Technical Report* 310 (1972): 1–11. Available at http://www.dfo-mpo.gc.ca/Library/28620.pdf.

Foulkes, T.J., and J.F. Caddy. "Towed Underwater Camera Vehicles for Fishery Resource Assessment." *Underwater Journal* 6 (1973): 110–16.

Foulkes, T.J., and D.J. Scarratt. "Design and Performance of 'TURP,' a Diver Controlled Towed Underwater Research Plane." *Fisheries Research Board of Canada Technical Report* 295 (1971): 1–11 + illustrations.

Fox, M. "Molly Kool, 93, a Pioneer of the Coastal Waters, Dies." *New York Times,* 3 March 2009, p. A25.

Frank, K.T., R.I. Perry, and K.F. Drinkwater. "Predicted Response of Northwest Atlantic Invertebrate and Fish Stocks to CO2-induced Climate Change." *Transactions of the American Fisheries Society* 119 (1990): 353–65.

Friars, G.W. *Breeding Atlantic Salmon: A Primer.* St Andrews, NB: Atlantic Salmon Federation, 1993.

Fridman, L., and P.J.G. Carrothers. *Calculations for Fishing Gear Design. FAO Fishing Manual Fn120.* Farnham, Surrey, England: Fishing News Books, 1987.

Fritz, C. "Experimental Cultures of Diatoms Occurring near St Andrews, N.B.." *Contributions to Canadian Biology 1918–1920* (1921): 63–8.

Fritz, C. "Plankton Diatoms, Their Distribution and Bathymetric Range in St Andrews Waters." *Contributions to Canadian Biology and Fisheries 1918–1920* (1921): 49–62.

Friedland, M.L. *The University of Toronto: A History*. Toronto: University of Toronto Press, 2002.

Frost, S.B. "Science Education in the 19th Century: The Natural History Society of Montreal, 1827–1925." *McGill Journal of Education* 17 (1982): 31–48.

Fry, F.E.J. "Effects of the Environment on Animal Activity." *University of Toronto Studies, Biology Series no. 55: Publications of the Ontario Fisheries Research Laboratory* 68 (1947): 1–62.

Ganong, W.F. "The Economic Mollusca of Acadia." *Bulletin of the Natural History Society of New Brunswick* 8 (1889): 1–116.

Gascoigne, J. *Joseph Banks and the English Enlightenment: Useful Knowledge and Polite Culture.* Cambridge: Cambridge University Press, 1994.

Gascoigne, J. *Science in the Service of Empire: Joseph Banks, the British State and the Uses of Science in the Age of Revolutions*. Cambridge: Cambridge University Press, 1998.

Gillett, M. "Carrie Derick (1862–1941) and the Chair of Botany at McGill." In *Despite the Odds: Essays on Canadian Women and Science*, ed. M.G. Ainley. Montreal: Véhicule Press, 1990. 74–87.

Gillett, M. *We Walked Very Warily: A History of Women at McGill*. Montreal: Eden Press Women's Publications, 1981.

Gingras, Y. *Physics and the Rise of Scientific Research in Canada*. Montreal and Kingston: McGill-Queen's University Press, 1991.

Golley, Benjamin. *A History of the Ecosystem Concept in Ecology: More than the Sum of the Parts*. New Haven: Yale University Press, 1993.

Good, G.A. "Between Two Empires: The Toronto Magnetic Observatory and American Science before Confederation." *Scientia Canadensis* 10 (1986): 34–52.

Good, G.A. "Toronto Magnetic Observatory and International Science ca. 1850." *Vistas in Astronomy* 28 (1985): 387–90.

Gordon Jr., D.C., P. Schwinghamer, T.W. Rowell, J. Prena, K. Gilkinson, W.P. Vass, D.L. McKeown, C. Bourbonnais, and K. MacIsaac. "Studies on the Impact of Mobile Fishing Gear on Benthic Habitat and Communities." Ottawa: Dept. of Fisheries and Oceans, 2003. Available at http://www2.mar.dfo-mpo.gc.ca/science/review/1996/Gordon/Gordon_e.html.

Gough, J. *Managing Canada's Fisheries: From Early Days to the Year 2000* Sillery,QC: Septentrion and Fisheries and Oceans Canada, 2006.

Graham, M. "Concepts of Conservation." In *Papers Presented at the International Technical Conference on the Conservation of the Living Resources of the Sea, Rome, 18 April to 10 May 1955*. New York: United Nations, 1956. 1–13.

Graham, M. *The Fish Gate*. London: Faber and Faber, 1943.
Graham, M. "Modern Theory of Exploiting a Fishery, and Application to North Sea Fishing." *Journal du Conseil* 10 (1935): 264–74.
Gran, H.H., and T. Braarud. "A Quantative Study of the Phytoplankton in the Bay of Fundy and the Gulf of Maine (Including Observations on Hydrography, Chemistry, and Turbidity)." *Journal of the Biological Board of Canada* 1 (1935): 279–467.
Graneli, E., B. Sundstrom, L. Edler, and D.M. Anderson, eds. *Toxic Marine Phytoplankton*. New York: Elsevier, 1990.
Grant, C.D. "Effects on Aquatic Insects of Forest Spraying with Phosphamidon in New Brunswick." *Journal of the Fisheries Research Board of Canada* 24 (1967): 823–32.
Greenberg, D.A. "The Effects of Tidal Power Development on the Physical Oceanography of the Bay of Fundy and Gulf of Maine." *Canada, Technical Reports of Fisheries and Aquatic Science* 1256 (1984): 349–69.
Greenberg, D.A., J.A. Shore, F.H. Page, and M. Dowd. "A Finite Element Circulation Model for Embayments with Drying Intertidal Areas and Its Application to the Quoddy Region of the Bay of Fundy." *Ocean Modelling* 10 (2005): 211–31.
Hachey, H.B. "The Circulation of Hudson Bay Water as Indicated by Drift Bottles." *Science* 82 (1935): 275–6.
Hachey, H.B. "The General Hydrography and Hydrodynamics of the Waters of the Hudson Bay Region." *Contributions to Canadian Biology*, new series, 7 (1931): 91–118.
Hachey, H.B. "Hydrographic Features of the Waters of Saint John Harbour." *Journal of the Fisheries Research Board of Canada* 4 (1939): 424–40.
Hachey, H.B. "Movements Resulting from Mixing of Stratified Waters." *Journal of the Biological Board of Canada* 1 (1934): 133–43.
Hachey, H.B. "Oceanography and Canadian Atlantic Waters." *Bulletin of the Fisheries Research Board of Canada* 134 (1961).
Hachey, H.B. "The Probable Effect of Tidal Power Development on Bay of Fundy Tides." *Journal of the Franklin Institute* 217 (1934): 747–56.
Hachey, H.B. "The Replacement of Bay of Fundy Waters." *Journal of the Biological Board of Canada* 1 (1934): 121–31.
Hachey, H.B. "Surface Water Temperatures of the Canadian Atlantic Coast." *Journal of the Fisheries Research Board of Canada* 4 (1939): 378–91.
Hachey, H.B. "Temporary Migrations of Gulf Stream Water on the Atlantic Seaboard." *Journal of the Fisheries Research Board of Canada* 4 (1939): 339–48.
Hachey, H.B. "Tidal Mixing in an Estuary." *Journal of the Biological Board of Canada* 1 (1935): 171–8.

Hachey, H.B. "Water Transports and Current Patterns for the Scotian Shelf." *Journal of the Fisheries Research Board of Canada* 7 (1947): 1–16.

Hachey, H.B. "The Waters of the Scotian Shelf." *Journal of the Fisheries Research Board of Canada* 5 (1942): 377–97.

Hachey, H.B., and H.J. McLellan. "Trends and Cycles in Surface Temperatures of the Canadian Atlantic." *Journal of the Biological Board of Canada* 7 (1948): 355–62.

Hall, S., S. Etheridge, D. Anderson, J. Kleindinst, M. Zhu, and Y. Zou, eds. *Harmful Algae Management and Mitigation*. Singapore: Asia-Pacific Economic Cooperation, 2004.

Halliday, R.G., and L.P. Fanning. "A History of Marine Fisheries Science in Atlantic Canada and Its Role in the Management of Fisheries." *Proceedings of the Nova Scotia Institute of Science* 43 (2006): 159–83.

Halliday, R.G., F.D. McCracken, A.W.H. Needler, and R.W. Trites. "A History of Canadian Fisheries Research in the Georges Bank Area of the Northwestern Atlantic." *Canadian Technological Report of Fisheries and Aquatic Science* 1550 (1987). Available at http://www.dfo-mpo.gc.ca/Library/101814.pdf.

Hansen, G.I. "Josephine Elizabeth Tilden (1869–1957)." In *Prominent Phycologists of the 20th Century*, ed. D. Garbary and M. Wynne. Hantsport, NS: Lancelot Press, 1996. 185–95.

Hargrave, B.T., ed. *The Handbook of Environmental Chemistry*, vol. 5, part M, *Environmental Effects of Marine Finfish Aquaculture*. Berlin: Springer-Verlag, 2005.

Hargrave, B.T., D.E. Duplisea, E. Pfeiffer, and D.J. Wildish. "Seasonal Changes in Benthic Fluxes of Dissolved Oxygen and Ammonia Associated with Marine Cultured Atlantic Salmon." *Marine Ecology Progress Series* 96 (1993): 249–57.

Hargrave, B.T., G.A. Phillips, L.I. Doucette, M.J. White, T.G. Milligan, D.J. Wildish, and R.E. Cranston. "Assessing Benthic Impacts of Organic Enrichment from Marine Aquaculture." *Water, Air & Soil Pollution* 99 (1997): 641–50.

Harris, M. *Lament for an Ocean: The Collapse of the Atlantic Cod Fishery. A True Crime Story*. Toronto: McClelland & Stewart, 1998.

Harris, R.S. *A History of Higher Education in Canada 1663–1960*. Toronto: University of Toronto Press, 1976.

Harrison, W.G. "Biological Oceanography in Canada (with Special Reference to Federal Government Science)." *Proceedings of the Nova Scotia Institute of Science* 43 (2006): 129–58.

Hart, D.R. "Yield- and Biomass-per-recruit Analysis for Rotational Fisheries, with an Application to the Atlantic Sea Scallop (*Placopecten magellanicus*)." *Fisheries Bulletin* 101 (2003): 44–57.

Hart, J.L. "Fisheries Research Board of Canada Biological Station, St Andrews, N.B., 1908–1958: Fifty Years of Research in Aquatic Biology." *Journal of the Fisheries Research Board of Canada* 15 (1958): 1127–61.

Hart, J.L., and D.L. McKernan. "International Passamaquoddy Fisheries Board Fisheries Investigations 1956–59: Introductory Account." *Journal of the Fisheries Research Board of Canada* 17:2 (1960): 127–31.

Harvey, M., E. Bourget, and N. Gagné. "Spat Settlement of Giant Scallop, *Placopecten magellanicus* (Gmelin 1791), and Other Bivalve Species on Artificial Filamentous Collectors Coated with Chitinous Material." *Aquaculture* 148 (1997): 277–98.

Hasegawa, Y., J.L. Martin, M.W. Giewat, and J.N. Rooney-Varga. "Microbial Community Diversity in the Phycosphere of Natural Populations of the Toxic Alga, *Alexandrium fundyense*." *Environmental Microbiology* 9 (2007): 3108–21.

Haslam, D.W. "The British Contribution to the Hydrography of Canada." In *From Leadline to Laser.* Centennial Conference of the Canadian Hydrographic Service. *Canadian Special Publication of Fisheries and Aquatic Science* 67 (1983): 20–35.

Haslam, D.W. "The British Contribution to the Hydrography of Canada." *International Hydrography Review* 61 (1984): 17–42.

Haya, K., L.E. Burridge, and B.D. Chang. "Environmental Impact of Chemical Wastes Produced by the Salmon Aquaculture Industry." *ICES Journal of Marine Science* 58 (2001): 492.

Haya, K., L.E. Burridge, I.M. Davies, and A. Ervik. "A Review and Assessment of Environmental Risk of Chemicals Used for the Treatment of Sea Lice Infestations of Cultured Salmon." In *The Handbook of Environmental Chemistry*, vol. 5, part M, *Water Pollution: Environmental Effects of Marine Finfish Aquaculture*, ed. B.T. Hargrave. Berlin: Springer-Verlag, 2005. 305–41.

Haya, K., J.L. Martin, S.L. Robinson, M.S. Khots, and J.D. Martin. "Does Uptake of *Alexandrium fundyense* Cysts Contribute to the Levels of PSP Toxins Found in the Giant Scallop, *Placopecten magellanicus*?" *Harmful Algae* 2 (2003): 75–81.

Haya, K., J.L. Martin, B.A. Waiwood, L.E. Burridge, J.M. Hungerford, and V. Zitko. "Identification of Paralytic Shellfish Toxins in Mackerel from Southwest Bay of Fundy, Canada." In *Toxic Marine Phytoplankton*, ed. E. Graneli, B. Sundstrom, L. Edler, and D.M. Anderson. New York: Elsevier, 1990. 350–5.

Haya, K., and A.J. Niemi, eds. *Proceedings of the 22nd Annual Aquatic Toxicity Workshop: October 2–4, 1995, St Andrews, NB. Canadian Technical Report of Fisheries and Aquatic Science* 2093 (1996).

Haya, K., D. Sephton, J. Martin, and T. Chopin. "Monitoring of Therapeutants and Phycotoxins in Kelps and Mussels Co-cultured with Atlantic Salmon in an Integrated Multi-trophic Aquaculture System." *Bulletin of the Aquaculture Association of Canada* 104:3 (2004): 29–34.

Haya, K., and B.A. Waiwood. "Acid pH and Chorionase Activity of Atlantic Salmon (*Salmo salar*) Eggs." *Bulletin of Environmental Contamination and Toxicology* 27 (1981): 7–12.

Haya, K., B.A. Waiwood, and L. Van Eeckhaute. "Disruption of Energy Metabolism and Smoltification during Exposure of Juvenile Atlantic Salmon (*Salmo salar*) to low pH." *Comparative Biochemistry and Physiology C: Toxicology and Pharmacology* 82 (1985): 323–9.

Hayes, F.R. *The Chaining of Prometheus: Evolution of a Power Structure for Canadian Science.* Toronto: University of Toronto Press, 1973.

Hays, Samuel P. *Conservation and the Gospel of Efficiency: The Progressive Conservation Movement 1890–1920*. Toronto: McClelland & Stewart, 1959.

Helmreich, S. *Alien Ocean: Anthropological Voyages in Microbial Seas*. Berkeley: University of California Press, 2009.

Henckel, I. "A Study of the Tide-pools on the West Coast of Vancouver Island." In *Postelsia: The Year Book of the Minnesota Seaside Station*. St Paul, MN: The Pioneer Press, 1906: 275–304. Available at http://archive.org/stream/postelsiayearboo1906minn#page/n5/mode/2up.

Hilborn, R., and C.J. Walters. *Quantitative Fisheries Stock Assessment*. New York: Chapman and Hall, 1992.

Hirt, P.W. *A Conspiracy of Optimism: Management of Forests since World War Two.* Lincoln: University of Nebraska Press, 1994.

Hjort, J. "Fluctuations in the Great Fisheries of Northern Europe, Viewed in the Light of Biological Research." *Rapports du Conseil Permanent Internationale pour l'Exploration de la Mer* 20 (1914): 1–228.

Hjort, J., G. Jahn, and P. Ottestad. "The Optimum Catch." *Hvalradets Skrifter* 7 (1933): 7–29.

Hoar, W.S., and D.J. Randall, eds. *Fish Physiology*, vol. 1, *Excretion, Ionic Regulation, and Metabolism*. New York: Academic Press, 1969.

Hoar, W.S., and D.J. Randall, eds. *Fish Physiology*, vol. 2, *The Endocrine System*. New York: Academic Press, 1969.

Hoar, W.S., and D.J. Randall, eds. *Fish Physiology*, vol. 3, *Reproduction and Growth, Bioluminescence, Pigments and Poisons*. New York: Academic Press, 1969.

Hoar, W.S., and D.J. Randall, eds. *Fish Physiology*, vol. 4, *The Nervous System, Circulation and Respiration*. New York: Academic Press, 1970.

Hoar, W.S., and D.J. Randall, eds. *Fish Physiology*, vol. 5, *Sensory Systems and Electric Organs*. New York: Academic Press, 1971.

Hoar, W.S., and D.J. Randall, eds. *Fish Physiology*, vol. 6, *Environmental Relations and Behavior*. New York: Academic Press, 1971.

Hoar, W.S., and D.J. Randall, eds. *Fish Physiology*, vol. 7, *Locomotion*. New York: Academic Press, 1978.

Holmberg, B., and R.L. Saunders. "The Effects of Pentachlorophenol on Swimming Performance and Oxygen Consumption in the American Eel (*Anguilla rostrata*)." *Rapports et Procès-verbaux des Réunions du Conseil International pour l'Exploration de la Mer* 174 (1979): 144–9.

Houston, S.,T. Ball, and M. Houston. *Eighteenth-century Naturalists of Hudson Bay*. Montreal and Kingston: McGill-Queen's University Press, 2003.

Hubbard, J. "Changing Regimes: Governments, Scientists and Fishermen and the Construction of Fisheries Policies in the North Atlantic 1850–2010." In *A History of the North Atlantic Fisheries*, vol. 2, *From the Mid-Nineteenth Century to the Present*, ed. D.J. Starkey, J.Th. Thór, and I. Heidbrink. Bremerhaven: Deutsches Schiffahrtsmuseum, 2012. 129–76.

Hubbard, J. "The Commission of Conservation and the Canadian Atlantic Fisheries." *Scientia Canadensis* 12: 1 (Spring-Summer 1988): 22–52.

Hubbard, J. *A Science on the Scales: The Rise of Canadian Atlantic Fisheries Biology, 1898–1939*. Toronto: University of Toronto Press, 2006.

Hubbard, J.M. "Home, Home, Sweet Home? A.G. Huntsman and the Homing Behaviour of Canadian Atlantic Salmon." *Acadiensis* 19(2) (1990): 40–71. Available at https://journals.lib.unb.ca/index.php/Acadiensis/article/view/11852/0.

Huntsman, A.G. *The Canadian Plaice*. Biological Board of Canada Bulletin no. 1 (1918).

Huntsman, A.G. "The Cause of Periodic Scarcity in Atlantic Salmon." *Transactions of the Royal Society of Canada*, series 3, vol. 31, sect. 5 (1937): 17–27.

Huntsman, A.G. "Circulation and Pollution of Water in and near Halifax Harbour." *Contributions to Canadian Biology*, new series, 2 (1924): 69–80.

Huntsman, A.G. "Cyclical Abundance and Birds versus Salmon." *Journal of the Fisheries Research Board of Canada* 24 (1941): 21–32.

Huntsman, A.G. "Edward Ernest Prince." *Canadian Field-Naturalist* 59 (1945): 1–3.

Huntsman, A.G. "The Effect of the Tide on the Distribution of the Fishes of the Canadian Atlantic Coast." *Transactions of the Royal Society of Canada* 12 (1918): 61–7.

Huntsman, A.G. "Fisheries Research in Canada." *Science* 98 (1943): 117–22.

Huntsman, A.G. "Fisheries Research in the Gulf of St Lawrence in 1917." *Canadian Fisherman* 14 (1918): 740–4.

Huntsman, A.G. "Fishing and Assessing Populations." In *A Symposium on Fish Populations Held at the Royal Ontario Museum of Zoology, Toronto, Canada, Jan. 10 and 11, 1947*, ed. Daniel Merriman. *Bulletin of the Bingham Oceanographic Collection, vol. 11*. New Haven, CT: Peabody Museum of Natural History, Yale University, 1948. 5–31.

Huntsman, A.G. "International Passamaquoddy Fishery Investigations." *Rapports et Procès Verbaux, Conseil Permanent International pour l'Exploration de la Mer* 13 (1938): 357–69.

Huntsman, A.G. "The Maritime Salmon of Canada." *Biological Board of Canada, Bulletin* 21 (1931): 15–17.

Huntsman, A.G. "North American Atlantic Salmon." *Rapports et Procès-Verbaux, Conseil Permanent International pour l'Exploration de la Mer* 101 (1938): 11–15.

Huntsman, A.G. "Progress in Fisheries Research in Canada." In *Fifty Years Retrospect, Canada, 1882–1932*. Royal Society of Canada. Toronto: Ryerson Press, 1932. 159–61.

Huntsman, A.G. "Résumé." In *The Wise Use of Our Resources: Papers for the Joint Session of Sections of the Royal Society of Canada, May 21, 1941*, ed. A.G. Huntsman. Ottawa: Royal Society of Canada, 1942. 33–8.

Huntsman, A.G. "Twenty-five Years of Canadian Fisheries Research." *Canadian Fisherman* 26:9 (1939): 95–8.

Huntsman, A.G. "The Scale Method of Calculating the Rate of Growth in Fishes." *Transactions of the Royal Society of Canada* 12:3 (1918): 47–52.

Huntsman, A.G., W.B. Bailey, and H.B. Hachey. "The General Oceanography of the Strait of Belle Isle." *Journal of the Fisheries Research Board of Canada* 11 (1954): 198–260.

Ide, F.P. "Effect of Forest Spraying with DDT on Aquatic Insects, Food of Salmon and Trout, in New Brunswick." *Proceedings of the Entomological Society of Ontario* 91 (1961): 39–40.

Ide, F.P. "Effect of Forest Spraying with DDT on Aquatic Insects of Salmon Streams." *Transactions of the American Fisheries Society* 86 (1957): 208–19.

Ide, F.P. "Effects of Forest Spraying with DDT on Aquatic Insects of Salmon Streams in New Brunswick." *Journal of the Fisheries Research Board of Canada* 24 (1967): 769–805.

Iles, T.D., and N. Tibbo. "Recent Events in Canadian Atlantic Herring Fisheries." *International Commission for the Northwest Atlantic Fisheries Redbook*, part 3 (1970): 134–47.

Jackson, F.S. "The Canadian Marine Biological Station." *Canadian Record of Science* 8:5 (1901): 308–14.

Jackson, J.R. "Earliest References to Age Determination of Fishes and Their Early Application to the Study of Fisheries." *Fisheries* 32:7 (2007): 321–8.

Jamieson, G.S., and A. Campbell, eds. *Proceedings of the North Pacific Symposium on Invertebrate Stock Assessment and Management. Canadian Special Publication of Fisheries and Aquatic Science* 125. Ottawa: NRC Research Press, 1988. Available at http://www.dfo-mpo.gc.ca/Library/223346.pdf.

Jamieson, G.S., and R.A. Chandler. "Paralytic Shellfish Poison in Sea Scallops (*Placopecten magellanicus*) in the West Atlantic." *Canadian Journal of Fisheries and Aquatic Science* 40 (1983): 313–18.

Jarrell, R.A. "The Influence of Irish Institutions upon the Organization and Diffusion of Science in Victorian Canada." *Scientia Canadensis* 9 (1985): 150–64.

Jarrell, R.A. "Science." In *The Canadian Encyclopedia*, vol. 3. Edmonton. Hurtig, 1985. 1653–55.

Jarrell, R.A. "Science Education at the University of New Brunswick in the Nineteenth Century." *Acadiensis* 2 (1973): 55–79.

Jarrell, R.A. "The Social Functions of the Scientific Society in 19th-century Canada." In *Critical Issues in the History of Canadian Science, Technology and Medicine*, ed. R.A. Jarrell and A.E. Roos. Thornhill and Ottawa, ON: HSTC Press, 1983. 31–44.

Jarrell, R.A., and Y. Gingras, eds. *Building Canadian Science: The Role of the National Research Council*. Thornhill, ON: Scientia Press, 1992.

Jarrell, R.A., and A.E. Roos, eds. *Critical Issues in the History of Canadian Science, Technology and Medicine*. Thornhill and Ottawa, ON: HSTC Press, 1983.

Johnstone, K. *The Aquatic Explorers: A History of the Fisheries Research Board of Canada*. Toronto: University of Toronto Press, 1977.

Jorgensen, C.B. *Biology of Suspension Feeding*. New York: Pergamon Press, 1966.

Josenhans, H.W., and J. Zevenhuizen. "Seafloor Dynamics on the Labrador Shelf." In *Ice Scour Bibliography*, ed. C.R. Goodwin, J.C. Finley. and L.M. Howard. Ottawa: Environmental Studies Revolving Funds, Report no. 010, 1985. 7.

Keenleyside, M.H.A. " Effects of Forest Spraying with DDT in New Brunswick on Food of Young Atlantic Salmon." *Journal of the Fisheries Research Board of Canada* 24 (1967): 807–22.

Keenleyside, M.H.A. "Effects of Spruce Budworm Control on Salmon and Other Fishes in New Brunswick." *Canadian Fish Culturist* 24:2 (1959): 17–22.

Ketchum, B.H., and D.J. Keen. "The Exchanges of Fresh and Salt Waters in the Bay of Fundy and in Passamaquoddy Bay." *Journal of the Fisheries Research Board of Canada* 10 (1953): 97–124.

Kerswill, C.J. "Effects of Water Circulation on the Growth of Quahaugs and Oysters." *Journal of the Fisheries Research Board of Canada* 7 (1949): 545–51.

Kerswill, C.J. "Relative Rates of Utilization by Commercial and Sport Fisheries of Atlantic Salmon (*Salmo salar*) from the Miramichi River, New Brunswick." *Journal of the Fisheries Research Board of Canada* 28:3 (1971): 351–63.

Kerswill, C.J. "Studies on Effects of Forest Spraying with Insecticides, 1952–63, on Fish and Aquatic Invertebrates in New Brunswick Streams: Introduction and Summary." *Journal of the Fisheries Research Board of Canada* 24 (1967): 701–8.

Kerswill, C.J. "Effects of Water Circulation on the Growth of Quahaugs and Oysters." *Journal of the Fisheries Research Board of Canada* 7 (1949): 545–51.

Kerswill, C.J., and H.E. Edwards. "Fish Losses after Forest Sprayings with Insecticides in New Brunswick, 1952–62, as Shown by Caged Specimens and Other Observations." *Journal of the Fisheries Research Board of Canada* 24 (1967): 709–29.

Kerswill, C.J., and P.F. Elson. "Preliminary Observations on Effects of 1954 DDT Spraying on Miramichi Salmon Stocks." *Fisheries Research Board of Canada Progress Reports (Atlantic)* 62 (1955): 17–24.

Kerswill, C.J., P.F. Elson, M.H.A. Keenleyside, and J.B. Sprague. "Effects on Young Salmon of Forest Spraying with DDT." In *Transactions of the 2nd Seminar on Biological Problems in Water Pollution, 1959. Technical Report W60–3.* Cincinnati, OH: Robert A. Taft Sanitary Engineering Center, 1960. 71.

Kirby-Smith, W.W. "Growth of the Bay Scallop: The Influence of Experimental Water Currents." *Journal of Experimental Marine Biology and Ecology* 8 (1972): 7–18.

Knight, A.P. "The Effects of Dynamite Explosions on Fish Life: A Preliminary Report." *Contributions to Canadian Biology 1902–1905* (1907): 21–30.

Knight, A.P. "The Effects of Polluted Water on Fish Life." *Contributions to Canadian Biology 1901* (1901): 9–18.

Kohlstedt, S.G., ed. *History of Women in the Sciences: Readings from Isis.* Chicago: University of Chicago Press, 1980.

Kosastsky, T. "Improving Epidemic Control: Lessons from the 1987 Toxic Mussels Affair." *Canadian Medical Association Journal* 147:12 (1992): 1769–72.

Kostylev, V.E., R.C. Courtney, G. Robert, and B.J. Todd. "Stock Evaluation of Giant Scallop (*Placopecten magellanicus*) Using High-resolution Acoustics for Seabed Mapping." *Fisheries Research* 60 (2003): 479–92.

Kruse, G.H., N. Bez, A. Booth, M. Dorn, S. Hills, R. Lipcius, D. Pelletier, C. Roy, S.J. Smith, and D. Witherell. *Spatial Processes and Management of Marine*

*Populations*. Fairbanks: University of Alaska Sea Grant College Program, AK-SG-01–02, 2001.

Lacroix, G.L. "Survival of Eggs and Alevins of Atlantic Salmon (*Salmo salar*) in Relation to the Chemistry of Interstitial Water in Redds in Some Acidic Streams of Atlantic Canada." *Canadian Journal of Fisheries and Aquatic Science* 42 (1985): 292–9.

Lacroix, G.L., and D. Knox. "Acidification Status of Rivers in Several Regions of Nova Scotia and Potential Impacts on Atlantic Salmon." *Canadian Technical Report of Fisheries and Aquatic Science* 2573 (2005): v + 71 p. Available at http://www.dfo-mpo.gc.ca/Library/287115.pdf.

Lamson, C., ed. *The Sea Has Many Voices: Ocean Policy for a Complex World*. Kingston, ON: McGill-Queen's University Press, 1994.

Land, C., and O. Pye. "Blinded by Science: The Invention of Scientific Forestry and Its Influence in the Mekong Region." *Watershed* 6:2 (2000–1): 25–34.

Lassus, P.D., D.J. Wildish, M. Bardouil, J.L. Martin, M. Bohec, and S. Bougrier. "Ecophysiological Study of Toxic *Alexandrium spp*. Effects on the Oyster, *Crassostrea gigas*. In *Harmful and Toxic Algal Blooms. Proceedings of the Seventh International Conference on Toxic Phytoplankton, Sendai, Japan, 12-16, 1995*, ed. T. Yasumoto, Y. Oshima, and Y. Fukuyo. Paris: Intergovernmental Oceanic Commission / UNESCO, 1996. 409–12.

Lauzier, L.M. "Bottom Residual Drift on the Continental Shelf Area of the Canadian Atlantic Coast." *Journal of the Fisheries Research Board of Canada* 24 (1967): 1845–59.

Lauzier, L.M. "Drift Bottle Observations in Northumberland Strait, Gulf of St Lawrence." *Journal of the Fisheries Research Board of Canada* 22 (1965): 353–68.

Lauzier, L.M. "Foreshadowing of surface water temperatures at St Andrews, N.B." *International Commission for the Northwest Atlantic Fisheries Special Publication* 6 (1966): 859–67.

Lauzier, L.M., and R.W. Trites. "The Deep Waters in the Laurentian Channel." *Journal of the Fisheries Research Board of Canada* 15 (1958): 1247–57.

Lavoie, D., and D. Hutchinson. "The U.S. Geological Survey Sea-going Women." *Oceanography* 18 (2005): 39–46.

Layton, L. "Mirror Image Twins: The Communities of Science and Technology in 19th Century America." *Technology and Culture* 12:4 (1971): 562–80.

Lee, Robert G. "The Classical Sustained Yield Concept: Concept and Philosophical Origins." In *Sustained Yield: Proceedings of a Symposium held April 27 and 28, 1982, Spokane, Washington*, ed. D.C. LeMaster, D.M. Baumgartner, and D. Adams. Pullman: Washington State University, 1982. 1–10.

Lee, Robert G. "Sustained Yield and Social Order." In *History of Sustained Yield Forestry: A Symposium*, ed. H.K. Steen. Portland, OR: Forestry History Society, 1984. 90–100.

Leim, A.H., and W.B. Scott. "Fishes of the Atlantic Coast of Canada." *Bulletin of the Fisheries Research Board of Canada* 155 (1966): 1–485.

Leim, A.H., S.N. Tibbo, L.R. Day, L. Lauzier, R.W. Trites, H.B. Hachey, and W.B. Bailey. "Report of the Atlantic Herring Investigation Committee." *Bulletin of the Fisheries Research Board of Canada* 111 (1957): 1–317.

LeMaster, D.C., D.M. Baumgartner, and D. Adams. *Sustained Yield: Proceedings of a Symposium held April 27 and 28, 1982, Spokane, Washington*. Pullman: Washington State University, 1982.

Levere, T.H. "The Most Select and the Most Democratic: A Century of Science in the Royal Society of Canada." *Scientia Canadensis* 20 (1998): 3–99.

Levere, T.H. "Science and the Canadian Arctic, 1818–76, from Sir John Ross to Sir George Strong Nares." *Arctic* (1988) 41: 127–37.

Levere, T.H. *Science and the Canadian Arctic. A Century of Exploration*, 1818–1918. Cambridge: Cambridge University Press, 1993.

Levere, T.H., and R.A. Jarrell. "General Introduction." In *A Curious Field-Book: Science and Society in Canadian History*, ed. T.H. Levere and R.A. Jarrell. Toronto: Oxford University Press, 1974. 1–24.

Li, M.F., G.S. Traxler, S. Clyburne, and J.E. Stewart. "Malpeque Disease: Isolation and Morphology of a *Labyrinthomyxa*-like Organism from Diseased Oysters." *International Council for Exploration of the Sea C.M.1980/F:15* (1980): 1–9.

Lightman, B. "'The Voices of Nature': Popularizing Victorian Science." In *Victorian Science in Context*, ed. B. Lightman. Chicago: University of Chicago Press, 1997. 187–211.

Lightman, B., ed. *Victorian Science in Context*. Chicago. University of Chicago Press, 1997.

Lillie, F.R. "The Woods Hole Marine Biological Laboratory." *Biological Bulletin* 174:1 (Suppl.) (1944): 1–284.

Lindsay, D. "The Historical Context: Science in Rupert's Land before 1859." In *The Modern Beginnings of Subarctic Ornithology: Correspondence to the Smithsonian Institution, 1856–1868*, ed. D. Lindsay. Winnipeg: Manitoba Record Society Publications 10, 1991. ix–xxx.

Lindsey, A.A. "The Harriman Alaska Expedition of 1899 Including the Identities of Those in the Staff Picture." *BioScience* 28 (1978): 383–6.

Loo, Tina. *States of Nature: Conserving Canada's Wildlife in the Twentieth Century*. Vancouver: UBC Press, 2006.

Loucks, R.H., R.W. Trites, K.E. Drinkwater, and D.J. Lawrence. "Summary of Physical, Biological, Socio-economic and Other Factors Relevant to Potential Oil Spills in the Passamaquoddy Region of the Bay of Fundy. Section 1: Physical Oceanographic Characteristics." *Fisheries Research Board of Canada Technical Report* 428 (1974): 1–59. Available at http://www.dfo-mpo.gc.ca/Library/23142.pdf.

Lowther, S. *The Relativity of Journeys*. Victoria: Trafford Publishing, 2001.

MacCrimmon, H.R., J.E. Stewart, and R. Brett. "Aquaculture in Canada: The Practice and the Promise." *Bulletin of the Fisheries Research Board of Canada* 188 (1974): 1–84; *Fisheries and Marine Service Bulletin* 188 (Ottawa: Department of the Environment, 1974): 1–84.

MacGregor, D.G., and H.J. McLellan. "Current Measurements in the Grand Manan Channel." *Journal of the Fisheries Research Board of Canada* 9 (1952): 213–22.

Maienschein, J. "Agassiz, Hyatt, Whitman, and the Birth of the Marine Biological Laboratory." *Biological Bulletin* 168 (Suppl.) (1985): 26–34.

Mann, K.H. and J.R.N. Lazier, *Dynamics of Marine Ecosystems: Biological–Physical Interactions in the Oceans* . Boston: Blackwell, 1991.

Martin, J.L., M.M. LeGresley, P.M. Strain, and P. Clement. "Phytoplankton Monitoring in the Southwest Bay of Fundy during 1993–96." *Canadian Technical Report of Fisheries and Aquatic Science* 2265 (1999): 1–132. Available at http://www.dfo-mpo.gc.ca/Library/232755.pdf.

Martin, J.L., K. Haya, L.E. Burridge, and D.J. Wildish. "*Nitzschia pseudodelicatissima* – A Sourceof Domoic Acid in the Bay of Fundy, Eastern Canada." *Marine Ecology Progress Series* 67 (1990): 177–82.

Martin, J.L, M.M. LeGresley, A. Hanke, and F.H. Page. "*Alexandrium fundyense*: Red Tides, PSP Shellfish Toxicity, Salmon Mortalities and Human Illnesses in 2003–04 – Before and After." In *Proceedings of the 12th International Conference on Harmful Algae*, ed. Ø. Moestrup. Copenhagen: International Society for the Study of Harmful Algae and the International Oceanic Commission / UNESCO, 2008. 206–8.

Martin, J.L., M.M. LeGresley, K. Haya, D.H. Sephton, L.E. Burridge, F.H. Page, and B.D. Chang. "Salmon Mortalities Associated with a Bloom of *Alexandrium fundyense* in 2003 and Subsequent Early Warning Approaches for Industry." *African Journal of Marine Science* 28 (2006). 431–4.

Martin, J.L., M.M. LeGresley, and P.M. Strain. "Phytoplankton Monitoring in the Western Isles Region of the Bay of Fundy during 1997–98." *Canadian Technical Report of Fisheries and Aquatic Science* 2349 (2001): 1–85.

Martin, J.L., M.M. LeGresley, and P.M. Strain. "Plankton monitoring in the Western Isles Region of the Bay of Fundy during 1999–2000." *Canadian Technical Reports of Fisheries and Aquatic Science* 2629 (2006): 1–88.

Martin, J.L., F.H. Page, A. Hanke, P.M. Strain, and M.M. LeGresley. "*Alexandrium fundyense* Vertical Distribution Patterns during 1982, 2001 and 2002 in the Bay of Fundy, Eastern Canada." *Deep-Sea Research II* 52 (2005): 2569–92.

Martin, J. L., F.H. Page, M.M. LeGresley, and D.J.A. Richard. "Phytoplankton Monitoring as a Management Tool: Timing of *Alexandrium* and *Pseudo-nitzschia* Blooms in the Bay of Fundy, Eastern Canada." In *Harmful Algae Management and Mitigation*, ed. S. Hall, S. Etheridge, D. Anderson, J. Kleindinst, M. Zhu, and Y. Zou. Singapore: Asia-Pacific Economic Cooperation, 2004. 136–40.

Martin, J.L., and D. Richard. "Shellfish Toxicity from the Bay of Fundy, Eastern Canada: 50 Years in Retrospect." In *Harmful and Toxic Algal Blooms. Proceedings of the Seventh International Conference on Toxic Phytoplankton, Sendai, Japan, 12–16, 1995*, ed. T. Yasumoto, Y. Oshima, and Y. Fukuyo. Paris: International Oceanic Commission / UNESCO, 1996. 3–6.

Martin, J.L., and A.W. White. "Distribution and Abundance of the Toxic Dinoflagellate *Gonyaulax excavata* in the Bay of Fundy." *Canadian Journal of Fisheries and Aquatic Science* 45 (1988): 1968–75.

Martin, J.L., A.W. White, and J.J. Sullivan. "Anatomical Distribution of Paralytic Shellfish Toxins in Soft-shell Clams." In *Toxic Marine Phytoplankton*, ed. E. Graneli, B. Sundstrom, L. Edler, and D.M. Anderson. New York: Elsevier, 1990. 379–84.

Martin, J.L., and D.J. Wildish. "Temporal and Spatial Dynamics of Alexandrium fundyense Cysts during 1981–83 and 1992 in the Bay of Fundy." In *Proceedings of the Fourth Canadian Workshop on Harmful Marine Algae*, ed. J.R. Forbes. *Canadian Technical Reports of Fisheries and Aquatic Science* 201 (1994). 22–4.

Mavor, J.W. "The Circulation of the Water in the Bay of Fundy. Part I: Introduction and Drift Bottle Experiments." *Contributions to Canadian Biology*, new series, 1 (1922): 103–24.

Mavor, J.W. "The Circulation of the Water in the Bay of Fundy. Part II: The Distribution of Temperature, Salinity and Density in 1919 and Movements of Water Which They Indicate in the Bay of Fundy." *Contributions to Canadian Biology*, new series, 1 (1923): 355–75.

Mavor, J.W. "Circulation of the Water in Bay of Fundy and Gulf of Maine." *Transactions of the American Fisheries Society* 50 (1921): 334–44.

Mavor, J.W. "The Course of the Gulf Stream in 1919–21 as Shown by Drift Bottles." *Science* 57 (1923): 14–15.

Mavor, J.W., E.H. Craigie, and J.D. Detweiler. "Investigation of the Bays of the Southern Coast of New Brunswick with a View to Their Use for Oyster Culture." *Contributions to Canadian Biology 1914–1915* (1916): 145–9.

Mavor, J.W. "Drift Bottles as Indicating a Superficial Circulation in the Gulf of Maine." *Science* 52 (1920): 442–3.

McGarvey, R., F.M. Serchuk, and I.A. McLaren. "Statistics of Reproduction and Early Life History Survival of the Georges Bank Sea Scallop (*Placopecten magellanicus*) Population." *Journal of the Northwest Atlantic Fisheries Science* 13 (1992): 83–99.

McGonigle, R.H. "Acute Catarrhal Enteritis of Salmonid Fingerlings." *Transactions of the American Fisheries Society* 70 (1940): 297–303.

McGonigle, R.H. "Some Effects of Temperature and Their Relation to Fish Culture." *Transactions of the American Fisheries Society* 62 (1932): 119–25.

McKay, G., ed. *The Proceedings of the Bras D'Or Lakes Aquaculture Conference, held in Sydney, Cape Breton, June 1975*. Sydney, NS: College of Cape Breton Press, 1976.

McKenzie, R. "Admiral Bayfield: Pioneer Nautical Surveyor." *Canadian Fisheries and Marine Service Special Publication* 32 (1976): 1–13.

McKeown, D.L., and D.E. Heffler. "Precision Navigation for Benthic Surveying and Sampling." In *Oceans '97 MTS/IEEE Conference Proceedings, Halifax, Nova Scotia, Canada, Oct. 6–9, 1997*, vol. 1. Dartmouth, NS: IEEE, 1997. 386–90.

McKeown, D.L., D.J. Wildish, and H.M. Akagi. "GAPS: Grab Acoustic Positioning System." *Canadian Technical Reports of Hydrography and Ocean Sciences*, Report 252 (2007). Available at http://www.dfo-mpo.gc.ca/Library/328342.pdf.

McLaughlin, P.A., and S. Gilchrist. "Women's Contributions to Carcinology." In *History of Carcinology*, ed. F. Truesdale and A.A. Balkema. Rotterdam: CRC Press, 1993. 165–206.

McLaughlin, P.A., and P.L. Illg, "Josephine F.L. Hart (1909–1993)." *Journal of Crustacean Biology* 14 (1994): 396–8.

McLaughlin, R.J., P.C. Young, R.B. Martin, and J. Parslow. "The Australian Scallop Dredge: Estimates of Catching Efficiency and Associated Indirect Fish Mortality." *Fisheries Research* 11 (1991): 1–24.

McLeese, D.W. "Behaviour of Lobsters Exposed to Bleached Kraft Mill Effluent." *Journal of the Fisheries Research Board of Canada* 27 (1970): 731–6.

McLeese, D.W. "Chemosensory Response of American Lobsters (*Homarus americanus*) in the Presence of Copper and Phosphamidon." *Journal of the Fisheries Research Board of Canada* 32 (1975): 2055–60.

McLeese, D.W. "Fenitrothion Toxicity to the Freshwater Crayfish, *Orconectes limosus*." *Bulletin of Environmental Contamination and Toxicology* 16 (1976): 411–16.

McLeese, D.W. "Olfactory Response and Fenitrothion Toxicity in American Lobsters (*Homarus americanus*)." *Journal of the Fisheries Research Board of Canada* 31 (1974): 1127–31.

McLeese, D.W. "Response of Lobsters, *Homarus americanus*, to Odor Solution in the Presence of Bleached Kraft Mill Effluent." *Journal of the Fisheries Research Board of Canada* 30 (1973): 279–82.

McLeese, D.W. "Toxicity of Phosphamidon to American Lobsters (*Homarus americanus*) held at 4 and 12° C." *Journal of the Fisheries Research Board of Canada* 31 (1974): 1556–8.

McLeese, D.W. "Uptake and Excretion of Cadmium by Marine Organisms from Sea Water with Cadmium at Low Concentrations: A Review." In "Cadmium Pollution of Belledune Harbour, New Brunswick, Canada," ed. J.F. Uthe and V. Zitko. *Canadian Technical Reports of Fisheries and Aquatic Science* 963 (1980). 55–63.

McLeese, D.W., and C.D. Metcalfe. "Toxicity of Mixtures of Phosphamidon and Methidathion to Lobsters (*Homarus americanus*)." *Chemosphere* 8 (1979): 59–62.

McLeese, D.W., C.D. Metcalfe, and V. Zitko. "Lethality of Permethrin, Cypermethrin and Fenvalerate to Salmon, Lobster and Shrimp." *Bulletin of Environmental Contamination and Toxicology* 25 (1980): 950–5.

McLeese, D.W., S. Ray, and L.E. Burridge. "Lack of Excretion of Cadmium from Lobsters." *Chemosphere* 10 (1981): 775–8.

McLeese, D.W., D.B. Sergeant, C.D. Metcalf, V. Zitko, and L.E. Burridge. "Uptake and Excretion of Aminocarb, nnylphenol, and Pesticide Diluent 585 by Mussels (*Mytilus edulis*)." *Bulletin of Environmental Contamination and Toxicology* 24 (1980): 575–81.

McLeese, D.W., and D.G. Wilder. "Lobster Storage and Shipment." *Bulletin of the Fisheries Research Board of Canada* 147 (1964): 1–69.

McLeese, D.W., V. Zitko, C.D. Metcalfe, and D.B. Sergeant. "Lethality of Aminocarb and the Components of the Aminocarb Formulation to Juvenile Atlantic Salmon, Marine Invertebrates and a Freshwater Clam." *Chemosphere* 9 (1980): 79–82.

McLellan, H.J. "Energy Considerations in the Bay of Fundy System." *Journal of the Fisheries Research Board of Canada* 15 (1958): 115–34.

McLellan, H.J. "On the Distinctness and Origin of the Slope Water off the Scotian Shelf and Its Easterly Flow South of the Grand Banks." *Journal of the Fisheries Research Board of Canada* 14 (1957): 213–39.

McLellan, H.J. "Temperature–salinity Relations and Mixing on the Scotian Shelf." *Journal of the Fisheries Research Board of Canada* 11 (1954): 419–30.

McLellan, H.J., L. Lauzier, and W.B. Bailey. "The Slope Water off the Scotian Shelf." *Journal of the Fisheries Research Board of Canada* 10 (1953): 155–76.

McMurrich, J.P. "Science in Canada." *The Week* 1:49 (1884): 776–7.

Medcof, J.C. 1985. "Life and Death with Gonyaulax: An Hstorical Perspective." In *Toxic Dinoflagellates*, ed. D.M. Anderson, A.W. White, and D.G. Baden. New York: Elsevier, 1985. 1–8.

Medcof, J.C. "Meat Yield from Digby Scallops of Different Sizes." *Fisheries Research Board of Canada Progress Report* 44 (1949): 6–9.

Medcof, J.C. "Modifications of Drags to Protect Small Scallops." *Journal of the Fisheries Research Board of Canada Atlantic Progress Report* 52 (1952): 9–14.

Medcof, J.C. "Oyster Farming in the Maritimes." *Bulletin of the Fisheries Research Board of Canada* 131 (1961): i–xi, 1–158.

Medcof, J.C., and Bourne, N. "Causes of Mortality of the Sea Scallop, *Placopecten magellanicus*." *Proceedings of the National Shellfisheries Association* 53 (1964): 33–50.

Medcof, J.C., A.H. Leim, A.B. Needler, A.W.H. Needler, J. Gibbard, and J. Naubert. "Paralytic Shellfish Poisoning on the Canadian Atlantic Coast." *Bulletin of the Fisheries Research Board of Canada* 75 (1947): 1–32.

Meehan, O.M. "The Canadian Hydrographic Service from the Time of Its Inception in 1883 to the End of the Second World War: Part 2." *Northern Mariner* 14:2 (2004): 159–297.

Meehan, O.M., ed., W. Glover with D. Gray. "The Canadian Hydrographic Service from the Time of Its Inception in 1883 to the End of the Second World War." *Northern Mariner, Special Edition*. Ottawa: Canadian Nautical Research Society, 2006.

Merriman, D., ed. *A Symposium on Fish Populations Held at the Royal Ontario Museum of Zoology, Toronto, Canada, Jan. 10 and 11, 1947. Bulletin of the Bingham Oceanographic Collection, vol. 11.* New Haven, CT: Peabody Museum of Natural History, Yale University, 1948.

Messieh, S.N., S.N. Tibbo, and L.M. Lauzier. "Distribution, Abundance and Growth of Larval Herring (*Clupea harengus L.*) in Bay of Fundy–Gulf of Maine Area." *Fisheries Research Board of Canada Technical Report* 277 (1971): 1–33. Available at http://www.dfo-mpo.gc.ca/Library/28628.pdf.

Mills, E.L. *Biological Oceanography: An Early History, 1870–1960*. London: Cornell University Press, 1989.

Mills, E.L. *The Fluid Envelope of Our Planet: How the Study of Ocean Currents Became a Science*. Toronto: University of Toronto Press, 2009.

Moestrup, Ø., ed. *Proceedings of the 12th International Conference on Harmful Algae*. Copenhagen: International Society for the Study of Harmful Algae and the International Oceanic Commission / UNESCO, 2008.

Mohn, R.K., G. Robert, and D.L. Roddick. "Research Sampling and Survey Design for Sea Scallops (*Placopecten magellanicus*) on Georges Bank." *Journal of Northwest Atlantic Fisheries Science* 7 (1987): 117–21.

Morgon, H.J., ed. *The Canadian Men and Women of the Times: A Handbook of Canadian Biography of Living Characters*. Toronto: William Briggs, 1912.

Mossop, B.K.E. "The Rate of Growth of the Sea Mussel (*Mytilus edulis L.*) at St Andrews, New Brunswick; Digby, Nova Scotia; and in Hudson Bay." *Transactions of the Royal Canadian Institute* 14 (1922): 3–22.

Mossop, B.K.E. "A Study of the Sea Mussel (*Mytilus edulis Linn.*)." *Contributions to Canadian Biology 1921–1922* (1922): 15–48.

Mounce, I. "Effect of Marked Changes in Specific Gravity upon the Amount of Phytoplankton in Departure Bay Waters." *Contributions to Canadian Biology and Fisheries*, new series, 1:1 (1922): 81–94.

Murray, J., and J. Hjort. *The Depths of the Ocean: A General Account of the Modern Science of Oceanography Based Largely on the Scientific Researches of the Norwegian Steamer* Michael Sars *in the North Atlantic.* London: Macmillan and Co., 1912.

Naidu, K.S. "Reproduction and Breeding Cycle of the Giant Scallop *Placopecten magellanicus* Gmelin in Port au Port Bay, Newfoundland." *Canadian Journal of Zoology* 48 (1970): 1003–12.

Neave, N.M., C.L. Dilworth, J.G. Eales, and R.L. Saunders. "Adjustment of Bouyancy in Atlantic Salmon Parr in Relation to Changing Water Velocity." *Journal of the Fisheries Research Board of Canada* 23 (1966): 1617–20.

Needler, A.B. "Paralytic Shellfish Poisoning and *Gonyaulax tamarensis*." *Journal of the Fisheries Research Board of Canada* 7 (1949): 490–504.

Needler, A.W.H. "The Oysters of Malpeque Bay." *Bulletin of the Fisheries Research Board of Canada* 22 (1931): 1–35.

Needler, A.W.H. "Prince, Edward Ernest." In *The Canadian Encyclopedia*, vol. 3. Edmonton: Hurtig, 1985. 1947.

Needler, A.W.H. "Reflections on the Fisheries Research Board." *Journal of the Fisheries Research Board of Canada* 31 (1974): 1283–4.

Needler, A.W.H. "The Seventy-fifth Anniversary of Two Canadian Biological Stations." *Canadian Journal of Fisheries and Aquatic Sciences* 41 (1984): 216–24.

Needler, A.W.H., and R.R. Logie. "Serious Mortalities in Prince Edward Island Oysters Caused by a Contagious Disease." *Transactions of the Royal Society of Canada*, series 3, 41:5 (1947): 73–89.

Nelson, J. "An Investigation of Oyster Propagation in Richmond Bay, P.E.I., during 1915." *Contributions to Canadian Biology 1915–1916* (1917): 53–78.

Nelson, R.W.P., K.M. Ellis, and J.N. Smith. "Environmental Monitoring Report for the Point Lepreau, N.B. Nuclear Generating Station – 1991 to 1994." *Canadian Technical Reports of Hydrography and Ocean Science* 211 (2001): 1–133. Available at http://www.dfo-mpo.gc.ca/Library/261283.pdf.

Newell, C.R., D.E. Campbell, and S.M. Gallagher. "Development of the Mussel Aquaculture Lease Site Model MUSMODC: A Field Program to Calibrate

Model Formulations." *Journal of Experimental Marine Biology and Ecology* 219 (1998): 143–69.

Newell, C.R., D.J. Wildish, and B.A. MacDonald. "The Effects of Velocity and Seston Concentration on the Exhalant Siphon Area and Valve Gape in the Mussel, *Mytilus edulis*." *Journal of Experimental Marine Biology and Ecology* 262 (2001): 91–111.

Norling, L. *Captain Ahab Had a Wife: New England Women and the Whale Fishery, 1720–1870*. Chapel Hill: University of North Carolina Press, 2000.

North American Council on Fishery Investigations. *Proceedings 1921–1930 (no. 1)*. Ottawa: Printer to the King's Most Excellent Majesty, 1932.

O'Brien, P. *Joseph Banks: A Life*. Chicago: University of Chicago Press, 1987.

Olla, B.L., J. Atema, R. Forward, J. Kittredge, R.J. Livingston, D.W. McLeese, D.C. Miller, W.B. Vernberg, P.G. Wells, and K. Wilson. "The Role of Behavior in Marine Pollution Monitoring. Behavior Panel Report." *Rapports et Procès-verbaux des Réunions du Conseil International pour l'Exploration de la Mer* 179 (1980): 174–81.

Orensanz, J.M., and G.S. Jamieson. "The Assessment and Management of Spatially Structured Stocks: An Overview of the North Pacific Symposium on Invertebrate Stock Assessment and Management." In *Proceedings of the North Pacific Symposium on Invertebrate Stock Assessment and Management*, ed. G.S. Jamieson and A. Campbell, *Canadian Special Publication of Fisheries and Aquatic Science* 125. Ottawa: NRC Research Press, 1998. 441–60.

Page, F.H., B.D. Chang, R.J. Losier, D.A. Greenberg, J.D. Chaffey, and P. McCurdy. "Water Circulation and Management of Infectious Salmon Anemia in the Salmon Aquaculture Industry of Southern Grand Manan Island, Bay of Fundy." *Canadian Technical Report of Fisheries and Aquatic Science* 2595 (2005): iii + 78 pp. Available at http://www.dfo-mpo.gc.ca/Library/315102.pdf.

Page, F.H., B.D. Chang, R.J. Losier, D.A. Greenberg, and P. McCurdy. "Water Circulation and Management of ISA in the Southwest New Brunswick Salmon Culture Industry." *Aquaculture Association of Canada Special Publication* 8 (2004): 64–8.

Page, F.H., B.D. Chang, and J.L. Martin. "A Century of Oceanographic Monitoring at the St Andrews Biological Station." *Gulf of Maine News (Regional Association for Research on the Gulf of Maine)* 7:2 (2000): 1–5.

Page, F.H., A. Hanke, J.L. Martin, M. LeGresley, B. Chang, and P. McCurdy. "Characteristics of *Alexandrium fundyense* Blooms That Affect Caged Salmon in the Bay of Fundy." *Aquaculture Association of Canada Special Publication* 9 (2005): 27–30.

Page, F.H., and R.J. Losier. "Temperature Variability during Canadian Bottom-trawl Summer Surveys Conducted in NAFO Division 4VWX, 1970–1992."

*International Council for the Exploration of the Sea Marine Science Symposium* 198 (1994): 323–31.

Page, F.H., R. Losier, P. McCurdy, D. Greenberg, J. Chaffey, and B. Chang. "Dissolved Oxygen and Salmon Cage Culture in the Southwestern New Brunswick Portion of the Bay of Fundy." In *Handbook of Environmental Chemistry*, vol. 5, part M, *Water Pollution Environmental Effects of Marine Finfish Aquaculture*, ed. B.T. Hargrave. Berlin: Springer-Verlag, 2005. 1–28.

Page, F.H., J.L. Martin, A. Hanke, and M.M. LeGresley. "The Relationship of *Alexandrium fundyense* to the Temporal and Spatial Pattern in Phytoplankton Community Structure within the Bay of Fundy, Eastern Canada." In *Harmful Algae 2002*, ed. K.A. Steidinger, J.H. Landsberg, C.R. Thomas, and G.A. Vargo. St Petersburg: Florida Fish and Wildlife Conservation Commission, Florida Institute of Oceanography, and International Oceanographic Commission / UNESCO, 2004. 92–4.

Page, F.H., J.L. Martin, A. Hanke, and M.M. LeGresley. "Temporal and Spatial Variability in the Characteristics of *Alexandrium fundyense* Blooms in the Coastal Zone of the Bay of Fundy, Eastern Canada." In *Harmful Algae 2004*, ed. G.C. Pitcher, T.A. Probyn, and H.M. Verheye. *African Journal of Marine Science* 28:2 (2006): 203–8.

Page, F.H., J. Piercey, B. Chang, B. MacDonald, and D. Greenberg. "An Introduction to the Oceanographic Aspects of Integrated Multi-Trophic Aquaculture." *Bulletin of the Aquaculture Association of Canada* 104:3 (2004): 35–43.

Page, F.H., M. Sinclair, C.E. Naimie, J.W. Loder, R.J. Losier, P.L. Berrien, and R.G. Lough. "Cod and Haddock Spawning on Georges Bank in Relation to Water Residence Times." *Fisheries Oceanography* 8 (1999): 212–26.

Page, F.H., and P.C. Smith. "Particle Dispersion in the Surface Layer off Southwest Nova Scotia: Description and Evaluation of a Model." *Canadian Journal of Fisheries and Aquatic Science* 46 (Suppl. 1) (1989): 21–43.

Page, F.H., R.L. Stephenson, and B.D. Chang. "A Framework for Addressing Aquaculture- Environment–fisheries Interactions in the Southwestern New Brunswick Portion of the Bay of Fundy." *Aquaculture Association of Canada Special Publication* 8 (2004): 69–72.

*Papers Presented at the International Technical Conference on the Conservation of the Living Resources of the Sea, April 18 to May 10, 1955, Rome.* New York: United Nations Publications, 1956.

Parsons, G.J., and M.J. Dadswell. "Effect of Stocking Density on Growth, Production, and Survival of the Giant Scallop, *Placopecten magellanicus*, Held in Intermediate Suspension Culture in Passamaquoddy Bay, New Brunswick." *Aquaculture* 103:3–4 (1992): 291–309.

Parsons, G.J., and M.J. Dadswell. "Evaluation of Intermediate Culture Techniques, Growth, and Survival of the Giant Scallop, *Placopecten magellanicus*, in Passamaquoddy Bay, New Brunswick." *Canadian Technical Report of Fisheries and Aquatic Sciences* 2012 (1994): 1–29. Available at http://www.dfo-mpo.gc.ca/Library/178925.pdf.

Parsons, G.J., S.M.C. Robinson, R.A. Chandler, L.A. Davidson, M. Lanteigne, and M.J. Dadswell. "Intra-annual and Long-term Patterns in the Reproductive Cycle of Giant Scallops *Placopecten magellanicus* (*Bivalvia: Pectinidae*) from Passamaquoddy Bay, New Brunswick, Canada." *Marine Ecology Progress Series* 80 (1992): 203–14.

Parsons, G.J., S.M.C. Robinson, and J.D. Martin. "Enhancement of a Giant Scallop Bed by Spat Naturally Released from a Scallop Aquaculture Site." *Bulletin of the Aquaculture Association of Canada* 94:2 (1993): 21–3.

Parsons, L.S. "Management of Marine Fisheries in Canada." *Canadian Bulletin of Fisheries and Aquatic Science* 225 (1993): 1–763.

Payre, R. "A European Progressive Era." *Contemporary European History* 11:3 (2002): 489–97.

Peck, R.M. "A Cruise for Rest and Recreation." *Audubon* 84:5 (1982): 86–99.

Penhallow, D.P. "Report on the Atlantic Biological Station of Canada, St Andrews, N.B., for 1908." *Contributions to Canadian Biology, 1906–10* (1912): 1–21.

Percy, J.A., P.G. Wells, and A.J. Evans, eds. *Bay of Fundy Issues: A Scientific Overview. Environment Canada – Atlantic Region, Occasional Report 8*. Dartmouth, NS, and Sackville, NB: Bay of Fundy Environmental Publications, 1997.

Perry, David A. "The Scientific Basis of Forestry." *Annual Review of Ecology and Systematics* 29 (1998): 435–66.

Perry, R.I., G.C. Harding, J.W. Loder, M.J. Tremblay, M.M. Sinclair, and K.F. Drinkwater. "Zooplankton Distributions at the Georges Bank Frontal System: Retention or Dispersion?" *Continental Shelf Research* 13 (1993): 357–83.

Perry, R.I., P.C.F. Hurley, P.C. Smith, J.A. Koslow, and R.O. Fournier. "Modelling the Initiation of Spring Phytoplankton Blooms: A Synthesis of Physical and Biological Interannual Variability off Southwest Nova Scotia, 1983–85." *Canadian Journal of Fisheries and Aquatic Science* 46 (Suppl. 1) (1989): 183–99.

Perry, R.I., and J.D. Neilson. "Vertical Distributions and Trophic Interactions of Age-0 Atlantic Cod and Haddock in Mixed and Stratified Waters of Georges Bank." *Marine Ecology Progress Series* 49 (1988): 199–214.

Perry, R.I., and S.J. Smith. "Identifying Habitat Associations of Marine Fishes Using Survey Data: An Application to the Northwest Atlantic." *Canadian Journal of Fisheries and Aquatic Science* 51 (1994): 589–602.

Peterson, R.H. "Effects on Fishes." In *Workshop on Long-range Transport for Air Pollution and Its Impacts on the Atlantic Region*. Halifax: Environment Canada Atmospheric Environmental Services (AES), Atlantic Region, 1979. 83–5.

Peterson, R.H. "Influence of Fenitrothion on Swimming Velocities of Brook Trout (*Salvelinus fontinalis*)." *Journal of the Fisheries Research Board of Canada* 31 (1974): 1757–62.

Peterson, R.H. "Temperature Selection of Atlantic Salmon (*Salmo salar*) and Brook Trout (*Salvelinus fontinalis*) as Influenced by various Chlorinated Hydrocarbons." *Journal of the Fisheries Research Board of Canada* 30 (1973): 1091–7.

Peterson, R.H. "Temperature Selection of Juvenile Atlantic Salmon (*Salmo salar*) as Influenced by Various Toxic Substances." *Journal of the Fisheries Research Board of Canada* 33 (1976): 1722–30.

Peterson, R.H., P.G. Daye, and J.L. Metcalfe. "Inhibition of Atlantic Salmon (*Salmo salar*) Hatching at Low pH." *Canadian Journal of Fisheries and Aquatic Science* 37 (1980): 770–4.

Peterson, R.H., G.F. Fletcher, S. Ray, and J. Doane. "Analysis of a Chlorinated Terphenyl (Aroclor 5460) and Its Deposition in Tissues of Cod (*Gadus morhua*)." *Bulletin of Environmental Contamination and Toxicology* 8 (1972): 52.

Peterson, R.H., and P. Harmon. "Swimming Ability of Pre-feeding Striped Bass Larvae." *Aquaculture International* 9 (2001): 361–6.

Peterson, R.H., D.J. Martin-Robichaud, and J. Power. "Toxicity of Potash Brines to Early Developmental Stages of Atlantic Salmon (*Salmo salar*)." *Bulletin of Environmental Contamination Toxicology* 41 (1988): 391–7.

Peterson, R.H., J.L. Metcalfe, and S. Ray. "The Effects of Cadmium on Yolk Utilization, Growth, and Survival of Atlantic Salmon Alevins and Newly Feeding Fry." *Archives of Environmental Contamination and Toxicology* 12 (1983): 37–44.

Peterson, R.H., and V. Zitko. "Variations in Insect Drift Associated with Operational and Experimental Contamination with Fenitrothion in New Brunswick." *Fisheries Research Board of Canada Technical Report* 439 (1974): 1–23.

Picard-Aitken, M., D. Campbell, and G. Côté. *Bibliometric Study in Support of Fisheries and Oceans Canada's International Science Strategy: Report to Fisheries and Oceans Canada by Science-Metrix*. Canada: Fisheries and Oceans Canada, 2009.

Pitcher, G.C., T.A. Probyn, and H.M. Verheye, eds. *Harmful Algae 2004. African Journal of Marine Science* 28:2 (2006).

Pixell G.H. "*Aggregata leandri*." *Quarterly Journal of Microscopical Science Series* 3:91 (1950): 465–7.

Pixell G.H. "Sporozoa of *Sipunculus*." *Quarterly Journal of Microscopical Science Series* 3:91 (1950): 469–76.

Pixell, H.M.L. "Polychaeta from the Pacific Coast of North America. Part. I. Serpulidae, with a Revised Table of Classification of the Genus *Spirobis*." *Proceedings of the Zoological Society of London* 82 (1912): 784–805.

Pixell, H.M.L. "Two New Species of the *Phoronidae* from Vancouver Island." *Quarterly Journal of Microscopical Science*, new series, 58 (1912): 257–84.

Poole, N.J., R.J. Parkes, and D.J. Wildish. "The Reactions of the Estuarine Ecosystem to Effluent from the Pulp and Paper Industry." *Helgoländ Wissenschaftliche Meeresuntersuchungen* 30 (1977): 622–32.

Poole, N.J., D.J. Wildish, and D.D. Kristmanson. "The Effects of the Pulp and Paper Industry on the Aquatic Environment." *CRC Critical Reviews in Environmental Control* 8 (1978): 153–95.

Posgay, J.A. "Population Assessment of the Georges Bank Sea Scallop Stocks." *International Council for the Exploration of the Sea Rapports et Procès-verbaux des Réunions* 175 (1979): 109–13.

Posgay, J.A. "Sea Scallop Investigations." In *Sixth Report on Investigations of the Shellfisheries of Massachusetts*. Commonwealth of Massachusetts: Department of Conservation Division of Marine Fisheries, 1953. 9–25.

*Postelsia: The Year Book of the Minnesota Seaside Station*. St Paul, MN: The Pioneer Press, 1906.

Poulson, E.M. "Conservation Problems in the Northwestern Atlantic." In *Papers Presented at the International Technical Conference on the Conservation of the Living Resources of the Sea, April 18 to May 10, 1955, Rome*. New York: United Nations Publications, 1956. 183–91.

Prakash, A. "Growth and Toxicity of a Marine Dinoflagellate, *Gonyaulax tamarensi*." *Journal of the Fisheries Research Board of Canada* 24 (1967): 1589–606.

Prakash, A. "Source of Paralytic Shellfish Toxin in the Bay of Fundy." *Journal of the Fisheries Research Board of Canada* 20 (1963): 983–96.

Prakash, A., and J.C. Medcof. "Hydrographic and Meteorological Factors Affecting Shellfish Toxicity at Head Harbour, New Brunswick." *Journal of the Fisheries Research Board of Canada* 19 (1962): 11–112.

Prakash, A., J.C. Medcof, and A.D. Tennant. "Paralytic Shellfish Poisoning in Eastern Canada." *Bulletin of the Fisheries Research Board of Canada* 177 (1971): 1–87.

Premdas, F., and J.M. Anderson. "The Uptake and Detoxification of C14-labelled DDT in Atlantic Salmon, *Salmo salar*." *Journal of the Fisheries Research Board of Canada* 20 (1963): 827–37.

Prince, E.E. "Marine Biological Station of Canada: Introductory Notes on Its Foundation, Aims and Work." *Contributions to Canadian Biology*, 1901: 1–8.

Prince, E.E. "Preface." In *Canada, Canadian Fisheries Expedition, 1914–1915: Investigations in the Gulf of St Lawrence and Atlantic Waters of Canada*. Ottawa: King's Printer, 1919. v–viii.

Prince, E.E. "Presidential Address: The Biological Investigation of Canadian Waters, with Special Reference to the Government Biological Stations."

*Transactions of the Royal Society of Canada*, 3rd series, 1907–8, vol. 1, *Section 4 (Geological and Biological Sciences)* (1908): 71–92.
Pycior, H.M., N.G. Slaack, and P.G. Abir-am, eds. *Creative Couples in the Sciences*. New Brunswick, NJ: Rutgers University Press, 1996.
Rand, G.M., and S.R. Petrocelli, eds. *Fundamentals of Aquatic Toxicology*. Washington, DC: Hemisphere Publishing Corp., 1985; 2nd ed., 1995.
Rathbun, M.J. "The Cancroid Crabs of America of the Families *Euryalidae, Portunidae, Atelecyclidae, Cancridae*, and *Xanthidae*." *U.S. National Museum Bulletin* 152 (1930): 1–593.
Rathbun, M.J. "Decapoda." *Canadian Atlantic Fauna* 10 (1929): 1–38.
Rathbun, M.J. "The Decapod Crustaceans of the Canadian Arctic Expedition 1913–18." In *Report of the Canadian Arctic Expedition 1913–18*, vol. 7, *Crustacea*. (A). Ottawa: F.A. Acland, 1919. 1–14a.
Rathbun, M.J. "The Grapsoid Crabs of America." *U.S. National Museum Bulletin* 97 (1918): 1–461.
Rathbun, M.J. "Oxystomatous and Allied Crabs of America." *U.S. National Museum Bulletin 166* (1937): 1–278.
Rathbun, M.J. "The Spider Crabs of America." *U.S. National Museum Bulletin* 129 (1925): 1–598.
Ray, S. "Bioaccumulation of Cadmium in Marine Organisms." *Experientia* 40 (1984): 14–23.
Ray, S. "Bioaccumulation of Lead in Atlantic Salmon (*Salmo salar*)." *Bulletin of Environmental Contamination and Toxicology* 19 (1978): 631–6.
Ray, S., and J. Coffin. "Ecological Effects of Cadmium Pollution in the Aquatic Environment: A Review." *Fisheries and Marine Service Technical Report* 734 (1977): 1–24.
Ray, S., D.W. McLeese, and C.D. Metcalfe. "Heavy Metals in Sediments and in Invertebrates from Three Coastal Areas in New Brunswick, Canada: A Natural Bioassay." *International Council for the Exploration of the Sea CIM 1979/E:29* (1979): 1–6.
Ray, S., D.W. McLeese, and D. Pezzack. "Chelation and Interelemental Effects on the Bioaccumulation of Heavy Metals by Marine Invertebrates." *Proceedings of the International Conference on the Management and Control of Heavy Metals in the Environment. London, U.K.* Edinburgh: CEP Consultants, 1979. 35–8.
Ray, S., and W. White. "*Equisetum arvense* – An Aquatic Vascular Plant as a Biological Monitor for Heavy Metal Pollution." *Chemosphere* 8 (1979): 125–8.
Ray, S., and W. White. "Selected Aquatic Plants as Indicator Species for Heavy Metal Pollution." *Journal of Environmental Science and Health* A11 (1976): 717–25.
Ray, S., and W.J. White. "Some Observations on Heavy Metal Concentration in Northeastern New Brunswick Estuarine Surficial Sediments." *Fisheries and*

*Marine Service Technical Report* 696 (1977): 1–33. Available at http://www.dfo-mpo.gc.ca/Library/58715.pdf.

Remington, J.E. "Katherine Jeanette Bush Peabody's Mysterious Zoologist." *Discovery* 12:3 (1977): 3–8.

Reidy, M.S. *Tides of History: Ocean Science and Her Majesty's Navy*. Chicago: University of Chicago Press, 2008.

Repetto, R. "A Natural Experiment in Fisheries Management." *Marine Policy* 25 (2001): 251–64.

Rhoads, D.C., and D.K. Young. "The Influence of Deposit-feeding Organisms on Sediment Stability and Community Trophic Structure." *Journal of Marine Research* 28 (1970): 150–78.

Richardson, H. *A Monograph on the Isopods of North America. Bulletin of the U.S. National Museum* 54 (1905) / Lochem, Netherlands: Antiquariat Junk, 1972: 1–727.

Robertson, A.D. "First Report on the "Barren Oyster Bottoms" Investigation, Richmond Bay, P.E.I." *Contributions to Canadian Biology, 1914–1915* (1916): 55–71.

Robinson, S.M.C. "A Review of the Biological Information Associated with Enhancing Scallop Production." *World Aquaculture* 24 (1993): 61–7.

Robinson, S.M.C. "Clam Enhancement Trials in the Bay of Fundy." *Department of Fisheries and Oceans Science Review, 1994–5* (1996): 1–16.

Robinson, S.M.C. "Shellfish Culture in the Bay of Fundy." *Aquaculture Association of Canada Special Publication* 2 (1997): 85–93.

Robinson, S.M.C. "The Shellfish Industry in the Gulf of Maine, Status and Possible Future Directions." *Gulf of Maine News* 1996 (Summer): 7–9.

Robinson, S.M.C., J.D. Castell, and E.J. Kennedy. "Developing Suitable Colour in the Gonads of Cultured Green Sea Urchins (*Strongylocentrotus droebachiensis*)." *Aquaculture* 206 (2002): 289–303.

Robinson, S.M.C., and T. Chopin. "Defining the Appropriate Regulatory and Policy Framework for the Development of Integrated Multi-trophic Aquaculture Practices: Summary of the Workshop and Issues for the Future." *Bulletin of the Aquaculture Association of Canada* 104:3 (2004): 73–82.

Rodgers, A.D. *Bernhard Eduard Fernow: A Story of North American Forestry*. Durham, NC: Forest History Society, 1991.

Roedel, P.M., ed. *Optimum Sustained Yield as a Concept in Fisheries Management*. Washington, DC: American Fisheries Society, 1975.

Rogers, J.E. *The Shell Book: A Popular Guide to a Knowledge of the Families of Living Mollusks, and an Aid to the Identification of Shells Native and Foreign*. Boston: Charles T. Brantford Co., 1908.

Rose, G.A. *Cod: The Ecological History of the North Atlantic Fisheries*. St John's, NL: Breakwater Press, 2007.

Rossiter, M.W. *Women Scientists in America: Struggles and Strategies to 1940.* Baltimore: Johns Hopkins University Press, 1982.

Rossiter, M.W. "'Women's Work' in Science, 1880–1910." In *History of Women in the Sciences: Readings from Isis*, ed. S.G. Kohlstedt. Chicago: University of Chicago Press, 1980. 287–304.

Rowell, T.W., P. Schwinghamer, M. Chin-Yee, K.D. Gilkinson, D.C. Gordon Jr., E. Hartgers, M. Hawryluk, D.L. McKeown, J. Prena, D.P. Reimer, G. Sonnichsen, G. Steeves, W.P. Vass, R. Vine, and P. Woo. "Grand Banks Otter Trawling Impact Experiment: III. Experimental Design and Methodology." *Canadian Technical Report of Fisheries and Aquatic Science* 2190 (1997): viii + 36 p. Available at http://www.dfo-mpo.gc.ca/Library/227460.pdf.

Rozwadowski, H. *The Sea Knows No Boundaries: A Century of Marine Science under ICES*. London: ICES and University of Washington Press, 2002.

Rozwadowski,, H., and D.K. Van Keuren, eds. *The Machine in Neptune's Garden: Historical Perspectives on Technology and the Marine Environment.* Sagamore Beach, MA: Science History Publications USA, 2004.

Salmond, J.A. *The Civilian Conservation Corps, 1933–42.* Durham, NC: Duke University Press, 1967.

Sandilands, R.W. "Hydrographic Charting and Oceanography on the West Coast of Canada from the Eighteenth Century to the Present Day." *Proceedings of the Royal Society of Edinburgh* 73 (1972): 75–83.

Sandström, W.J. "The Hydrodynamics of Canadian Atlantic Waters." In *Canadian Fisheries Expedition, 1914–1915, in the Gulf of St Lawrence and Atlantic Waters of Canada.* Ottawa: Canadian Department of the Naval Service, 1919. 221–347.

Saunders, R.L. "Adjustment of Buoyancy in Young Atlantic Salmon and Brook Trout by Changes in Swim Bladder Volume." *Journal of the Fisheries Research Board of Canada* 22 (1965): 335–52.

Saunders, R.L. "Heated Effluent for the Rearing of Fry – For Farming and for Release." In *Harvesting Polluted Waters*, ed. O. Devik. New York: Plenum Press, 1976. 213–36.

Saunders, R.L. "Seasonal Pattern of Return of Atlantic Salmon in the Northwest Miramichi River, New Brunswick." *Journal of the Fisheries Research Board of Canada* 24:1 (1967): 21–32.

Saunders, R.L. "The Thermal Biology of Atlantic Salmon: Influence of Temperature on Salmon Culture with Particular Reference to Constraints Imposed by Low Temperature." *Report, The Institute of Freshwater Research* (Drottningholm) 63 (1986): 77–90.

Saunders, R.L. "Winterkill! The Reality of Lethal Winter Sea Temperature in East Coast Salmon Farming." *Bulletin of the Aquaculture Association of Canada* 1 (1987): 36–40.

Saunders, R.L., and K.R. Allen. "Effects of Tagging and Fin-Clipping on the Survival and Growth of Atlantic Salmon between Smolt and Adult Stages." *Journal of the Fisheries Research Board of Canada* 24 (1967): 2595–611.

Saunders R.L., and E.B. Henderson. "Atlantic Herring as a Dietary Component for Culture of Atlantic Salmon." *Aquaculture* 3:4 (1974): 369–85.

Saunders, R.L., and E.B. Henderson. "Changes in Gill ATPase Activity and Smolt Status of Atlantic Salmon (*Salmo solar*)." *Journal of the Fisheries Research Board of Canada* 35 (1978): 1542–6

Saunders, R.L., and E.B. Henderson. "Growth of Atlantic Salmon Smolts and Post-smolts in Relation to Salinity, Temperature, and Diet." *Fisheries Research Board of Canada Technical Report* 149 (1969): 1–74. Available at http://www.dfo-mpo.gc.ca/Library/24053.pdf.

Saunders, R.L., and E.B. Henderson. "Influence of Photoperiod on Smolt Development and Growth of Atlantic Salmon (*Salmo solar*)." *Journal of the Fisheries Research Board of Canada* 27 (1970): 1295–311.

Saunders, R.L., E.B. Henderson, B.D. Glebe, and E.J. Loudenslager. "Evidence of a Major Environmental Component in the Determination of the Grilse: Larger Salmon Ratio in Atlantic Salmon (*Salmo salar*)." *Aquaculture* 33 (1983): 107–18.

Saunders, R.L., E.B. Henderson, and P.R. Harmon. "Effects of Photoperiod on Juvenile Growth and Smolting of Atlantic Salmon and Subsequent Survival and Growth in Sea Cages." *Aquaculture* 45 (1985): 55–66.

Saunders, H.E., and C.W. Hubbard. "The Circulating Water Channel of the David W. Taylor Model Basin." *Society of Naval Architects and Marine Engineers, Transactions* 52 (1944): 325–64.

Saunders, R.L., and M.N. Kutty. "Swimming Performance of Young Atlantic *Salmon* (*Salmo salar*) as Affected by Ambient Oxygen Concentration." *Journal of the Fisheries Research Board of Canada* 30 (1973): 223–7.

Saunders, R.L., B.C. Muise, and E.B. Henderson. "Mortality of Salmonids Cultured at Low Temperature in Sea Water." *Aquaculture* 5 (1975): 243–52.

Saunders, R.L., and J.B. Sprague. "Copper-Zinc Mining Pollution on a Spawning Migration of Atlantic Salmon." *Water Research* 1 (1967): 419–32.

Scarratt, D.J. "An Artificial Reef for Lobsters, *Homarus americanus*." *Journal of the Fisheries Research Board of Canada* 24 (1968): 2683–90.

Scarratt, D.J. "The Effects of Raking Irish Moss (*Chondrus crispus*) on Lobsters in Prince Edward Island." *Helgoländ Wissenschaftliche Meersuntersuchungen* 24 (1973): 415–24.

Scarratt, D.J. "Impact of Spills and Clean-up Technology on Living Natural Resources and Resource-based Industry." In "Summary of Physical, Biological, Socio-economic and Other Factors Relevant to Potential Oil

Spills in the Passamaquoddy Region of the Bay of Fundy." *Fisheries Research Board of Canada Technical Report* 428 (1974): 141–58. Available at http://www.dfo-mpo.gc.ca/Library/23142.pdf.

Scarratt, D.J. "Lobster Larvae off Pictou, Nova Scotia, Not Affected by Bleached Kraft Mill Effluent." *Journal of the Fisheries Research Board of Canada* 26 (1969): 1931–4.

Scarratt, D.J., and G.E. Raine. "Avoidance of Low Salinity by Newly Hatched Lobster Larvae." *Journal of the Fisheries Research Board of Canada* 24 (1967): 1403–6.

Scarratt, D.J., J.B. Sprague, D.G. Wilder, V. Zitko, and J.M. Anderson. "Some Biological and Chemical Investigations of a Major Winter Oil Spill on the Canadian East Coast." *International Council for the Exploration of the Sea C.M. 1970/E:14* (1970): 1–7.

Scarratt, D.J., and V. Zitko. "Bunker C Oil in Sediments and Benthic Animals from Shallow Depths in Chedabucto Bay, N.S." *Journal of the Fisheries Research Board of Canada* 29 (1972): 1347–50.

Scarratt, D.J., and V. Zitko. "Sublittoral Sediment and Benthos Sampling and Littoral Observations in Chedabucto Bay, April 1973." In *Proceedings of Conference on "Oil and the Canadian Environment" May 16, 1973*, ed. D. MacKay and W. Harrison. Toronto: Institute of Environmental Science and Engineering, University of Toronto, 1975. 78–9.

Schlichting, H. *Boundary-Layer Theory*. New York: McGraw-Hill, 1955.

Schmitt, W.L. "Mary J. Rathbun 1860–1943." *Crustaceana* 24 (1973): 283–97.

Schwach, V. "An Eye into the Sea: The Early Development of Fisheries Acoustics in Norway, 1935–1969." In *The Machine in Neptune's Garden: Historical Perspectives on Technology and the Marine Environment*, ed. H.M. Rozwadowski and D.K. van Keuren. Sagamore Beach, MA: Science History Publications, 2004. 211–42.

Schwach, V., and J.M. Hubbard. "Johan Hjort and the Birth of Fisheries Biology: The Construction and Transfer of Knowledge, Approaches and Attitudes, Norway and Canada, 1890–1920." *Studia Atlantica* 13 (2010): 20–39.

Scott, J.C. *Seeing Like a State: How Certain Schemes to Improve the Human Condition Have Failed*. New Haven: Yale University Press, 1998.

Scott, R.B. *People of the Southwest Coast of Vancouver Island: A History of the Southwest Coast*. Victoria, BC: Morris Printing Co., 1974.

Scott, W.B., and M.G. Scott. "Atlantic Fishes of Canada." *Canadian Bulletin of Fisheries and Aquatic Science* 219 (1988): 1–731.

Seijo, J.-C., and J.F. Caddy. "Port Location for Inshore Fleets Affects the Sustainability of Coastal Source-sink Resources: Implications for Spatial Management of Populations." *Fisheries Research* 91 (2008): 336–48.

Seijo, J-C., J.F. Caddy, and J. Euan. "SPATIAL: Space-time Dynamics in Marine Fisheries. A Bioeconomic Software Package for Sedentary Species." FAO Computerized Information Series (Fisheries), 6 (1994): 116 pp. (plus discs).

Sephton, D.H., K. Haya, J.L. Martin, M.M. LeGresley, and F.H. Page. "Paralytic Shellfish Toxins in Zooplankton, Mussels, Lobsters and Caged Atlantic Salmon, *Salmo salar*, during a Bloom of *Alexandrium fundyense* off Grand Manan Island, in the Bay of Fundy." *Harmful Algae* 6 (2007): 745–58.

Sette, O.E. "Studies on the Pacific Pilchard or Sardine (*Sardinops caerula*). I. Structure of a Research Program to Determine How Fishing Determines the Resource." United States Fish and Wildlife Service, Special Scientific Report, 19 (1943): 1–210. Available at http://spo.nmfs.noaa.gov/SSRF/SSRF15.pdf.

Sheets-Pyenson, S. "Stones and Bones and Skeletons: The Origins and Early Development of the Peter Redpath Museum (1882–1912)." *McGill Journal of Education* 17:1 (1982): 49–62.

Shteir, A.B. "Elegant Recreations? Configuring Science Writing for Women." In *Victorian Science in Context*, ed. B. Lightman. Chicago. University of Chicago Press, 1997. 236–55.

Skud, B.E. "Revised Estimates of Halibut Abundance and the Thompson-Burkenroad Debate." Scientific Report no. 56. Seattle, WA: International Halibut Commission, 1975. 1–36. Available at http://iphc.int/publications/scirep/SciReport0056.pdf.

Smallwood, J.R., ed. "Nancy Frost Button." In *Encyclopedia of Newfoundland and Labrador*, vol. 2. St John's: Newfoundland Book Publishers Ltd., 1981. 428–9.

Smith, M.W. "Physical and Biological Conditions in Heavily Fertilized Waters." *Journal of the Biological Board of Canada* 1 (1934): 67–93.

Smith, P.F., ed. *Underwater Photography: Scientific and Engineering Applications*. New York: Van Nostrand Reinhold, 1984.

Smith, S.J., and F.H. Page. "Associations between Atlantic Cod (*Gadus morhua*) and Hydrographic Variables: Implications for the Management of the 4VsW Cod Stock." *International Council for the Exploration of the Sea Journal of Marine Science* 53 (1996): 597–614.

Smith, S.J., and P. Rago. "Biological Reference Points for Sea Scallops (*Placopecten magellanicus*): The Benefits and Costs of Being Nearly Sessile." *Canadian Journal of Fisheries and Aquatic Science* 61 (2004): 1338–54.

Smith, S.J., and G. Robert. "Getting More out of Your Survey Information: An Application to Georges Bank Scallops (*Placopecten magellanicus*)." In *North Pacific Symposium on Invertebrate Stock Assessment and Management*, ed. G.S. Jamieson and A. Campbell. Special publication of *Fisheries and Aquatic Science* 125 (1998). 3–13.

Smith, S.J., and G. Robert. "Scallops, Sampling and the Law." *American Statistical Association Proceedings (1991), Statistics and Environment Section* (1992): 102–9.

Smith, S.J., E.L. Kenchington, M.J. Lundy, G. Robert, and D. Roddick. "Spatially Specific Growth Rates for Sea Scallops (*Placopecten magellanicus*)." In *Spatial Processes and Management of Marine Populations*, ed. G.H. Kruse, N. Bez, A. Booth, M. Dorn, S. Hills, R. Lipcius, D. Pelletier, C. Roy, S.J. Smith, and D. Witherell. Fairbanks: University of Alaska Sea Grant College Program, AK-SG-01–02, 2001. 211–31.

Smith, S.J., D. McKeown, M. Lundy, D. Gordon, J. Anderson, M. Strong, and M. Power. "TOWCAM – Towed Camera Array for Video/still Benthic Surveys. *Journal of Shellfish Research* 25 (2006): 308–9.

Smith, T.D. *Scaling Fisheries: The Science of Measuring the Effects of Fishing, 1855–1955.* Cambridge: Cambridge University Press, 1994.

Solberg, B., and A. Svendsrud. "Development of Forest Research in Norway since 1927: Some Issues." *Forestry* 70:4 (1997): 359–66.

Sollas, W.J. "John Joseph Frederick Whiteaves, LL.D (McGill), F.R.S. Canada (1835–1909)." *Quarterly Journal of the Geological Society of London* 66 (1910): xlix–l.

Sprague, J.B. "The ABC's of Pollutant Bioassay Using Fish." In *Biological Methods for the Assessment of Water Quality*, ed. J. Cairns and K.L. Dickson. American Society for Testing and Materials Special Technical Publication 528. Philadelphia, PA: ASTM, 1973. 6–30.

Sprague, J.B. "Apparatus Used for Studying Avoidance of Pollutants by Young Atlantic Salmon." In *Biological Problems in Water Pollution: Third Seminar, 1962. Robert A. Taft Sanitary Engineering Center*, ed. C.M. Tarzwell. Cincinnati, OH: US Department of Health, Education and Welfare, Public Health Service, Division of Water Supply and Pollution Control, 1965. 315.

Sprague, J.B. "Apparent DDT Tolerance in an Aquatic Insect Disproved by Test." *Canadian Entomology* 100 (1968): 279–84.

Sprague, J.B. "Avoidance of Copper-Zinc Solutions by Young Salmon in the Laboratory." *Journal of the Water Pollution Control Federation* 36 (1964): 990–1004.

Sprague, J.B. "Avoidance Reactions of Rainbow Trout to Zinc Sulphate Solutions." *Water Research* 2 (1968): 367–72.

Sprague, J.B. "Effects of Sublethal Concentrations of Zinc and Copper on Migration of Atlantic Salmon." In *Biological Problems in Water Pollution: Third Seminar, 1962. Robert A. Taft Sanitary Engineering Center*, ed. C.M. Tarzwell. Cincinnati, OH: US Department of Health, Education and Welfare, Public Health Service, Division of Water Supply and Pollution Control, 1965. 332–3.

Sprague, J.B. "Factors That Modify Toxicity." In *Fundamentals of Aquatic Toxicology*, ed. G.M. Rand and S.R. Petrocelli. Washington, DC: Hemisphere Publishing Corp., 1985; 2nd ed., 1995. 124–63.

Sprague, J.B. "Highly Alkaline Water Caused by Asbestos-cement Pipeline." *Progress in Fish-Culture* 26 (1964): 111–14.

Sprague, J.B. "An Informal Look at the Parents of Canadian Aquatic Toxicology." In *Proceedings of the 22nd Annual Aquatic Toxicity Workshop: October 2–4, 1995, St Andrews, NB*, ed. K. Haya and A.J. Niemi. *Canadian Technical Report of Fisheries and Aquatic Science* 2093 (1996). 2–14.

Sprague, J.B. "Lethal Concentrations of Copper and Zinc for Young Atlantic Salmon." *Journal of the Fisheries Research Board of Canada* 21 (1964): 17–26.

Sprague, J.B. "Measurement of Pollutant Toxicity to Fish. I: Bioassay Methods for Acute Toxicity." *Water Research* 3 (1969): 793–821.

Sprague, J.B. "Measurement of Pollutant Toxicity to Fish. II. Utilizing and Applying Bioassay Results." *Water Research* 4 (1970): 3–32.

Sprague, J.B. "Measurement of Pollutant Toxicity to Fish. III. Sublethal Effects and 'Safe' Concentrations." *Water Research* 5 (1971): 245–66.

Sprague, J.B. "Perspective on a Career: Changing Approaches to Water Pollution Evaluation." *Marine Pollution Bulletin* 25 (1992): 6–13.

Sprague, J.B. "Promising Anti-pollutant: Chelating Agent NTA Protects Fish from Copper and Zinc." *Nature* 220 (1968): 1345–6.

Sprague, J.B. "Resistance of Four Freshwater Crustaceans to Lethal High Temperature and Low Oxygen." *Journal of the Fisheries Research Board of Canada* 20 (1963): 387–415.

Sprague, J.B., and W.G. Carson. "Toxicity Tests with Oil Dispersants in Connection with Oil Spill at Chedabucto Bay, Nova Scotia." *Fisheries Research Board of Canada Technical Report* 201 (1970): 1–30.

Sprague, J.B., P.F. Elson, and R.L. Saunders. "Sublethal Copper-Zinc Pollution in a Salmon River: A Field and Laboratory Study." *International Journal of Air and Water Pollution* 9 (1965): 531–43.

Stark, S.J. *Female Tars: Women aboard Ship in the Age of Sail*. Annapolis, MD: Naval Institute Press, 1996.

Stasko, A.B. "Progress of Migrating Atlantic Salmon (*Salmo salar*) along an Estuary, Observed by Ultrasonic Tracking." *Journal of Fish Biology* 7:3 (1975): 329–38.

Steen, H.K., ed. *History of Sustained Yield Forestry: A Symposium*. Portland, OR: Forestry History Society, 1984.

Steidinger, K.A., J.H. Landsberg, C.R. Thomas, and G.A. Vargo, eds. *Harmful Algae 2002*. St Petersburg, FL: Florida Fish and Wildlife Conservation

Commission, Florida Institute of Oceanography, and International Oceanographic Commission / UNESCO, 2004.
Stephenson, R.L., D.E. Lane, D.G. Aldous, and R. Nowak. "Management of the 4WX Atlantic Herring (*Clupea harengus*) Fishery: An Evaluation of Recent Events." *Canadian Journal of Fisheries and Aquatic Science* 50 (1993): 2742–57.
Stevenson, J.A., and Dickie, L.M. "Annual Growth Rings and Rate of Growth of the Giant Scallop, *Placopecten magellanicus* (Gmelin) in the Digby area of the Bay of Fundy." *Journal of the Fisheries Research Board of Canada* 11: 660–71.
Stevenson, J.C. "Edith and Cyril Berkeley: An Appreciation." *Journal of the Fisheries Research Board of Canada* 28 (1971): 1360–4.
Stevenson, J.C. *The Fisheries Research Board of Canada, 1898–1973: The First 75 years*. Fisheries Research Board of Canada Miscellaneous Special Publication 20. Toronto: University of Toronto Press, 1973.
Stokesbury, K.D.E., B.P. Harris, M.C. Marino II, and J.I. Nogueira. "Estimation of Sea Scallop Abundance Using a Video Survey in Off-shore US Waters." *Journal of Shellfish Research* 23 (2004): 33–40.
Stewart, J.E., and J.D. Castell. "Various Aspects of Culturing the American Lobster (*Homarus americanus*)." In Proceedings of the FAO Conference on Aquaculture, Kyoto, Japan 1976. FIR: AQ/Cont/76/E: 11; and in *Advances in Aquaculture*, ed. T.V.R. Pillay and W.A. Dill. Farnham, Surrey, Eng.: Fishing News Books, 1979. 314–19.
Stewart, J.E., and A. Safer. "A Retrospective: Three Quarters of a Century at the Halifax Fisheries Research Laboratory." *Proceedings of the Nova Scotia Institute of Science* 43 (2005): 19–44. Available at nsis.chebucto.org/wp-content/uploads/2013/04/Stewart_Safer_lo.pdf.
Stortz, P., and E.L. Panayotidis, eds. *Historical Identities: The Professoriate in Canada*. Toronto: University of Toronto Press, 2005.
Strain, P.M., D.J. Wildish, and P.A. Yeats. "The Application of Simple Models of Nutrient Loading and Oxygen Demand to the Management of a Marine Tidal Inlet." *Marine Pollution Bulletin* 30 (1995): 253–61.
Strong, M.B., and P. Lawton. "URCHIN–Manually-deployed Geo-referenced Video System for Underwater Reconnaissance and Coastal Habitat Inventory." *Canadian Technical Reports of Fisheries and Aquatic Science* 2553 (2004): iv + 28 pp.
Stroud, R.H. "Introductory Remarks." In *Optimum Sustained Yield as a Concept in Fisheries Management*, ed. P.M. Roedel. Washington: American Fisheries Society, 1975. 1–4.
Studhalter, R.A. "Early History of Crossdating." *Tree Ring Bulletin* 21 (1956): 31–5.

Suffet, I.H., ed. *Advances in Environmental Science and Technology*, vol. 8, *Fate of Pollutants in the Air and Water Environments.* New York: Wiley-Interscience, 1977.

Sullivan, C.M. "Bivalve Larvae of Malpeque Bay, PEI." *Bulletin of the Fisheries Research Board of Canada* 77 (1948): 1–30.

Suomala, J.B., ed. *Meeting on Hydroacoustical Methods for the Estimation of Marine Fish Populations, 25–29 June, 1979*. Cambridge, MA: Charles Stark Draper Laboratory, 1981.

Sutterlin, A.M., E.B. Henderson, S.P. Merrill, R.L. Saunders, and A.A. MacKay. "Salmonid Rearing Trials at Deer Island, New Brunswick, with Some Projections on Economic Viability." *Canadian Technical Report of Fisheries and Aquatic Sciences* 1011 (1981): 1–32.

Sutterlin, A.M., and S.P. Merrill. "Norwegian Salmonid Farming." *Fisheries and Marine Service Technical Report* 779 (1978): 1–47.

Sutterlin, A.M., R.L. Saunders, E.B. Henderson, and P.R. Harmon. "The Homing of Atlantic Salmon (*Salmo salar*) to a Marine Site." *Canadian Technical Reports of Fisheries and Aquatic Science* 1058 (1982): 1–6.

Symons, P.E.K. "Behaviour and Growth of Juvenile Atlantic Salmon *Salmo salar* and Three Competitors at Two Stream Velocities." *Journal of the Fisheries Research Board of Canada* 33 (1976): 2766–73.

Symons, P.E.K. "Behaviour of Young Atlantic Salmon (*Salmo salar*) Exposed to or Force-fed Fenitrothion, an Organophosphate Insecticide." *Journal of the Fisheries Research Board of Canada* 30 (1973): 651–5.

Tait, B.J., and L.F. Ku. "A Historical Review of Tidal, Current and Water Level Surveying in the Canadian Hydrographic Service." In *From Leadline to Laser*. Centennial Conference of the Canadian Hydrographic Service. *Canadian Special Publication of Fisheries and Aquatic Science* 67. Ottawa: Fisheries and Oceans, Scientific Information and Publications Branch (1983). 51–61.

Tarzwell, C.M., ed. *Biological Problems in Water Pollution: Third Seminar, 1962. Robert A. Taft Sanitary Engineering Center*. Cincinnati, OH: US Department of Health, Education and Welfare, Public Health Service, Division of Water Supply and Pollution Control, 1965.

Thiessen, A.D. "The Founding of the Toronto Magnetic Observatory and the Canadian Meteorological Service." *Journal of the Royal Astronomical Society of Canada* 34 (1940): 308–48.

Thomas, C. "Moira Dunbar: Woman Scientist the Navy Refused to Take on Board." *The Guardian*, 12 January 2000.

Thomas, M. "The Beginnings of Canadian Meteorology." Toronto: ECW Press, 1991.

Tibbo, S.N. "Herring – the 'Golden Goose' of the Sea." *Fisheries Research Board of Canada Biological Station, St Andrews, NB, General Series Circular* 55 (1970): 1–5.

Tibbo, S.N., and L.M. Lauzier. "Larval Swordfish (*Xiphias gladius*) from Three Localities in the Western Atlantic." *Journal of the Fisheries Research Board of Canada* 26 (1969): 3248–51.

Tizard, T.H., H.N. Moseley, J.Y. Buchanan, and J. Murray. *Narrative of the Cruise of H.M.S. Challenger with a General Account of the Scientific Results of the Expedition. Report on the Scientific Results of the Voyage of H.M.S. Challenger.* Narrative vol. 1. London: Her Majesty's Government 1885: 1–509. Available at http://archimer.ifremer.fr/doc/1885/publication-4746.pdf.

*Transactions of the 2nd Seminar on Biological Problems in Water Pollution, 1959. Technical Report* W60–3. Cincinnati, OH: Robert A. Taft Sanitary Engineering Center, 1960.

Tremblay, M.J., W. Loder, F.E. Werner, C.E. Naimie, F.H. Page, and M.M. Sinclair. "Drift of Sea Scallop Larvae *Placopecten magellanicus* on Georges Bank: A Model Study of Roles of Mean Advection, Larval Behaviour and Larval Origin." *Deep-Sea Research, Part II, Topical Studies in Oceanography* 41 (1994): 7–49.

Tremblay, M.J., and M.M. Sinclair. "Planktonic Sea Scallop Larvae (*Placopecten magellanicus*) in the Georges Bank Region: Broadscale Distribution in Relation to Physical Oceanography." *Canadian Journal of Fisheries and Aquatic Science* 49 (1992): 1597–615.

Trippel, E.A. "The First Marine Biological Station in Canada: 100 Years of Scientific Research at St Andrews." *Canadian Journal of Fisheries and Aquatic Science* 56 (1999): 2495–507.

Trites, R.W. "Comments on Flushing Time for Passamaquoddy Bay and Wind-generated Waves in the Quoddy Region." *Fisheries Research Board of Canada Technical Report* 901 (1979): 3–7.

Trites, R.W. "Probable Effects of Proposed Passamaquoddy Power Project on Oceanographic Conditions." *Journal of the Fisheries Research Board of Canada* 18 (1961): 163–201.

Trites, R.W. "Temperature and Salinity in the Quoddy Region of the Bay of Fundy." *Journal of the Fisheries Research Board of Canada* 19 (1962): 975–8.

Trites, R.W., and R.E. Banks. "Circulation on the Scotian Shelf as Indicated by Drift Bottles." *Journal of the Fisheries Research Board of Canada* 15 (1958): 79–89.

Trites, R.W., and D.G. MacGregor. "Flow of Water in the Passages of Passamaquoddy Bay Measured by the Electromagnetic Method." *Journal of the Fisheries Research Board of Canada* 19 (1962): 895–919.

Tvedt, H.B., T.J. Benfey, D.J. Martin-Robichaud, C. McGowan, and M. Reith. "Gynogenesis and Sex Determination in Atlantic Halibut (*Hippoglossus hippoglossus*)." *Aquaculture* 252:2–4 (2006): 573–83.

Uthe, J.F., and V. Zitko. "An Overview of Marine Environmental Quality Issues on the Atlantic Coast of Canada." In *Canadian Conference on Marine Environmental Quality: Proceedings, Halifax, NS, 29 February to 3 March, 1988*, ed. P.G. Wells and J. Gratwick. Halifax, NS: International Institute for Transportation and Ocean Policy Studies, 1988. 199–205.

Vachon, A. "Hydrography in Passamaquoddy Bay and Vicinity, New Brunswick." *Contributions to Canadian Biology 1917–1918* (1918): 295–328.

Vogel, S. *Life in Moving Fluids: The Physical Biology of Flow*. Boston: Willard Grant Press, 1981.

Volterra, V. "Variations and Fluctuations of the Number of Individuals in Animal Species Living Together." *Journal du Conseil International pour l'Exploration de la Mer* 3 (1926): 3–51.

Waddy, S.L., and D.E. Aiken. "Lobster (*Homarus americanus*) Culture and Resource Enhancement: The Canadian Experience." *Canadian Industry Report of Fisheries and Aquatic Science* 244 (1998): 9–18.

Waddy, S.L., and D.E. Aiken. "Broodstock Management for Year-round Production of Larvae for Culture of the American Lobster." *Canadian Technical Report of Fisheries and Aquatic Science* 1272 (1984): 1–14.

Waddy, S.L., and D.E. Aiken. "Temperature Regulation of Reproduction in Female American Lobsters (*Homarus americanus*)." *ICES Marine Science Symposia* 199 (1995): 54–60.

Waddy, S.L., D.E. Aiken, and D.P.V. De Kleijn. "Control of Growth and Reproduction." In *Biology of the Lobster* Homarus americanus, ed. J.R. Factor. Toronto: Academic Press, 1995. 217–66.

Waddy, S.L., L.E. Burridge, M.N. Hamilton, S.M. Mercer, D.E. Aiken, and K. Haya. "Emamectin Benzoate Induces Molting in American Lobster, *Homarus americanus*." *Canadian Journal of Fisheries and Aquatic Science* 59 (2002): 1096–9.

Waddy, S.L., M.N. Hamilton, L.E. Burridge, S.M. Mercer, D.E. Aiken, and K. Haya. "Molting Response of Female Lobsters (*Homarus americanus*) to Emamectin Benzoate Varies with Reproductive Stage." In *Aquaculture Canada 2002. Proceedings of the Contributed Papers of the19th Annual Meeting of the Aquaculture Association of Canada, Charlottetown, PEI*, ed. C.I. Hendry. *Aquaculture Association of Canada Special Publication* 6 (2003): 75–7.

Waddy, S.L., S.M. Mercer, M.N. Hamilton-Gibson, D.E. Aiken, and L.E. Burridge. "Feeding Response of Female American Lobsters, *Homarus americanus*, to SLICE®-medicated Salmon Feed." *Aquaculture* 269 (2007): 123–9.

Waddy, S.L., V.A. Merritt, M.N. Hamilton-Gibson, D.E. Aiken, and L.E. Burridge. "Relationship between Dose of Emamectin Benzoate and Molting Response of Ovigerous American Lobsters (*Homarus americanus*)." *Ecotoxicology and Environmental Safety* 67 (2007): 95–9.

Waggoner, Theodore R. "Community Stability as a Forest Management Objective." *Journal of Forestry* 97 (1997): 710–14.

Waiser, W.A. "Canada on Display: Towards a National Museum, 1881–1911." In *Critical Issues in the History of Canadian Science, Technology and Medicine*, ed. R.A. Jarrell and A.E. Roos. Thornhill and Ottawa: HSTC Press, 1983. 167–77.

Waiser, W.A. *The Field-Naturalist: John Macoun, the Geological Survey, and Natural Science*. Toronto: University of Toronto Press, 1989.

Waiser, W.A. "The Government Explorer in Canada, 1870–1914." In *North American Exploration*. Vol. 3, *A Continent Comprehended*, ed. J.L. Allen. Lincoln: University of Nebraska Press, 1997. 412–60, 593–9.

Waiwood, K.G. "Haddock." *Bulletin of the Aquaculture Association of Canada* 94:1 (1994): 16–21.

Waiwood K.G., and M.I. Buzeta. "Reproductive Biology of Southwest Scotian Shelf Haddock (*Melanogrammus aeglefinus*)." *Canadian Journal of Fisheries and Aquatic Science* 46 (1989): 153–70.

Waiwood, K.G., K.G. Howes, and J. Reid. "Halibut Aquaculture Research at the St Andrews Biological Station." In *Science Review 1992 & '93 of the Bedford Institute of Oceanography, Halifax Fisheries Research Laboratory, and the St Andrews Biological Station*, ed. A. Fiander. Dartmouth, NS: Department of Fisheries and Oceans, Scotia-Fundy Region, 1994. 43–6.

Waldichuk, M. "The Nature and Extent of Marine Contamination Caused by Land-based Sources in Canada." In *Canadian Conference on Marine Environmental Quality: Proceedings. Halifax, NS, 29 February to 3 March, 1988*, ed. P.G. Wells and J. Gratwick. Halifax: International Institute for Transportation and Ocean Policy Studies, Dalhousie University, 1988. 75–135.

Wallace, D.H. "Keynote Address." In *Optimum Sustained Yield as a Concept in Fisheries Management*, ed. P.M. Roedel. Washington, DC: American Fisheries Society, 1975. 5–8.

Warner, P.C., and M.S. Ewing. "Wading in the Water: Women Aquatic Biologists Coping with Clothing, 1877–1945." *BioScience* 97 (2002): 98–104.

Watson, E.E. "Mixing and Residual Currents in the Tidal Waters as Illustrated in the Bay of Fundy." *Journal of the Fisheries Research Board of Canada* 2 (1936): 141–208.

Wells, P.G. "Effects of Venezuelan Crude Oil on Young Stages of the American Lobster, *Homerus americanus*." Guelph, ON: University of Guelph doctoral dissertation, 1976. 1–274.

Wells, P.G., P.D. Keizer, J.L. Martin, P.A. Yeats, K.M. Ellis, and D.W. Johnston. "The Chemical Environment of the Bay of Fundy." In *Bay of Fundy Issues: A Scientific Overview. Workshop Proceedings, Wolfville, N.S., Jan. 29–Feb. 1, 1996*, ed. J.A. Percy, P.G. Wells, and A. Evans. Dartmouth, NS, and Sackville, NB: Environment Canada Atlantic Region Occasional Report no. 8, 1997. 37–61.

Wells, P.G., and S.J. Rolston, eds. *Health of Our Oceans: A Status Report on Canadian Marine Environmental Quality*. 2nd ed. Ottawa, ON, and Dartmouth NS: Environment Canada, Conservation and Protection, 1991.

Wells, P.G., and J.B. Sprague. "Effects of Crude Oil on American Lobster (*Homarus americanus*) Larvae in the Laboratory." *Journal of the Fisheries Research Board of Canada* 33 (1976): 1604–14.

Werner, F.E., F.H. Page, D.R. Lynch, J.W. Loder, R.G. Lough, R.I. Perry, D.A. Greenberg, and M.M. Sinclair. "Influences of Mean Advection and Simple Behavior on the Distribution of Cod and Haddock Early Life Stages on Georges Bank." *Fisheries Oceanography* 2 (1993): 43–64.

White, A.W. "Dinoflagellate Toxins in Phytoplankton and Zooplankton Fractions during a Bloom of *Gonyaulax excavate*." In *Toxic Dinoflagellate Blooms*, ed. D.L. Taylor and H. Seliger. North Holland: Elsevier, 1979. 381–4.

White, A.W. "Dinoflagellate Toxins as Probable Cause of an Atlantic Herring (*Clupea harangus harangus*) Kill, and Pteropods as Apparent Vector." *Journal of the Fisheries Research Board of Canada* 34 (1977): 2421–4.

White, A.W. "Growth Inhibition Caused by Turbulence in the Toxic Marine Dinoflagellate *Gonyaulax excavate*." *Journal of the Fisheries Research Board of Canada* 33 (1976): 2598–602.

White, A.W. "Marine Zooplankton Can Accumulate and Retain Dinoflagellate Toxins and Cause Fish Kills." *Limnology and Oceanography* 26 (1981): 104–10.

White, A.W. "Recurrence of Kills of Atlantic Herring (*Clupea harengus harengus*) Caused by Dinoflagellate Toxins Transferred through Herbivorous Zooplankton." *Canadian Journal of Fisheries and Aquatic Science* 37 (1980): 2262–5.

White, A.W. "Relationships of Environmental Factors to Toxic Dinoflagellate Blooms in the Bay of Fundy." *Rapports et Procès-verbaux des Réunions du Conseil International pour l'Exploration de la Mer* 187 (1987): 38–46.

White, A.W. "Sensitivity of Marine Fishes to Toxins from the Red-tide Dinoflagellate *Gonyaulax excavata* and Implications for Fish Kills." *Marine Biology* 3 (1981): 255–60.

White, A.W., and C.M. Lewis. "Resting Cysts of the Toxic, Red Tide Dinoflagellate *Gonyaulax excavata* in Bay of Fundy Sediments." *Journal of the Fisheries Research Board of Canada* 39 (1982): 1185–94.

White, A.W., and L. Maranda. "Paralytic Toxins in the Dinoflagellate *Gonyaulax excavata* and in Shellfish." *Journal of the Fisheries Research Board of Canada* 35 (1980): 397–402.

White, A.W., J.L. Martin, M. LeGresley, and W.J. Blogoslawski. "Inability of Ozone to Detoxify Paralytic Shellfish Toxins in Soft-shell Clams." In *Toxic Dinoflagellates*, ed. D.M. Anderson, A.W. White, and D.G. Baden. New York: Elsevier, 1985. 473–8.

White, H.C. "The Eastern Belted Kingfisher in the Maritime Provinces." *Bulletin of the Fisheries Research Board of Canada* 97 (1953): 1–44.

White, H.C. "The Food of Kingfishers and Mergansers on the Margaree River, Nova Scotia." *Journal of the Biological Board of Canada* 2 (1936): 299–309.

White, H.C. "Local Feeding of Kingfishers and Mergansers." *Journal of the Biological Board of Canada* 3 (1937): 323–38.

White, H.C. "Some Facts and Theories concerning the Atlantic Salmon." *Transactions of the American Fisheries Society* 64 (1934): 360–2.

Whiteaves, J.F. *Catalogue of the Marine Invertebrata of Eastern Canada. Geological Survey of Canada Special Report* 722. Ottawa: S.E. Dawson, 1901.

Whitfield, C.M., and R.A. Jarrell. "Lefroy, Sir John Henry." *Dictionary of Canadian Biography* 11 (2000): 508–10.

Wilder, D.G. "The Relative Toxicity of Certain Metals to Lobsters." *Journal of the Fisheries Research Board of Canada* 8 (1952): 486–7.

Wildish, D.J. "Coastal Zone Management and the Pulp and Paper Industry." *Pulp and Paper Canada*, June 1983: 1145–8.

Wildish, D.J. "The Toxicity of Polychlorinated Biphenyls (PCB) to *Gammarus oceanicus*." *Bulletin of Environmental Contamination and Toxicology* 5 (1970): 202–4.

Wildish, D.J., H.M. Akagi, and N. Hamilton. "Effect of Velocity on Horse Mussel Initial Feeding Behaviour." *Canadian Technical Report of Fisheries and Aquatic Science* 2325 (2000): 1–34. Available at http://www.dfo-mpo.gc.ca/Library/250056.pdf.

Wildish, D.J., H.M. Akagi, and N. Hamilton. "Sedimentary Changes at a Bay of Fundy Salmon Farm Associated with Site Fallowing." *Bulletin of the Aquaculture Association of Canada* 101 (2001): 49–54.

Wildish, D.J., H.M. Akagi, N. Hamilton, and B.T. Hargrave. "A Recommended Method for Monitoring Sediments to Detect Organic Enrichment from Mariculture in the Bay of Fundy." *Canadian Technical Report of Fisheries and Aquatic Science* 2286 (1999): 1–31. Available at http://www.dfo-mpo.gc.ca/Library/238355.pdf.

Wildish, D.J., H.M. Akagi, D.L. McKeown, and G.W. Pohle. "Pockmarks Influence Benthic Communities in Passamaquoddy Bay, Bay of Fundy, Canada." *Marine Ecology Progress Series* 357 (2008): 51–66.

Wildish, D.J., H.M. Akagi, and N.J. Poole. "Avoidance by Herring of Dissolved Components in Pulp Mill Effluents." *Bulletin of Environmental Contamination and Toxicology* 18 (1977): 521–5.

Wildish, D.J., B.T. Hargrave, and G. Pohle, "Cost Effective Monitoring of Organic Enrichment Resulting from Salmon Mariculture." *International Council for the Exploration of the Sea Journal of Marine Science* 58 (2001): 469–76.

Wildish, D.J., P.D. Keizer, A.J. Wilson, and J.L. Martin. "Seasonal Changes of Dissolved Oxygen and Plant Nutrients in Seawater near Salmonid Net Pens in the Macrotidal Bay of Fundy." *Canadian Journal of Fisheries and Aquatic Science* 50 (1993): 303–11.

Wildish, D.J., and D.D. Kristmanson. *Benthic Suspension Feeders and Flow*. New York: Cambridge University Press, 1997.

Wildish, D.J., and D.D. Kristmanson. "Control of Suspension-feeding Bivalve Production by Current Speed." *Helgoländer Wissenschaftliche Meeresuntersuchungen* 39 (1985): 237–43.

Wildish, D.J., and D.D. Kristmanson. "Growth Response of Giant Scallops to Periodicity of Flow." *Marine Ecology Progress Series* 42 (1988): 163–9.

Wildish, D.J., and D.D. Kristmanson. "Importance to Mussels of the Benthic Boundary Layer." *Canadian Journal of Fisheries and Aquatic Science* 41 (1984): 1618–25.

Wildish, D.J., and D.D. Kristmanson. "Tidal Energy and Sublittoral Macrobenthic Animals in Estuaries." *Journal of the Fisheries Research Board of Canada* 36 (1979): 1197–206.

Wildish, D.J., D.D. Kristmanson, R.L. Hoar, A.M. DeCoste, S.D. McCormick, and A.W. White. "Giant Scallop Feeding and Growth Responses to Flow." *Journal of Experimental Marine Biology and Ecology* 113 (1987): 207–20.

Wildish, D.J., D.D. Kristmanson, and S.M.C. Robinson. "Does Skimming Flow Reduce Growth in Horse Mussels?" *Journal of Experimental Marine Biology and Ecology* 358 (2008): 33–8.

Wildish, D.J., D.D. Kristmanson, and A.M. Saulnier. "Interactive Effect of Velocity and Seston Concentration on Giant Scallop Feeding Inhibition." *Journal of Experimental Marine Biology and Ecology* 155 (1992): 161–8.

Wildish, D.J., P. Lassus, J.L. Martin, A.M. Saulnier, and M. Bardouil. "Effect of the PSP-causing Dinoflagellate, *Alexandrium sp.*, on the Initial Feeding Response of Crassostrea gigas." *Aquatic Living Resources* 11 (1998): 35–43.

Wildish, D.J., and N.A. Lister. "Biological Effects of Fenitrothion in the Diet of Brook Trout." *Bulletin of Environmental Contamination and Toxicology* 10 (1973): 233–9.

Wildish, D.J., and J.L. Martin. "Determining the Potential Harm of Marine Phytoplankton to Finfish Aquaculture Resources of the Bay of Fundy." *Fisken og Harvet* 13 (1994): 115–26.

Wildish, D.J., J.L. Martin, A.J. Wilson, and M. Ringuette. "Environmental Monitoring of the Bay of Fundy Salmonid Mariculture Industry during 1988–89." *Canadian Technical Report of Fisheries and Aquatic Science* 1760 (1990): 1–123. Available at http://www.dfo-mpo.gc.ca/Library/116576.pdf.

Wildish, D.J., and M.P. Miyares. "Filtration Rate of Blue Mussels as a Function of Flow Velocity: Preliminary Experiments." *Journal of Experimental Marine Biology and Ecology* 142 (1990): 213–19.

Wildish, D.J., D.L. Peer, and D.A. Greenberg. "Benthic Macrofauna Production in the Bay of Fundy and Possible Effects of a Tidal Power Barrage at Economy Point – Cape Tenny." *Canadian Journal of Fisheries and Aquatic Science* 43 (1986): 2410–7.

Wildish, D.J., and G.W. Pohle. "Benthic Macrofaunal Changes Resulting from Finfish Mariculture." In *The Handbook of Environmental Chemistry*, vol. 5, part M, *Water Pollution: Environmental Effects of Marine Finfish Aquaculture*, ed. B.T. Hargrave. Berlin: Springer-Verlag, 2005. 275–304.

Wildish, D.J., N.J. Poole, and D.D. Kristmanson. "The Effect of Anaerobiosis on Measurement of Sulfite Pulp Mill Effluent Concentration in Estuarine Water by U.V. Spectrophotometry." *Bulletin of Environmental Contamination and Toxicology* 16 (1976): 208–13.

Wildish, D.J., N.J. Poole, and D.D. Kristmanson. "Pulp Mill Pollution in L'Etang Estuary: A Case History and Clean-up Alternatives." *Fisheries and Marine Services Technical Report* 884 (1979): 1–6. Available at http://www.dfo-mpo.gc.ca/Library/26232.pdf.

Wildish, D.J., N.J. Poole, and D.D. Kristmanson. "Temporal Changes of Sublittoral Macrofauna in L'Etang Inlet Caused by Sulfite Pulp Mill Pollution." *Fisheries and Marine Services Technical Report* 718, (1977): 1–13. Available at http://www.dfo-mpo.gc.ca/Library/18296.pdf.

Wildish, D.J., and A.M. Saulnier. "Hydrodynamic Control of Filtration in *Placopecten magellanicus*." *Journal of Experimental Marine Biology and Ecology* (1993): 65–82.

Wildish, D.J., and M.L.H. Thomas. "Effects of Dredging and Dumping on Benthos of Saint John Harbour, Canada." *Marine Environmental Research* 15 (1985): 45–57.

Wildish, D.J., A.J. Wilson, A.J., W.W. Young-Lai, A.M. DeCoste, D.E. Aiken, and J.M. Martin. "Biological and Economic Feasibility of Four Growout Methods for the Culture of Giant Scallops in the Bay of Fundy." *Canadian Technical Report of Fisheries and Aquatic Science* 1658 (1988): 1–21.

Wildish, D.J., V. Zitko, H.M. Akagi, and A.J. Wilson. "Sedimentary Anoxia Caused by Salmonid Mariculture Wastes in the Bay of Fundy and Its Effects on Dissolved Oxygen in Seawater." In *Proceedings of Canada-Norway Finfish Aquaculture Workshop, Sept. 11–14, 1989*, ed. R.L. Saunders. *Canadian Technical Report of Fisheries and Aquatic Science* 1761 (1990). 11–18.

Wilks, B. *Browsing Science Research at the Federal Level in Canada: History, Research Activities, and Publications*. Toronto: University of Toronto Press, 2004.

Williams, K.B. "From Civilian Planktonologist to Navy Oceanographer: Mary Sears in World War II." In *The Machine in Neptune's Garden: Historical Perspectives on Technology and the Marine Environment*, ed. H.M. Rozwadowski and D.K. van Keuren. Sagamore Beach, MA: Watson Publishing International, 2004. 243–72.

Wolff, M. "Population Dynamics of the Peruvian Scallop *Argopecten purpuratus* during the El Nino Phenomena 1983." *Canadian Journal of Fisheries and Aquatic Science* 44 (1987): 1684–91.

Wright, J.L.C., R.K. Boyd, A.S.W. de Freitas, M. Falk, R.A. Foxall, W.D. Jamieson, M.V. Laycock, A.W. McCulloch, A.G. McInnes, P. Odense, V.P. Pathak, M.A. Quilliam, M.A. Ragan, P.G. Sim, P. Thibault, J.A. Walter, M. Gilgan, D.J.A. Richard, and D. Dewar. "Identification of Domoic Acid, a Neuroexcitatory Amino Acid, in Toxic Mussels from Eastern Prince Edward Island." *Canadian Journal of Chemistry* 67 (1989): 481–90.

Yasumoto, T., Y. Oshima, and Y. Fukuyo, eds. *Harmful and Toxic Algal Blooms. Proceedings of the Seventh International Conference on Toxic Phytoplankton, Sendai, Japan, 12–16, 1995.* Paris: International Oceanic Commission / UNESCO, 1996.

Young-Lai, W.W., and D.E. Aiken."Biology and Culture of the Giant Scallop, *Placopecten magellanicus*: A Review." *Canadian Technical Report of Fisheries and Aquatic Science* 1478 (1986): 1–21.

Zaslow, M. *Reading the Rocks: The Story of the Geological Survey of Canada 1842–1972*. Toronto: Macmillan of Canada, 1975.

Zeller, S. "Darwin Meets the Engineers: Scientizing the Forest at McGill University, 1890–1910." *Environmental History* 6:3 (2001): 428–50.

Zeller, S. *Inventing Canada: Early Victorian Science and the Idea of a Transcontinental Nation*. Toronto: University of Toronto Press, 1987.

Zeller, S. *Land of Promise, Promised Land: The Culture of Victorian Science in Canada*. Historical Booklet no. 56. Ottawa: Canadian Historical Association, 1996.

Zeller, S. "Nature's Gullivers and Crusoes: The Scientific Exploration of North America, 1800–1870." In *North American Exploration*, vol. 3, *A Continent Comprehended*, ed. J.L. Allen. Lincoln: University of Nebraska Press, 1997.

Zitko, V. "Determination of Residual Fuel Oil Contamination of Aquatic Animals." *Bulletin of Environmental Contamination and Toxicology* 5 (1971): 559–64.

Zitko, V. "Fifteen Years of Environmental Research." *Marine Pollution Bulletin* 10 (1979): 100–3.

Zitko, V. "Polychlorinated biphenyls and Organochlorine Pesticides in Some Freshwater and Marine Fishes." *Bulletin of Environmental Contamination and Toxicology* 6 (1971): 464–70.

Zitko, V. "Uptake and Excretion of Chlorinated and Brominated Hydrocarbons by Fish." *Fisheries and Marine Service Technical Report* 737 (1977). Available at http://www.dfo-mpo.gc.ca/Library/18314.pdf.

Zitko, V., D.E. Aiken, S.N. Tibbo, K.W.T. Besch, and J.M. Anderson. "Toxicity of Yellow Phosphorus to Herring (*Clupea harengus*), Atlantic Salmon (*Salmo salar*), Lobster (*Homarus americanus*), and Beach Flea (*Gammarus oceanicus*)." *Journal of the Fisheries Research Board of Canada* 27 (1970): 21–9.

Zitko, V., and E. Arsenault. "Chlorinated Paraffins: Properties, Uses, and Pollution Potential." *Fisheries and Marine Service Technical Report* 491 (1974): 1–38. Available at http://www.dfo-mpo.gc.ca/Library/22633.pdf.

Zitko, V., and E. Arsenault. "Fate of High-Molecular Weight Chlorinated Paraffins in the Aquatic Environment." *American Chemical Society, Division of Environmental Chemistry* 15 (1975): 174–6.

Zitko, V., and E. Arsenault. "Fate of High-Molecular Weight Chlorinated Paraffins in the Aquatic Environment." In *Advances in Environmental Science and Technology*, vol. 8, *Fate of Pollutants in the Air and Water Environments*, ed. I.H. Suffet. New York: Wiley-Interscience, 1977. 409–18.

Zitko, V., and W.V. Carson. "Accumulation of Thallium in Clams and Mussels." *Bulletin of Environmental Contamination and Toxicology* 14 (1975): 530–3.

Zitko, V., W.V. Carson, and W.G. Carson. "Thallium: Occurrence in the Environment and Toxicity to Fish." *Bulletin of Environmental Contamination and Toxicology* 13 (1975): 23–30.

Zitko, V., B.J. Finlayson, D.J. Wildish, J.M. Anderson, and A.C. Kohler. "Methylmercury in Freshwater and Marine Fishes in New Brunswick, in the Bay of Fundy, and on the Nova Scotia Banks." *Journal of the Fisheries Research Board of Canada* 28 (1971): 1285–91.

Zitko, V., and D.J. Wildish. "The First Marine Biological Station in Canada: Highlights of Environmental Research at St Andrews." *Water Quality Research Journal of Canada* 35 (2000): 809–17.

# Contributors

**Mary Needler Arai** is a professor emerita at the Department of Biological Sciences, University of Calgary, and a senior volunteer investigator, Pacific Biological Station, Nanaimo, BC. She is a third-generation woman marine biologist, and grew up in St Andrews. She is an honorary member the Canadian Society of Zoologists and an honorary Doctor of Science of the University of New Brunswick. Her publications mainly concern coelenterate biology, including *A Functional Biology of Scyphozoa*, published by Chapman and Hall in 1997.

**John F. Caddy** joined the St Andrews Biological Station in 1966 as the scientist responsible for methods of assessing and managing Maritime stocks of the scallop *Placopecten magellanicus* and, later, other molluscan stocks. He focused on the dynamics of shellfish and crustacean populations, using diving, underwater photography, and submersibles to monitor stocks. As editor, his book *Marine Invertebrate Fisheries* brought together global studies on the exploitation of marine invertebrate populations. After moving to Halifax in 1977 as service chief of the Invertebrates and Marine Plants Division of the St Andrews station, he was recruited by FAO Rome, where he retired 15 years ago as chief of the (then) Marine Resources Service.

**Blythe D. Chang** has been a biologist at the St Andrews Biological Station since 1989. Prior to that, he worked for Fisheries and Oceans Canada on the Pacific coast. Since 2003 he has been in the station's Coastal Oceanography and Ecosystem Research Section, led by Fred Page. His work within this group has focused on oceanographic and environmental research in the coastal zone of the Bay of Fundy, particularly on aquaculture–environment interactions.

**Robert H. Cook** was director of the St Andrews Biological Station and the regional chief of the Fisheries and Environmental Sciences Division from 1977 until 1999. During his tenure at St Andrews, Bob Cook played a key role in salmon aquaculture development in the region, nationally, and internationally. He left St Andrews in 1993 to become the director of aquaculture coordination at the regional headquarters in Halifax. He retired in 1998.

**Timothy James Foulkes** studied mechanical engineering at McGill University from 1955 to 1958 and then at Ryerson University in Toronto, where he graduated as an engineering technologist in 1961. Following a year working with mechanical standards at Canadian Aviation Electronics in Montreal, he moved to New Zealand, where he worked for the Department of Scientific and Industrial Research until 1964. Returning to Canada he joined John Carrothers at the St Andrews Biological Station to work on fishing gear engineering research. He left St Andrews in 1987 to resolve technical difficulties with underwater sampling gear at the Northwest Atlantic Fisheries Centre in St John's, NL. He retired in 1995 and moved back to his home in Bayside, NB.

**Jennifer Hubbard** is an associate professor of the history of science and technology at Ryerson University. Her *A Science on the Scales: The Rise of Canadian Atlantic Fisheries Biology, 1898–1939* was published by the University of Toronto Press in 2006. She has recently published articles on the political and environmental role and impact of fisheries science in the *ICES Journal of Marine Science*, *Isis*, and *Environmental History*.

**Jennifer L. Martin** has been a biologist at the St Andrews Biological Station since 1977, where she has studied phytoplankton ecology with particular emphasis on harmful algal blooms and their effects on shellfish, finfish, and other vertebrate consumers. After retirement in 2014 she has continued these studies as an emeritus research scientist.

**Eric L. Mills**, after a career as a biological oceanographer, is professor emeritus of the history of science in the Department of Oceanography at Dalhousie University and a member of the History of Science and Technology Programme at the University of King's College, Halifax. He works on the history of 19th-century natural history and of 20th- and 21st-century biological and physical oceanography. His publications include *Biological Oceanography: An Early History, 1870–1960* (1989/2012) and *The Fluid Envelope of Our Planet: How the Study of Ocean Currents Became a Science* (2009/11).

**Fred H. Page** has been a research scientist at the St Andrews Biological Station since 1991, and is the head of the station's Coastal Oceanography and Ecosystem Research Section. He is a bio-physical oceanographer specializing in the investigation of linkages between the physical characteristics and processes of the coastal and shelf seas and their living resources. His research is directed towards developing an understanding of the hydrography and water circulation within south-west New Brunswick and other coastal areas of Atlantic Canada, with an emphasis on issues related to the salmon aquaculture industry and integrated coastal zone management.

**Richard H. Peterson** was a research scientist at the St. Andrews Biological Station beginning in 1970. For 15 years he investigated the possible influence of acid precipitation on embryonic development of salmonids, and on diversity of stream fishes and invertebrates. For the next 15 years, he investigated methods used in aquaculture of salmonids and striped bass. He was an emeritus research scientist until 2008, during which time he completed publication of his aquaculture studies.

**Shawn M.C. Robinson** has been a research scientist with Fisheries and Oceans Canada at the St Andrews Biological Station since 1988, and is an adjunct professor at the University of New Brunswick. His research interests are in applied ecological research of blue mussels, sea scallops, sea urchins, sea cucumbers, soft-shell clams, worms, sea lice, and marine bacteria. He is studying natural ecological processes and how these animals utilize their environment so that more sustainable culture techniques can be developed, notably by integrated multitrophic aquaculture.

**Robert L. Stephenson** has been a research scientist at the St Andrews Biological Station since 1984, and was director from 2005–9. His research was on the ecology, assessment, and management of Atlantic herring, and more broadly on issues related to fisheries resource evaluation, fisheries management science, and integrated management. He is currently visiting research professor at the University of New Brunswick, where he is principal investigator of the Canadian Fisheries Research Network – an NSERC-funded network linking academics, industry, and government in collaborative fisheries research across Canada.

**Peter G. Wells** is a senior research fellow and adjunct professor at Dalhousie University, Halifax, studying marine ecosystem health, ecotoxicology, and the use and influence of marine environmental information at the science-policy interface. After graduate studies in toxicology at the St Andrews Biological Station, he worked as a marine environmental scientist with Environment Canada, retiring in 2006. He is the author or editor of *Controlling Chemical Hazards*, *Exxon Valdez Oil Spill: Fate and Effects in Alaskan Waters*, *The Rio de la Plata: An Environmental Overview*, and *Microscale Testing in Aquatic Toxicology*.

**David J. Wildish** is an emeritus research scientist at the St Andrews Biological Station and a research associate of the Huntsman Marine Science Centre, both in St Andrews, NB. He published *Benthic Suspension Feeders and Flow* with David Kristmanson in 1997 and over 200 primary, technical, and manuscript reports. The latter include studies on a wide range of marine benthic ecological topics, for which he was awarded a D.Sc. by examination from the University of London, UK, in 1992.

# Index

Lightning Source UK Ltd.
Milton Keynes UK
UKHW041101260919
350490UK00001B/195/P